Study Guide

Biology
The Dynamic Science

Peter J. Russell

Stephan L. Wolfe

Paul E. Hertz

Cecie Starr

Beverly McMillan

Prepared by

Carolyn Bunde
Idaho State University

Mark Sheridan
North Dakota State University

William Kroll
Loyola University Chicago

Jyoti Wagle
Houston Community College

THOMSON
BROOKS/COLE

Australia • Brazil • Canada • Mexico • Singapore • Spain • United Kingdom • United States

Printed in the United States of America

1 2 3 4 5 6 7 11 10 09 08 07

ISBN-13: 978-0-534-40321-8
ISBN-10: 0-534-40321-2

Thomson Higher Education
10 Davis Drive
Belmont, CA 94002-3098
USA

For more information about our products, contact us at:
Thomson Learning Academic Resource Center
1-800-423-0563

For permission to use material from this text or product, submit a request online at **http://www.thomsonrights.com.**
Any additional questions about permissions can be submitted by email to **thomsonrights@thomson.com.**

CONTENTS

1 Introduction to Biological Concepts and Research 1

UNIT ONE MOLECULES AND CELLS

2 Life, Chemistry, and Water 9
3 Biological Molecules: The Carbon Compounds of Life 16
4 Energy, Enzymes, and Biological Reactions 25
5 The Cell: An Overview 35
6 Membranes and Transport 44
7 Cell Communication 52
8 Harvesting Chemical Energy: Cellular Respiration 58
9 Photosynthesis 69
10 Cell Division and Mitosis 82

UNIT TWO GENETICS

11 Meiosis: The Cellular Basis of Sexual Reproduction 89
12 Mendel, Genes, and Inheritance 96
13 Genes, Chromosomes, and Human Genetics 102
14 DNA Structure, Replication, and Organization 108
15 From DNA to Protein 116
16 Control of Gene Expression 132
17 Bacterial and Viral Genetics 144
18 DNA Technologies and Genomics 155

UNIT THREE EVOLUTIONARY BIOLOGY

19 The Development of Evolutionary Thought 163
20 Microevolution: Genetic Changes within Populations 169
21 Speciation 175
22 Paleobiology and Macroevolution 182
23 Systematic Biology: Phylogeny and Classification 189

UNIT FOUR BIODIVERSITY

24 The Origin of Life 195
25 Prokaryotes and Viruses 205
26 Protists 215
27 Plants 223
28 Fungi 232
29 Animal Phylogeny, Acoelomates, and Protostomes 240
30 Deuterostomes: Vertebrates and Their Closest Relatives 247

UNIT FIVE PLANT STRUCTURE AND FUNCTION

31 The Plant Body 254
32 Transport in Plants 261
33 Plant Nutrition 266
34 Reproduction and Development in Flowering Plants 273
35 Control of Plant Growth and Development 281

UNIT SIX ANIMAL STRUTURE AND FUNCTION

36 Introduction to Animal Organization and Physiology 288
37 Information Flow and the Neuron 294
38 Nervous Systems 302
39 Sensory Systems 309
40 The Endocrine System 318
41 Muscles, Bones, and Body Movements 326
42 The Circulatory System 332
43 Defenses against Disease 339
44 Gas Exchange: The Respiratory System 347
45 Animal Nutrition 355
46 Regulating the Internal Environment 361
47 Animal Reproduction 369
48 Animal Development 375

UNIT SEVEN ECOLOGY AND BEHAVIOR

49 Population Ecology 381
50 Population Interactions and Community Ecology 389
51 Ecosystems 397
52 The Biosphere 402
53 Biodiversity and Conservation Biology 408
54 The Physiology and Genetics of Animal Behavior 413
55 The Ecology and Evolution of Animal Behavior 418

ANSWERS 425

STRUCTURE OF THIS STUDENT WORKBOOK

The following outline shows how each chapter in this student workbook is organized.

Chapter Number ⇒	2
Chapter Title ⇒	**LIFE, CHEMISTRY, AND WATER**
Chapter Highlights ⇒	A quick overview to the subject matter of the chapter in a bulleted list.
Study Strategies ⇒	This is a bulleted list of tips to help you use the chapter to study to the best of your ability. In certain chapters, there are topic maps that can be used to study material as well
Interactive Exercises ⇒	The interactive exercises are organized by sections with page references. There are sets of questions and exercises under each section. Some options of exercises follow: fill in the blank section review, building vocabulary with prefixes and suffixes, complete the table, short answer, choice, matching, sequence, identification, labeling, completion, true/false, and problems
Self-Quiz ⇒	This is a set of questions designed to provide a quick assessment of how well you understand the information from the chapter.
Integrating and Applying Key Concepts ⇒	These represent "big-picture" or "real-life" applications of the concepts presented in the chapter.
Answers to Interactive Exercises and Self-Quiz ⇒	Answers for all the Interactive Exercises and the Self-Quiz can be found at the end of the student workbook. The answers are arranged by chapter number and main heading

1

INTRODUCTION TO BIOLOGICAL CONCEPTS AND RESEARCH

CHAPTER HIGHLIGHTS

- Living systems a) are organized in a hierarchy, b) contain chemical instructions that govern their structure and function, c) engage in metabolic activities, d) have energy flows and cycle matter through them, e) compensate for changes in the external environment, f) reproduce and undergo development, and g) change from one generation to the next.
- Populations evolve over time; evolution has produced the great diversity of life on earth.
- Biologists classify organisms in a hierarchical system.
- Biological knowledge results for basic and applied research.

STUDY STRATEGIES

- This chapter contains what appears to be a lot of different concepts. However, the various topics form a cohesive picture of what characterizes living systems and how we go about studying them.
- DO NOT try to go through the chapter in one sitting. Take one section at a time. Start with the text by first looking at the figures and their legends. Then read the text, carefully noting the meaning of boldface terms. Finally, work through the companion section(s) in the study guide.
- Repeat this process for each section—text, then study guide, until you have completed the chapter.
- Test your overall understanding by taking the self-test.

INTERACTIVE EXERCISES

Why It Matters [pp. 1-2]

Section Review

Biology is the study of (1)____.

1.1 What is Life? Characteristics of Living Systems [pp. 2–7]

Section Review

Living systems have several characteristics. First, living systems are organized into a(n) (2)____. A(n) (3)____ is the lowest level of organization. Many single cells, such as bacteria, exist as (4)____. Plants and animals are (5)____. A(n) (6)____ is a group of unicellular or multicellular organisms. The collection of all the populations of different organisms living in the same place form a(n) (7)____. A(n) (8)____ consists of the community and the nonliving factors with which it interacts. The collection of all of the earth's ecosystems make up the (9)____. Each level of organization possesses a unique set of characteristics called (10)____. Second, living systems contain chemical instructions that govern their structure and function. The presence of (11)_____ distinguishes living systems from nonliving matter. (12)____ is copied from DNA and directs the production of (13)____. Third, living organisms engage in (14)____, the ability to transform and use energy. Fourth, energy flows and matter cycles through living organisms. Photosynthetic organisms are (15)____, whereas animals are (16)____. (17)____ feed on the remains of dead organisms. Fifth, living organisms compensate for changes in the (18)____. (19)____ is a steady internal condition maintained by responses that compensate for changes in the external environment. Sixth, living organisms reproduce and undergo development. (20)____ is the process in which parents produce offspring. The transmission of DNA from one generation to the next is called (21)____. (22)____ is a series of programmed events that to a fertilized egg becoming an adult. The (23)____ of an organism is the sequential stages through which individuals develop, grow, maintain themselves, and reproduce. Seventh (and last), populations of living organisms change from one generation to the next. Such change over time is called (24)____.

Building Vocabulary

Prefixes	Meaning
bio-	life
eco-	"home"
homeo-	similar, constant
proto-	first, earliest form of

Suffix	Meaning
-ology	study of
-stasis	equilibrium caused by opposing forces
-zoa	animal

Prefix	Suffix	Definition
_____	____	25. The study of life
_____	____	26. The study of groups of organisms interacting with on another and their non-living environment
_____	____	27. maintaining a constant internal environment
_____	system	28. a community along with its nonliving environment
_____	sphere	29. the collection of the earth's ecosystems
_____	____	30. animal-like unicellular organisms

Complete the Table

	Level	Feature
31.	cell	
32.		individual made up of interdependent cells
33.		group of organisms of the same kind that live together in the same place
34.	community	
35.		the community and the nonliving factors with which it interacts

Short Answer

36. Define biological magnification. ______________________________

____

Choice

For each of the following statements, choose the most appropriate term from the list below.

a. diversity   b. stability

37. ____ The degree to which the populations of a community remain the same.

38. ____ The number and types of populations in a community.

For each of the following statements, choose the most appropriate term from the list below.

a. primary producer   b. consumer   c. decomposer

39. ____ Organisms that convert sunlight to chemical energy which they use to make complex molecules, such as sugars, from simple molecules such as carbon dioxide and water.

40. ____ Organisms that feed on the complex molecules made by producers.

41. ____ Organisms that feed on the remains of dead organisms.

Matching

Match each of the following terms with its correct definition.

42. ____ RNA — A. molecules which distinguish living systems from nonliving systems and serve as the hereditary material

43. ____ life cycle — B. molecules that direct the production of protein

44. ____ DNA — C. molecules that carry out most of the activities of life, including the synthesis of all other biological molecules

45. ____ protein — D. sequential stages through which they develop, grow, maintain themselves, and reproduce

46. ____ inheritance — E. a series of programmed changes that enable a fertilized egg to become an adult

47. ____ reproduction — F. the process through which parents produce offspring

48. ____ development — G. the transmission of DNA from one generation to the next

1.2 Biological Evolution [pp. 7–9]

1.3 Biodiversity [pp. 9–13]

Section Review

Populations of organisms change through time as a result of changes in the frequency of traits encoded by genes. In (49)____, only certain individuals are permitted to mate so as to increase the frequency of desired traits in domesticated species. The selection of traits in nature is (50)____. The origin and inheritance of new variations arise from the structure and variability of DNA. DNA is organized into structural units called (51)____. (52)____ are random changes in DNA. Favorable mutations may produce (53)____ that improve survival and reproduction under particular environmental conditions.

The millions of kinds of living organisms on earth are classified into a hierarchy of categories. (54)____ is the fundamental grouping in which individuals are closely related in structure, biochemistry, and behavior. Each species is assigned a two-part (55)____. (56)____ is a group of similar species that have a common ancestor. Related genera are organized into a(n) (57)____, related families into the same (58)____, and related orders are in the same (59)____. Related classes are grouped into a(n) (60)____, and related phyla are assigned to a(n) (61)____. The most inclusive grouping is the (62)____, of which there are three recognized: (63)____, (64)____, and (65)____. Archaea and Bacteria are both (66)____. Eukarya are described as (67)____ because their DNA is enclosed in a(n) (68)____, which along with other specialized compartments are called (69)____. The domain Eukarya is divided into four kingdoms: (70)____, (71)____, (72)____, and (73)____.

Matching

Match each of the following terms with its correct definition.

74. ____ genes
75. ____ adaptations
76. ____ mutations
77. ____ natural selection
78. ____ artificial selection

A. selection of desired traits through breeding practices conducted by humans

B. occurs when certain characteristics allow some organisms to survive better and reproduce more than other organisms in the same population, and results in the successful characteristics becoming more common in the population in later generations.

C. functional unit of DNA

D. random changes in the structure, number, or arrangement of DNA molecules

E. characteristics that help an organism survive/reproduce in a particular environment

Choice

For each of the following characteristics, choose the most appropriate term from the list below.

a. prokaryote b. eukaryote

79. ____ Has cells that possess nuclei and other membrane-bound organelles

80. ____ Has cells that do not possess nuclei or other membrane-bound organelles

For each of the following characteristics, choose the most appropriate term from the list below.

a. plants b. fungi c. animals

81. ____ Eukaryotic; multicellular; consumers; generally motile

82. ____ Eukaryotic; multicellular; generally carryout photosynthesis; includes conifers and mosses

83. ____ Eukaryotic; unicellular or multicellular; do not carryout photosynthesis; includes yeasts and molds

Sequence

Arrange the following taxonomic terms in the correct order from the least inclusive (starting with number 84) to the most inclusive (ending with number 91).

84. ____ A. family

85. ____ B. phylum

86. ____ C. class

87. ____ D. species

88. ____ E. order

89. ____ F. genus

90. ____ G. domain

91. ____ H. kingdom

1.4 Biological Research [pp. 13–19]

Section Review

What we know about living systems is the product of (92)____. The (93)____ is an approach in which scientists make observations, develop working explanations, and test those explanations. The search for explanations without a specific practical goal in mind is termed (94)____. (95)____ has a goal of solving specific practical problems. (96)____ is information that describes basic structures or processes, whereas (97)____ is information that results from manipulation of the system under study. A(n) (98)____ is a working explanation for observations. A hypothesis yields (99)____ about what the researcher expects to happen that can tested experimentally. (100)____ are other possible explanations that should be considered when designing experiments. A(n) (101)____is a treatment group that is not experimentally manipulated. The (102)____is the one factor that is manipulated. Experiments are generally conducted in (103)____, in which multiple samples receive the same experimental or control treatment. A(n) (104) ____ is a statement of what a scientist would observe if the hypothesis being tested is wrong. (105)____ are useful to study because of characteristics such as rapid development or short life cycle. The manipulation of organisms to produce useful products is (106)____. A(n) (107)____ is supported by exhaustive experimentation and is not likely to be contradicted by future research.

Choice

For each of the following characteristics, choose the most appropriate term from the list below.

a. applied research b. basic research c. biotechnology

108. ____ Advances collective understanding of living systems

109. ____ The study of how illness spreads from animals to humans or through human population is an ____ example

110. ____ The production of human insulin by bacteria to treat diabetes is an example

For each of the following characteristics, choose the most appropriate term from the list below.

a. observational data b. experimental data

111. ____ Information that involves detailed descriptions of structures or processes

112. ____ Information derived from careful manipulation of the system under study

Matching

Match each of the following characteristics with its correct definition.

113. ____ control A. A working explanation of observed facts

114. ____ hypothesis B. A statement arising from an hypothesis that can be tested experimentally

115. ____ prediction C. Other explanations for observations

116. ____ null hypothesis D. The treatment group that is not manipulated in an experiment

117. ____ replicate E. The single factor that is varied in an experiment

118. ____ alternate hypothesis F. Multiple samples/subjects of a given treatment in an experiment

119. ____ experimental variable G. A statement of what would be observed if the tested hypothesis is wrong

Short Answer

120. Describe a model organism. ____

____

121. Describe a scientific theory. ____

____

SELF-TEST

Use the following information to answer questions 1–3. A biologist interested in growth notes that there are fish of different sizes in an aquarium and that smaller fish have lower levels of plasma insulin-like growth factor (IGF) than larger fish. She then designs an experiment in which fish of the same size are given once daily injections of saline or 100ng IGF/g body weight. Water temperature, light conditions, and type and amount of found were the same for all fish. [p. 14]

1. Based on the initial observations, what is the most likely hypothesis that is being tested with the experiment?

a. Small fish have reduced IGF compared to large fish.

b. IGF regulates fish growth.

c. The growth of fish is not controlled.

d. The growth of fish depends on the amount of food eaten.

2. The results of the above experiment yield

a. experimental data.

b. observational data.

c. scientific facts.

d. new biotechnology.

3. The experimental variable in the above experiment is ____.

a. IGF level

b. body weight

c. food intake

d. water temperature

4. Emergent properties of populations include ________. [p. 3]

a. birth rate/death rate

b. stability and diversity

c. ability to learn

d. being alive

5. The biomolecule that carries out most of the activities of life, including the synthesis of other biomolecule types, is [p. 3]

a. protein.

b. DNA.

c. RNA.

d. carbohydrate.

e. lipid.

6. Organisms that use energy from sunlight to synthesize complex molecules from CO_2 and H_2O are [p. 5]

a. primary producers.

b. consumers.

c. decomposers.

d. heterotrophs.

7. The maintenance of internal conditions by organisms in the face of changing external environment is [p. 6]

a. response.
b. adaptation.
c. homeostatis.
d. tolerance range.

8. Heritable variations that allow some individuals to compete more successfully in their environment and pass on favorable characteristics to their offspring, resulting in the favorable trait becoming more common in the population over time is [p. 8]

a. artificial selection.
b. mutation.
c. natural selection.
d. bottleneck.

9. The species name for modern humans: [p. 11]

a. *Homo.*
b. *sapiens.*
c. *Homo sapiens.*
d. Homidae.

10. This kingdom contains organisms that are multicellular consumers that are generally motile. [p. 13]

a. Fungi
b. Plantae
c. Protoctista
d. Animalia

11. Prokaryotic organisms [pp. 11]

a. are all bacteria.
b. lack nuclear envelops.
c. are all producers.
d. are comprised of two kingdoms.

INTEGRATING AND APPLYING KEY CONCEPTS

1. Explain how evolution contributes to the diversity of living organisms.
2. Explain how the process of reproduction underlies the continuity of life.

2

LIFE, CHEMISTRY, AND WATER

CHAPTER HIGHLIGHTS

- Matter is defined, including organizational levels of atoms, elements, molecules, and compounds.
- Chemical bonds are characterized using the atomic structure of elements.
- Water's properties are explained by understanding its unique atomic structure.
- Acids', bases', and buffers' structure and reactivity are explained.

STUDY STRATEGIES

First, focus on major levels of organization of matter—elements, atoms, molecules, and compounds.

Exam the atomic structure of elements and the underlying reasons for chemical bonds between atoms.

Next, make comparisons between the various types of chemical bonds and the stability or reactivity of the resulting molecules and compounds.

Evaluate the properties of water and associate these properties with water's chemical composition, hydrogen bonding, and reactivity characteristics.

TOPIC MAP

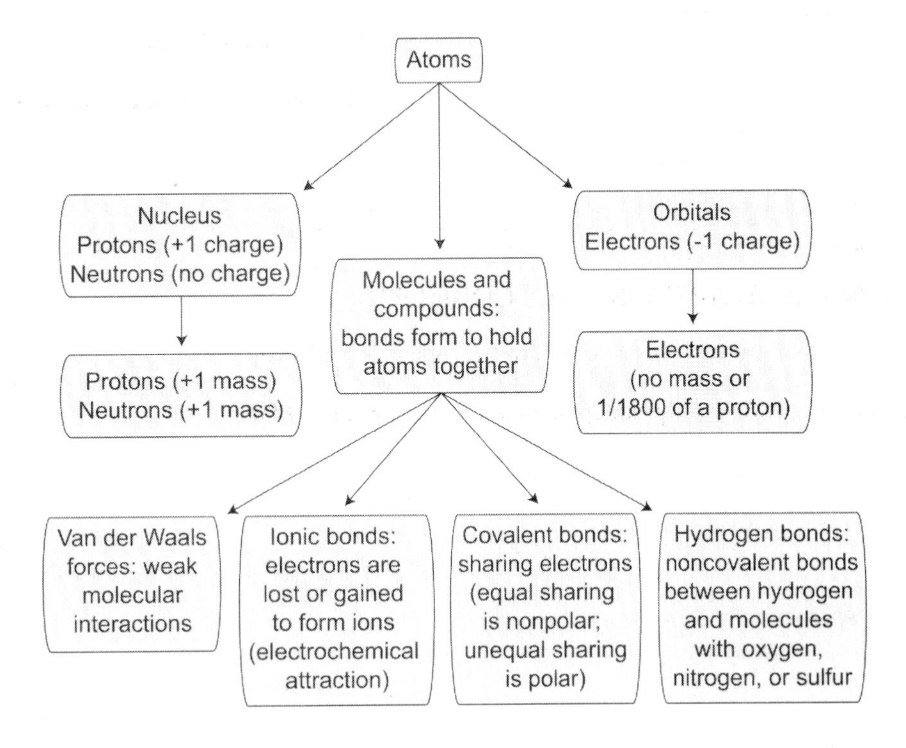

INTERACTIVE EXERCISES

Why It Matters [pp. 21-22]

Section Review (Fill-in-the-Blanks)

All life, (1) _____, (2) _____, and other organisms, are composed of various (3) _____ (the smallest unit of a given element) held together by (4) _____ _____. Understanding the properties and behavior of living organisms, or (5) _____, is based on the structure and reactivity of (6) _____ _____.

2.1 The Organization of Matter: Elements and Atoms [pp. 22-23]

Section Review (Fill-in-the-Blanks)

All (7) _____ are composed of (8) _____. Elements are (9) _____ substances, which (10) _____ be broken or divided into simpler substances. Living organisms are primarily composed of (11) _____, (12) _____, (13) _____, and (14) _____. The smallest unit of an element which retains the properties of that element is a(n) (15) _____. Each element is identified by a unique (16) _____. Characteristics of elements are listed in a periodic table. Some of the information listed in the periodic table, the (17) _____ _____ and (18) _____ _____ can provide a significant amount of information about the atoms of any given element. Two or more atoms combine in unique ratios to form (19) _____. Molecules composed of two or more different atoms are known as (20) _____.

2.2 Atomic Structure [pp. 23-28]

Matching

Match each of the following components of an atom with its correct definition.

21. _____ Proton
22. _____ Neutron
23. _____ Electron

A. Negatively charged particle
B. Positively charged particle
C. Uncharged particle

Match each of the following components of an atom with its location.

24. _____ Proton
25. _____ Neutron
26. _____ Electron

A. Nucleus
B. Energy levels or Shells

Choice

For each of the following statements, choose the most appropriate definition from the list below.

A. atomic number B. atomic mass C. 1s orbital D. 2s orbital E. 2p orbital F. isotope

27. _____ Total number of particles in the nucleus
28. _____ Number of protons in the nucleus of an atom
29. _____ Atomic number is the same, but the number of neutrons vary
30. _____ 1st energy level that only holds 2 electrons
31. _____ 2nd energy level of 3 dumbbell-shaped electron paths, each holding 2 electrons
32. _____ 2nd energy level that is spherical, holding 2 electrons

Section Review (Fill-in-the-Blanks)

The 1st energy shell has 1 orbital. The 1s orbital holds (33) _____ electrons. The 2nd energy shell has (34) _____ different types of orbitals, the (35) _____ and the (36) _____. The 2s orbital holds (37) _____ electrons. There are (38) _____ dumbbell-shaped 2p orbitals, each holding (39) _____ electrons. In summary, the 1st energy shell can hold a maximum of (40) _____ electrons, while the 2nd energy shell can hold a maximum of (41) _____ electrons.

Labeling

Identify each numbered part of the following illustration.

Figure A

42. ______________

43. ______________

44. ______________

45. ______________

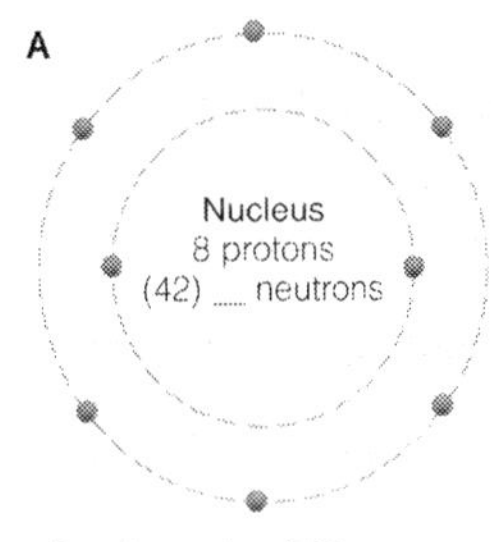

Figure B

46. ______________

47. ______________

48. ______________

49. ______________

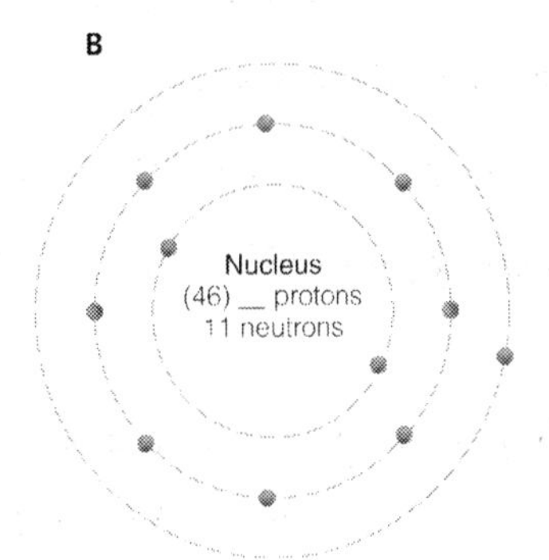

Figure C

50. ______________

51. ______________

52. ______________

53. ______________

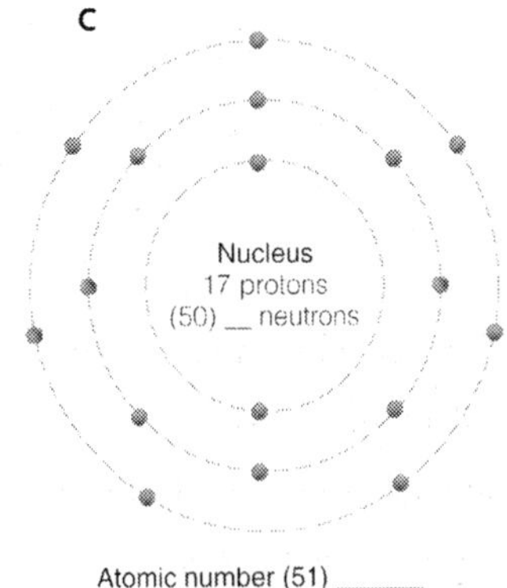

2.3 Chemical Bonds [pp. 28-32]

Section Review (Fill-in-the-Blanks)

Given the figures A, B, and C above, Fig A, (54) _____ has (55) _____ electrons in the outermost shell, Fig B, (56) _____ has (57) _____ electrons in the outermost shell, and Fig C, (58) _____, has (59) _____ electrons in the outermost shell. Which two of the above elements are most likely to form a bond? (60) _____ and (61) _____ will form a(n) (62) _____ bond. Since, there is only (63)_____ electron(s) in the outermost shell of (64) _____, it will be lost to form a(n) (65) _____ ion, and the outermost shell of (66) _____ will gain the electron to form a(n) (67) _____ ion. The opposite (68) _____ of the ions will attract to form a(n) (69) _____ _____. The resulting bond is formed by the (70) _____ attraction between the (71) _____ ion and the (72) _____ ion.

The other 3 major chemical linkages that occur between reactive elements are (73) _____ and (74) _____ bonds and (75) _____ _____ _____ forces. If the electrons are shared between the two elements, instead of a complete loss and gain of electrons, a (76) _____ bond forms. Sharing of (77) _______ between the two elements can be (78) _____ or (79) _____, depending on the (80) _______ between the bonded atoms. If the bond is a result of equal sharing, the bond is considered a(n) (81) _____ _____, while unequal sharing of electrons results in a(n) (82) _____ _____ bond. The covalent bonds in the water molecule are (83) _____. Molecules with similar bonds are (84) _____ to each other, while molecules with differing bonds are (85) _____. (86) _____ molecules which are attracted to water are (87) _____ and (88) _____ molecules which are repealed by water are (89) _____.

2.4 Hydrogen Bonds and the Properties of Water [pp. 32-36]

Short Answer

90. Explain the importance of hydrogen bonds in the water lattice. ____

____.

91. Why do lipid molecules with a polar end and a nonpolar end form a bilayer in water? ____________________

____

Labeling

92. Identify each lettered part of the following illustration.

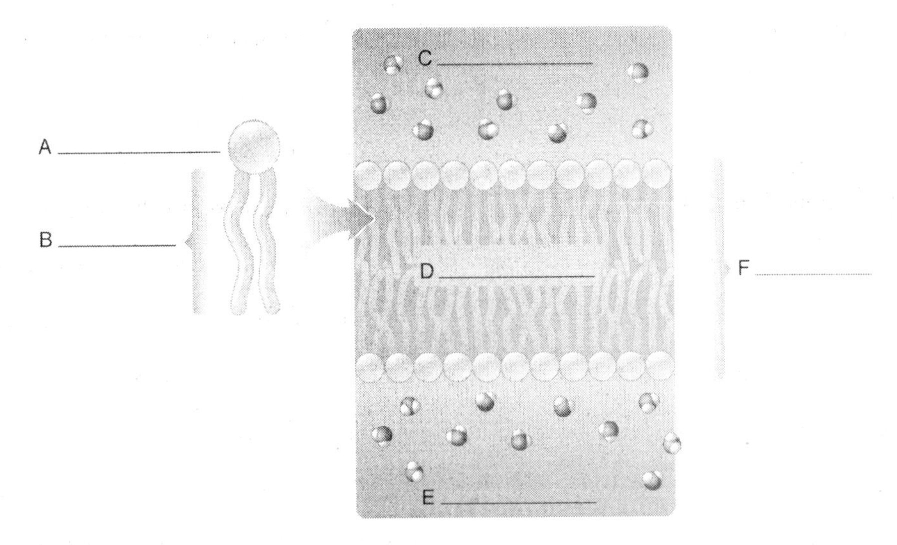

The polar end of this phospholipid is the __A__ group. The nonpolar ends of this phospholipid are the __B__ chains. When phospholipids are placed in a __C and E__ solution such as water, a(n) __F__ forms. The __C and E__ solution interacts with the __A__ group, also considered to be hydrophilic, while the inner portion of the membrane is composed of the nonpolar __B__ chains and excludes water or is __D__.

Matching

Match each of the following with its correct definition.

93.	____	Solute	A. layer of water that reduces molecule attraction
94.	____	Solution	B. liquid portion of a solution, typically water
95.	____	Solvent	C. suspension of solute and solvent
96.	____	Hydration layer	D. number of ions/molecules of solute per unit solvent
97.	____	Concentration	E. ions or molecules dissolved in water

True/False

If the statement is true, write a "T" in the blank. If the statement is false, make it correct by changing the underlined word(s) and writing the correct word(s) in the answer blank.

98. ____________ Water has a high boiling point, the hydrogen-bond lattice makes it <u>difficult</u> for individual water molecules to escape or break loose.

99. ____________ The <u>specific heat</u> of water is the amount of heat necessary to turn liquid water into a gas.

100. ____________ The hydrogen-bond lattice makes it difficult for water molecules to separate. This is known as <u>adhesion</u>.

101. ____________ A long column of water goes from the roots to the leaves in tubes found in a redwood tree. This can occur by <u>adhesion</u> of the water molecules to the surface of the tube by forming a hydrogen-bond lattice.

102. ____________ The <u>surface tension</u>, attraction of water molecules at the interface between water and air, allows water striders to "walk on water."

2.5 Water Ionization and Acids, Bases, and Buffers [pp. 36-39]

Section Review (Fill-in-the-Blanks)

When pure water dissociates, equal concentration of (103) _____ ions and (104) _____ ions are formed. If additional substances are added to the water which releases additional hydrogen ions (or protons), the solution would be considered a(n) (105) _____. When the substances added to water accepts protons, the solution is then a(n) (106) _____. A(n) (107) _____ prevents changes in pH by either (108) _____ or (109) _____ (110) _____ ions.

Choice

Given the pH scale 0-14, identify if the following substances are acids or bases

111. pH 3.2 _____ A. Acid

112. pH 11.8 _____ B. Base

113. pH 7.3 _____ C. Neutral

114. pH 6.8 _____

115. pH 7.0 _____

SELF-TEST

1. Which component of an atom determines the specific type of atom? [p. 24]
   a. isotope
   b. neutron
   c. electron
   d. proton

2. Radioactive isotopes are useful in dating and as tracers because unstable _______ decay at a uniform rate. [p. 25]
   a. neutrons
   b. protons
   c. electrons
   d. both protons and electrons

3. Electrons in the orbitals closest to the nucleus are more stable than electrons in the outer orbitals. This is because the overall charge of the nucleus is _______. [p. 25]
   a. neutral
   b. positive
   c. negative
   d. none of these

4. Hydrogen bonds can form between atoms of two molecules that have a ________ and ________ end. [pp. 28-30]
   a. positive and negative
   b. positive and neutral
   c. negative and neutral
   d. none of these

5. Water is a polar solution; molecules with ______ bonds would most likely dissolve in water. [pp. 28-30]

a. nonpolar covalent

b. ionic

c. polar covalent

d. both ionic and polar covalent

6. Ice floats because the hydrogen-bond lattice has spaces that are __________. [p. 32]

a. closer together

b. farther apart

c. the same distance as in the liquid state

d. a combination of all of these

7. If the molarity of a solution increases, the number of solute particles is _________. [p. 33]

a. increased

b. decreased

c. unchanged

d. not related to molarity

8. The property of water called ________ allows fish to survive in a pond that is covered with ice. [p. 35]

a. heat of vaporization

b. surface tension

c. high specific heat

d. adhesion

9. Enzymes, proteins which breakdown chemical bonds, are pH sensitive. Enzyme Z is most active in an acidic environment. Which pH would cause a decrease in enzyme Z activity? [p. 36]

a. 3.2

b. 4.8

c. 9.3

d. 5.8

10. With respect to question #9, if a solution with enzyme Z had a _______, the pH would not be expected to change greatly, if either acids or bases were added. [p. 37]

a. buffer

b. weak acid

c. weak base

d. strong base

INTEGRATING AND APPLYING KEY CONCEPTS

1. Silicon (Si), like carbon, has four electrons in the outermost orbital. Predict the possibility and results of biological molecules that are silicon based instead of carbon based.

2. Hydrogen bonds are weak bonds that form and break rapidly. Explain the effect of temperature and pH changes on the stability of hydrogen bonds and the resulting effect on biological molecules.

3

BIOLOGICAL MOLECULES: THE CARBON COMPOUNDS OF LIFE

CHAPTER HIGHLIGHTS

- Overview of major carbon-based molecules that are essential to living organisms.
- Identification and role of functional groups in biological molecules are explained.
- Common reactions to make or break bonds associated with biological molecules are addressed.
- Presentation of primary types of carbohydrates, lipids, proteins, and nucleic acids.
- Basic structures and functions of carbohydrates, lipids, proteins, and nucleic acids are presented.

STUDY STRATEGIES

First, focus on the structure and role of carbon as well as the functional groups in biological molecules.
Understand the structural building blocks of each major type of biological molecule.
Next, examine the structure as well as the functional groups of each major type of biological molecule.
Be able to relate the structural properties of each major type of molecule to its functional characteristics.

TOPIC MAP

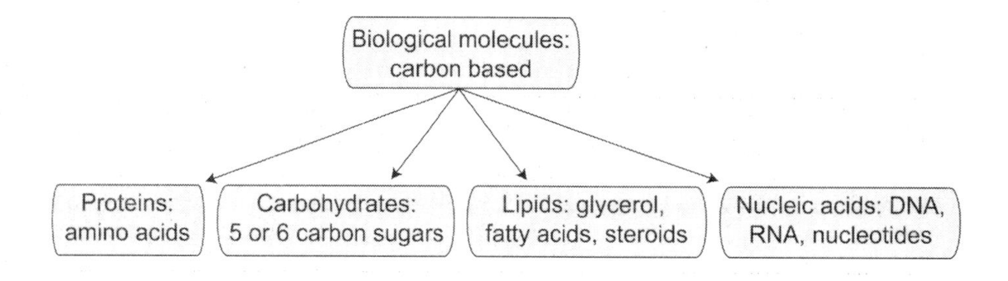

INTERACTIVE EXERCISES

Why It Matters [pp. 41-42]

Section Review (Fill-in-the-Blanks)

All living organisms are composed of (1) _____ compounds. A simple molecule, atmospheric (2) _____ _____, is used by plants to produce different (3) _____ compounds that can be used as an energy source for all other (4) _____. The process fixing CO_2 by plants, (5) _____ is essential to all living organisms.

3.1 Carbon Bonding [pp. 42-43]

3.2 Functional Groups in Biological Molecules [pp. 43-45]

Section Review (Fill-in-the-Blanks)

Molecules or compounds with carbon are known as (6) _______, while those without (7) ______ are known as (8) ______. Carbon has (9) ______ ______ in the outer orbital. When bonds are formed with other atoms, the electrons are equally shared, so these bonds are (10) ______ ______ bonds. Due to the structure of carbon, long chains of (11) ______ form, often with branches or as ring structures. (12) ______ is a common atom that binds to carbon, resulting in (13) ______. If you recall from chapter 2, the most common elements in living organisms are (14) ______, (15) ______, (16) ______, and (17) ______. These elements are major components of the biological molecules, (18) ______, (19), ______, (20) ______, and (21) ______ ______.

Identification

Identify the functional group or class of molecule.

Functional Group	Name	Class of Molecule
OH	22. ________	Alcohols, Water
H N H	23. ________	24. ________
O C // OH	Carboxyl	25. ________
O^- O═P—O^- O	26. ________	DNA or RNA, Phospholipids
C═O C	Carbonyl	27. ________

Section Review (Fill-in-the-Blanks)

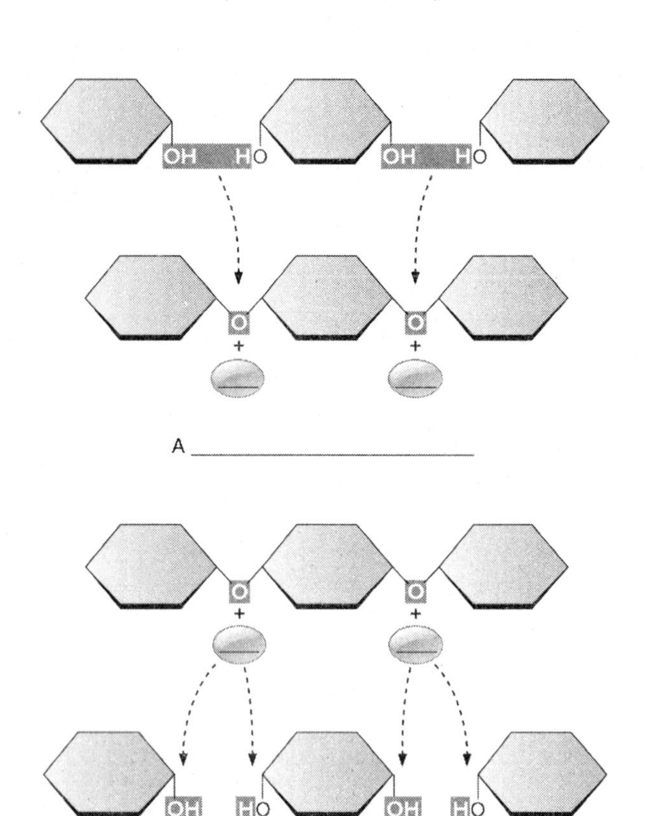

In figure A, (28) _____ _____ reaction, the (29) _____ group is lost from one molecule and the (30) _____ from another molecule and remaining (31) _____ is part of the bond between the two molecules. In summary, a bond is formed between two molecules, the (32) _____ portion of the reaction and (33) _____ is removed, the (34) _______ portion of the reaction. In figure B, (35) _____ reaction, a bond is broken between two molecules. (36) _____ is split and (37) _____ group is added to one molecule and (38) _____ is added to the other molecule. In summary, a bond is broken between two molecules by a (39) _____ reaction—"water splitting". These reactions are not random. In future chapters you will learn that specific enzymes are necessary for these reactions to occur.

3.3 Carbohydrates [pp. 45-50]

Building Vocabulary

Prefixes	Meaning
mono-	one
poly-	much, many

Prefix	Suffix	Definition
_____	-saccharum	40. Multiple sugar units
_____	-saccharum	41. A single sugar unit

Section Review (Fill-in-the-Blanks)

Carbohydrates have (42) _____ atoms that are hydrated. The ratio of carbon to (43) ____ and (44) _____ is typically (45) _____:_____:_____. (46) ______ is the process by which water is removed by (47) _____ _____ and monosaccharides are joined to form (48) _____.

Short Answer

49. Define *isomer* and explain the difference between optical and structural isomers. ______________________________

____

50. Explain the primary difference between glycogen and cellulose. Which of these polysaccharides is most easily digested by animals? Explain why. ____

____

3.4 Lipids [pp. 50-55]

Section Review (Fill-in-the-Blanks)

There are 3 major groups of lipids: (51) _____, (52) _____, and (53) _____. Neutral lipids are composed of one molecule of (54) _____ and (55) _____ (56) _____ _____chains forming (57)_____. Fatty acids contain a (58) _____ group. A(n) (59) _____ _____ reaction bonds the (60) _____ end of the fatty acid to the (61) _____ end of glycerol. Fatty acids can be (62) _____ with (63) _____ or (64) _____, which contain (65) _____ bonds. If one double bond is present, the fatty acid is (66) _____, and (67) _____ if two or more double bonds are present.

Short Answer

68. Explain why some fats are liquids, while others are solids. ______________________________

____

69. Explain the primary difference between a triglycerides and phospholipids ______________________________

____

Identification

70. In the diagram (from chapter 2), identify the layer or nature of the layer.

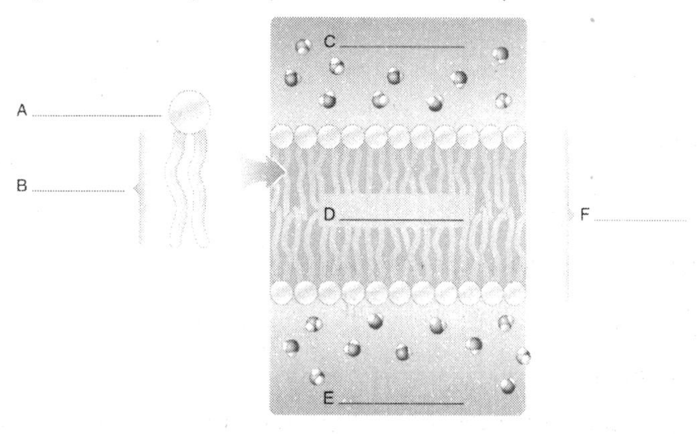

A. Identify the functional group.

B. Identify the two chains; assume one has one double bond and the other has two double bonds.

C & E. The environment is _____ or _____.

D. The environment is _____ or _____.

F. When phospholipids are put into a polar solution, this is likely to form.

Short Answer

71. Given the diagram (#70), explain why a membrane is likely to form with phospholipids in a polar solution.____

____

72. Predict what would happen if phospholipids were placed in a nonpolar solution. ________________________

____

73. Explain why cholesterol is necessary to living organisms.____

____

3.5 Proteins [pp. 55-64]

Section Review (Fill-in-the-Blanks)

The building blocks or structural units of (74) _____ are (75) _____ _____. The basic molecular plan of an amino acid is a carbon with a(n) (76) _____ group, a(n) (77) _____, a(n) (78) _____ and a hydrogen. The (79) _____ end has the (80) _____ group, while the (81) _____ end has the (82) _____ group. The bond between two amino acids is called a(n) (83) _____ bond, formed by a(n) (84) _____ _____ reaction. A chain of more than three amino acids is called a(n) (85) _____.

Choice

For each of the following statements, choose the most appropriate protein structural level from the list below.

A. Primary B. Secondary C. Tertiary D. Quaternary

86. ____ Typically either an alpha helix or a beta sheet
87. ____ Two or more polypeptide chains
88. ____ Folding of the primary sequence into a three-dimensional arrangement
89. ____ Sequence of amino acids, like an alphabet

Short Answer

90. Which weak bond is very important in the secondary and tertiary structure of the protein? ______________

____

91. Explain denaturation and several likely causes. ____

____

Matching

Match each of the following type with its appropriate function.

92. _____ Enzymatic — A. Provide regulatory signals between cells; insulin is an example.

93. _____ Structural — B. Initiates cellular responses by binding molecules to cell surface or the internal environment.

94. _____ Receptor — C. Provides support to cells.

95. _____ Hormones — D. Involved in cellular movements; actin and myosin are examples.

96. _____ Motile — E. Increases movement across the cell membrane, typically ions

97. _____ Transport — F. Increases the rate of biological reactions

3.6 Nucleotides and Nucleic Acids [pp. 64-68]

Section Review (Fill-in-the-Blanks)

The two major types of (98) _____ _____ are (99) _____ _____ or (100)_____ and (101) _____ _____ or (102) _____. The basic unit or monomer is the (103) _____. A(n) (104) _____ has three components, a(n) (105) _____ _____, a(n) (106) _____ _____ sugar, and a(n) (107) _____ group. Nitrogenous bases are of (108) _____ major types, (109) _____ and (110) _____. The backbone of nucleic acids are repeating units of (111) _____ and (112) _____, which are connected by a(n) (113) _____ bond. DNA is composed of (114) _____ strands of nucleotides, while RNA is (115) _____ stranded. The chains of DNA are (116) _____ to each other.

Choice

For each of the following statements, choose the most appropriate nitrogenous base

A. Pyrimidine B. Purine

117. _____ Two carbon rings

118. _____ One carbon ring

119. _____ Three different types

120. _____ Two different types

A. Uracil B. Adenine C. Thymine D. Guanine E. Cytosine F. Two G. Three

121. _____ Pyrimidine base found only in DNA

122. _____ Purine base that binds with Guanine

123. _____ The purine nitrogenous bases

124. _____ The pyrimidine nitrogenous bases

125. _____ Binds with Adenine in DNA

126. _____ Binds with Adenine in RNA

127. _____ Number of hydrogen bonds between Adenine and its complementary base in DNA

128. _____ Number of hydrogen bonds between Cytosine and its complementary base

Complete the Table

129. Complete the table; nitrogenous bases should be aligned in complementary pairs.

Nucleic Acid	Strands	Sugar	Purines	Pyrimidines
DNA	A. ______	B. ______	C. ______ E. ______	D. ______ F. ______
RNA	G. ______	H. ______	I. ______ K. ______	J. ______ L. ______

Summary—Completion

Identify the biological molecule for each of the building block units.

Structural Building Block	*Biological Molecule*	*Structural Building Block*	*Biological Molecule*
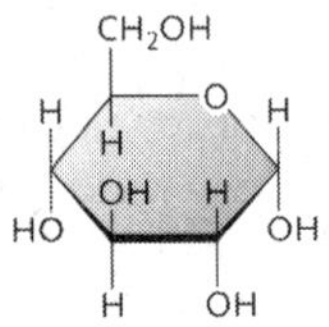	130. __________	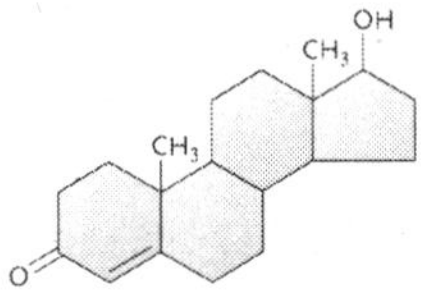	133. __________
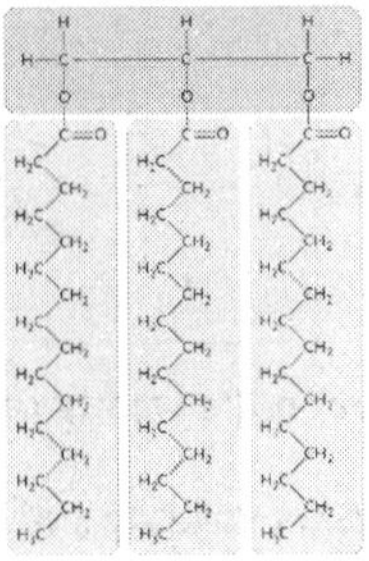	131. __________	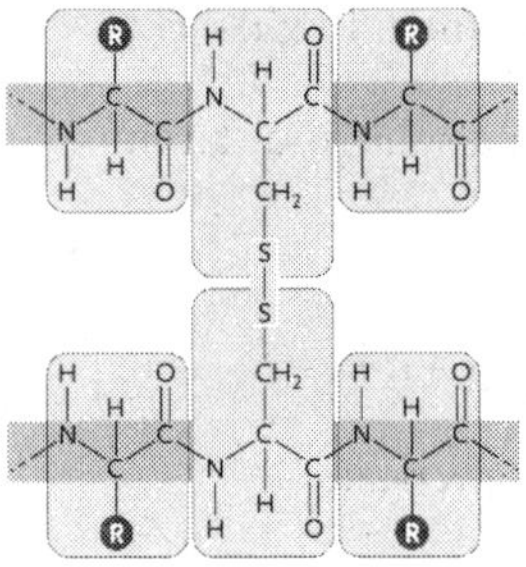	134. __________
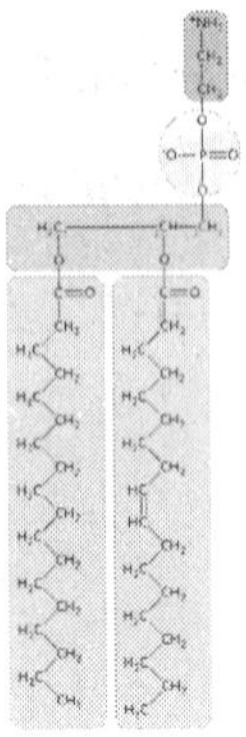	132. __________	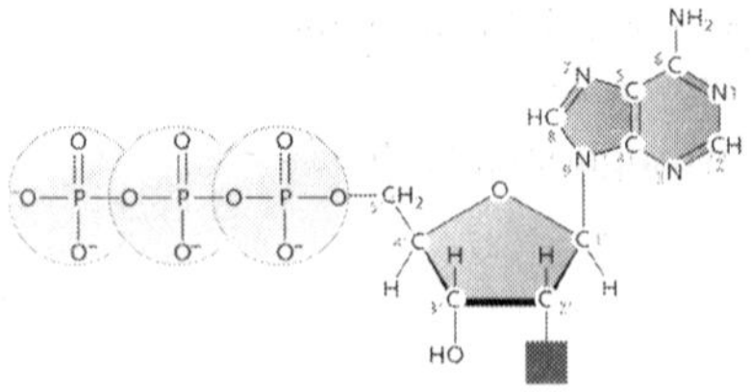	135. __________

SELF-TEST

1. In a dehydration synthesis reaction, water is _______. [p. 43]

a. produced

b. used

c. both used and produced

d. not involved in the reaction

2. Which functional group plays an important role in the hydrophilic portion of the cell membrane? [pp. 40-41]

a. Phosphate

b. Amino

c. Carboxyl

d. Hydroxyl

3. Horses and cows are large animals; they consume a tremendous amount of grass and hay—what humans would consider diet food. Predict what is different about these animals that is different from humans. [pp. 48-50]

a. GI tract doesn't contain any microbes

b. GI tract microbes consume glycogen

c. They never sleep, so they eat more

d. GI tract microbe enzymes breakdown cellulose

4. Which lipid group is hydrophobic? [p. 51-54]

a. phospholipids

b. triglycerides

c. steroids

d. both triglycerides and steroids

5. For a fat to be polyunsaturated, the number of double bonds is ________. [p. 50]

a. 0

b. 1

c. 2 or more

d. 0, a phosphate group is necessary

6. To which biological molecule group does chlorophyll belong? [p. 55]

a. Proteins

b. Carbohydrates

c. Lipids

d. Nucleic Acids

7. Which functional group(s) play a key role in the formation of a peptide bond? [p. 57]

a. hydroxyl

b. carboxyl

c. sulfhydryl

d. phosphate

8. If the primary structure were in the wrong order, predict the effect on the tertiary structure of a protein. [pp. 57-62]

a. nothing, no effect on structure or function

b. altered function due to abnormal folding

c. secondary structure would be altered

d. either b or c could occur

9. A highly specialized region of a protein is called a ________. [pp. 62-63]

a. zipper

b. motif

c. chaperone

d. domain

10. Given the building blocks (structural units), which biological molecule has the potential for the most variation? [pp. 65-68]

a. Proteins

b. Carbohydrates

c. Lipids

d. Nucleic Acids

11. If DNA and RNA were both exposed to high temperatures, which molecule would be most effected with respect to complementary base paring? [pp. 64-69]

a. DNA

b. RNA

12. With respect to question #11, which bond type is most likely to break? [pp. 64-69]

a. phosphodiester

b. peptide

c. polyunsaturated

d. hydrogen

INTEGRATING AND APPLYING KEY CONCEPTS

1. Proteins are the primary biological molecule that is involved tissue rejection of transplants. Predict a way to modify the protein to prevent rejection without affecting function.

2. If you wanted to determine if dehydration synthesis or hydrolysis occurred, what would you measure and why? (Assume you are able to measure all components of the reactions.)

4

ENERGY, ENZYMES, AND BIOLOGICAL REACTIONS

CHAPTER HIGHLIGHTS

- Energy exists in two interchangeable forms: Kinetic and potential energy:
    - Living organisms convert potential energy into kinetic energy during metabolic reactions.
- Laws of thermodynamics explain energy flow.
- Metabolic reactions can be divided into:
    - Catabolic reactions that breakdown chemicals to release energy (exergonic reactions).
    - Anabolic reactions that use energy (endergonic reactions) to create larger chemicals.
- Energy released during exergonic reactions is stored in ATP molecules.
- Living things use ATP as an immediate source of energy to support cellular functions.
- Enzymes:
    - Are biological catalysts that speed chemical reactions.
    - Reduce activation energy required to start a reaction.
    - Catalyze a specific reaction, have a specific substrate and specific product(s).
    - Have names that end in -ase.
    - Are mostly proteins (RNA catalysts are called ribozymes).
    - Are not changed during a reaction; therefore used repeatedly and made in very small numbers.
    - Need cofactors and coenzymes to catalyze (many of them, but not all).
    - Are affected by temperature and pH that change their 3D structure
    - Are often regulated by inhibitors and feed back mechanisms

STUDY STRATEGIES

Remember the goal of this chapter is to understand how energy and enzymes are involved in chemical reactions.
Concentrate on the key concepts of thermodynamics—energy release and conversions.
Focus on biological reactions that make or break chemicals while absorbing or releasing energy.
Learn about ATP—how it is made and supports cellular reactions.
Understand enzymes—their chemical nature, how they catalyze chemical reactions, and their regulation.

INTERACTIVE EXERCISES

Why It Matters [pp. 75–76]

Section Review (Fill-in-the-Blanks)

Chemical reactions that support living things are referred to as (1) ______________. These reactions are catalyzed by (2) ______________. Some of these reactions require (3) ______________ or release to support cellular functions.

4.1 Energy, Life, and the Laws of Thermodynamics [pp. 76–79]

Section Review (Fill-in-the-Blanks)

(4) ______________ is the capacity to do work. Heat, chemical, electrical, mechanical, and radiation are different forms of (5) ______________. Visible light, infrared, ultraviolet, gamma rays, and X rays are all different types of (6) ______________ ______________. Energy exists in two interchangeable forms. (7) ______________ energy is the energy of motion such as in the moving objects, atoms, electrons, electricity, sound, thermal, and radiation waves. (8) ______________ energy is the stored energy such as chemical energy, nuclear energy, and energy in objects due to their position.

Living things break down sugar to release its (9) ______________ energy to convert it into (10) ______________ energy, which can then be used for cellular functions. Chemical reactions that break down larger chemicals are referred to as (11) ______________ reactions. These reactions often release energy and therefore are also called (12) ______________ reactions. On the other hand, chemical reactions that build larger chemicals are called (13) ______________ reactions. These reaction often require energy and are also called (14) ______________ reactions.

(15) ______________ is the study of energy flow that allows for the understanding of biological reactions. In a reaction and flow of energy, the molecules involved are called the (16) ______________ and everything else is referred to the (17) ______________. Living things can have closed (18) ______________ where molecules do not exchange energy with their surroundings although most reactions are (19) ______________ ______________ where molecules exchange freely exchange energy with the surroundings.

According to the first law of thermodynamics, energy is interchangeable, but it can never be (20) ______________ or (21) ______________. The total amount of energy in the universe remains (22) ______________. The (23) ______________ is the source of energy for all living things. Plants, through the process of (24) ______________, absorb kinetic light energy from this source and convert it to chemical energy that is stored in carbohydrates and fats as (25) ______________. All living things (including plants themselves) metabolize these chemicals to release stored (26) ______________ energy partly as (27) ______________ energy to support their functions. The remaining energy is released back into the environment as (28) ______________ energy.

According to the second law of thermodynamics, as changes take place in a system, it leads to increased disorder or (29) ______________.

A chemical reaction that takes place without outside help is called a(n) (30) ______________ reaction. (31) ________ energy is the energy available to do work, such as cellular functions, growth and development. The change in this energy (ΔG) can be positive or negative. When the change is (32) ____________, the reaction requires the input of energy and is called endergonic. If the change is (33) ____________, the reaction releases energy. The reaction will occur spontaneously and is called (34) ____________. Example: Hydrolysis of sucrose to form two monosaccharides has (35) ____________ ΔG, which means that this split is spontaneous. However, combining of two monosaccharides to form sucrose has (36) ______________ ΔG, which means that this synthesis would require input of energy.

Matching

Match each of the following terms with its correct definition.

37.	____	Free energy	A. Reactions that breakdown larger chemicals into smaller chemicals
38.	____	Metabolism	B. Refers to stored energy
39.	____	Endergonic	C. Refers to the energy available to do work
40.	____	Energy	D. It is the ability to do work
41.	____	Catabolism	E. Refers to the energy due to motion of the object
42.	____	Kinetic energy	F. Reactions that build larger chemicals
43.	____	ΔG	G. Change in free energy
44.	____	Anabolism	H. Refers to reactions that release energy
45.	____	Exergonic	I. Reactions that take place without outside help
46.	____	Potential energy	J. Refers to all chemical reactions that take place in living things
47.	____	Spontaneous reactions	K. Refers to reactions that require energy

Complete the Table

Complete the following table by giving a specific example to explain each of the opposite terms.

48. Catabolism:	49. Anabolism:
50. Kinetic energy:	51. Potential energy:
52. Endergonic reactions:	53. Exergonic reaction:
54. Positive ΔG:	55. Negative ΔG:

4.2 How Living Organisms Couple Reactions to Make Synthesis Spontaneous [pp. 79–81]

4.3 Thermodynamics and Reversible Reactions [pp. 81–82]

Section Review (Fill-in-the-Blanks)

(56) ____________ reactions that take place in cells, build larger chemicals and often require energy, (57) ____________ reactions. They have a(n) (58) ____________ ΔG and do not occur spontaneously. Therefore, they often connect with (59) ____________ reactions that release energy and have a high negative ΔG. The net affect of these endergonic and exergonic reactions is a(n) (60) ____________ ΔG. These combined reactions are called (61) ____________ reactions.

ATP ________ acts as the (62) ____________ agent to support the anabolic reactions. ATP consists of five-carbon sugar (63) ____________, a nitrogenous base (64) ____________, and three (65) ____________ groups. When ATP splits, large amounts of (66) ____________ energy is released; ADP ____________ and inorganic (67) ____________ are formed.

In a cell, ATP splits to transfer phosphate group to the substrate of an endergonic reaction, increasing the (68) ____________ energy of the molecule; the molecule is now "energized." This step that is referred to as (69) ____________. These coupled reactions are now spontaneous.

Making of ATP from ADP is a(n) (70) ____________ reaction and does not occur spontaneously. This reaction is coupled with breakdown of glucose by the cell, which is a(n) (71) ____________ reaction. As a result, (72) ____________ energy in glucose is transferred to ADP in order to make (73) ________. Energy in ATP is then transferred to chemicals by attaching (74) ____________ group.

Many reactions reach a balance point when both the reactants and the products are present in the solution and have reached the maximum entropy. The reaction has reached a(n) (75) ____________ point. The reaction can go forward or backward at this point and is now called a(n) (76) ____________ reaction.

Some metabolic reactions in living cells do not reach (77) ____________ point because they are part of a chain reactions where the (78) ____________ get used by the next reaction.

Matching

Match each of the following terms with its correct definition.

79. ____ Phosphorylation
80. ____ ATP
81. ____ Coupled reactions
82. ____ Equilibrium point
83. ____ Reversible reaction

A. Primary coupling agent in all living things
B. Where an exergonic reaction is connected to an endergonic reaction
C. Addition of phosphate group to a molecule
D. Reaction can go forward or backward
E. A balance of reactants and products in a solution

Short Answer

84. What are coupled reactions? ____

____

____

85. Give a short description of ATP structure and its role in coupled reactions. ______________

____

____

4.4 The Role of Enzymes in Biological Reactions [pp. 82–86]

4.5 Conditions and Factors Affecting Enzyme Activity [pp. 86–89]

4.6 RNA-based Biological Catalysts: Ribozymes [pp. 89-90]

Section Review (Fill-in-the-Blanks)

Most metabolic reactions are too slow. Enzymes (86) ____________ the rate of a chemical reaction by decreasing the (87) ___________ energy of the reaction. They do not undergo changes during the reaction and therefore are referred to as (88) ____________. They are mostly (89) ____________ although some RNA molecules can also catalyze a chemical reaction. A substrate/reactant binds to the (90) __________ site of the enzyme. Living things produces many different types of enzymes—one for each type of chemical reaction—known as the enzymatic (91) _____________. Their names end in (92) -______.

Enzymes are found inside the cell, as part of the cell membranes, and in body fluids. Some enzymes require inorganic factors called (93) ______________. Others require organic chemicals called (94) ____________, which are often derived from vitamins.

The enzyme binds with the reactant or the (95) ______________ to form a temporary, unstable state called the (96) ____________ state. There are three mechanisms to describe the role of an enzyme in most chemical reactions:

A. Enzymes use their (97) __________ site to increase the chances of bringing the (98) _____________ closer together.

B. Enzymes use their (99) __________ site to orient the (100) _____________ correctly to increase the chance of forming the (101) ____________ state.

C. Enzymes provide the (102) ___________ the appropriate ionic environment for forming the (103) ____________ state.

Enzymes are regulated by factors such as (104) ______ and (105) ______________. Typically, each enzyme has a range for each factor.

Temperature affects an enzymatic reaction by increasing the kinetic motion of the reactants, thereby increasing the (106) ________________ frequency of the molecules. It also affects the (107) ________________ structure of the enzyme, making it more active or even (108) ____________ it when extreme levels are reached.

The pH also affects the reaction by affecting the (109) ________________ structure of the enzyme.

Substrate level affects the rate of enzyme reaction by increasing the frequency of (110) ____________ of the substrate with the enzyme. Once the level of the substrate reaches (111) ________________, the rate is not affected.

Chemicals that stop enzymatic reactions are called (112) ________________. Those chemicals that somewhat resemble the substrate molecule and occupy the (113) ___________ site are referred to as (114) ________________ ________________. Those chemicals that bind to a location other than the (115) ___________ site, thereby affecting the (116) ________________ structure of the enzyme and its (117) __________ site, are called (118) ________________ ______________. Certain poisons, toxins and antibiotics act as enzymatic (118 inhibitors) ________________ for important chemical reactions. Chemical reactions are often regulated by the changing concentration of (119) ________________.

Some enzymes are regulated by inhibitors and activators that bind to the a site other than the (120) ___________ site, a mechanism called the (121) ________________ mechanism. The inhibitor is often the end-product of a chain reaction and therefore this mechanism is also called the (122) ________________ ________________ mechanism.

Some enzymes are regulated by attaching a phosphate functional group to an inactive enzyme—(123) ________________; and detaching the phosphate—(124) ________________. Enzymes involved are (125) ___________ _____________ and (126) ___________ ________________, respectively.

Ribozymes are (127) __________ molecules that catalyze chemical reactions. Ribosomes without the (128) ______________ can catalyze the assembly of amino acids to form (129)__________.

Matching

Match each of the following scientists with their contribution to cellular respiration.

130. ____ Thomas Cech and Sidney Altman — A. Determined the sequence of reaction in glycolysis

131. ____ ____ Jacques Monod, J.P.Changeux, and J. Wyman — B. Established that RNA can catalyze chemical reactions as ribozymes

Match each of the following chemicals and enzymes with their correct definition.

132. ____ protein kinase — A. RNA molecules that catalyze chemical reactions.

133. ____ competitive inhibitor — B. Substances that speed up chemical reactions without changing themselves.

134. ____ allosteric regulation — C. Organic chemicals that help enzymes.

135. _____ feedback inhibition

136. _____ protein phosphatase

137. _____ noncompetitive inhibitor

138. _____ ribozyme

139. _____ cofactors

140. _____ catalyst

141. _____ activation energy

142. _____ active site

143. _____ coenzyme

144. _____ enzyme specificity

145. _____ transition state

D. Chemicals that resemble the substrate and block the active site.

E. When the end-product acts as the inhibitor of a chain reaction.

F. A temporary state of enzyme binding with the substrate.

G. Energy required to start a reaction.

H. The location on an enzyme where the substrate binds.

I. Where an enzyme is regulated by an activator or an inhibitor binding to the enzyme at a site away from the active site.

J. An enzyme that detaches the phosphate group to an enzyme to change its activity.

K. Refers to enzymes catalyzing only a specific chemical reaction.

L. Chemicals that bind to a location on the enzyme other than the active site in order to inhibit the chemical reaction.

M. Inorganic chemicals that help chemical reactions.

N. An enzyme that attaches a phosphate group to an enzyme to change its activity.

Complete the Table

146. There are three mechanisms that contribute to the formation of the transition state. Complete the following table with specific information about these mechanisms.

A.
B.
C.

147. Chemical pathways are regulated by different types of inhibitors and mechanisms. Complete the following table explain each mechanism.

Inhibitor	*Explanation*
A. Competitive Inhibitor	
B. Noncompetitive Inhibitor	
C. Allosteric Regulation	
D. Feedback Inhibition	

True/False

Mark if the statement is true or false. If the statement is false, justify your answer in the line below each statement.

148. ______________ All enzymes are active only at 37°C.

____

149. ______________ Enzymatic inhibitors bind at the active site.

____

150. ______________ Ribozymes are RNA molecules that catalyze chemical reactions.

____

SELF-TEST

1. Enzymes are ____________ molecules. [p. 80]
   a. carbohydrate
   b. protein
   c. lipid
   d. RNA

2. All of the following are forms of radiant energy except [p. 72]
   a. ultraviolet.
   b. X rays.
   c. heat.
   d. visible light.

3. In which of the following form is the energy stored in glucose? [p. 72]
   a. potential energy
   b. radiation energy
   c. kinetic energy
   d. electrical energy

4. Which of the following describes disorder in a system? [p. 74]
   a. endergonic
   b. anabolism
   c. free energy
   d. entropy

5. All of the following would be true for a chemical reaction has a negative ΔG except: [p. 75]

a. The reaction will be spontaneous.
b. The reaction releases free energy.
c. The reaction will be endergonic.

6. A place on the enzyme where a substrate binds is called the [p. 80]

a. regulatory site.
b. allosteric site.
c. active site.
d. substrate site.

7. All of the following are true for an enzyme except: [p. 79]

a. They catalyze a specific reaction.
b. They do undergo changes during a reaction.
c. They temporarily combine with the substrate.
d. They increase activation energy for the reaction.

8. Which of the following has a shape similar to the substrate for an enzyme? [p. 84]

a. competitive inhibitor
b. allosteric inhibitor
c. noncompetitive inhibitor
d. coenzyme

9. Which of the following is an inorganic enzyme helper? [p. 80]

a. coenzyme
b. inhibitor
c. cofactor
d. allosteric activator

10. Which of the following is an organic enzyme helper? [p. 80]

a. coenzyme
b. inhibitor
c. cofactor
d. allosteric activator

11. Ribozymes are _______________ molecules. [p. 86]

a. carbohydrate
b. protein
c. lipid
d. RNA

12. Which of the following correctly describes a catalyst? [p. 79]

a. It increases the activation energy of a reaction.
b. It undergoes a chemical change during a reaction.
c. It speeds up a chemical reaction.
d. It is always an inorganic molecule.

13. All of the following are factors that affect enzyme activity except [p. 82]

a. temperature.
b. pH.
c. substrate concentration.
d. ATP.

14. All of the following are parts of ATP except [p. 76]

a. ribose.

b. 3 phosphates.

c. adenine.

d. thymine.

15. All of the following are true for a coupled reactions except: [p. 75]

a. They have combined exergonic and endergonic reactions.

b. They have an overall positive ΔG.

c. They are spontaneous.

d. They release free energy.

INTEGRATING AND APPLYING KEY CONCEPTS

1. Penicillin is an antibiotic. Discuss how this drug acts and relates to the topics in this chapter.
2. Explain why ATP is called the primary coupling agent.
3. Discuss the evolutionary significance of Ribozymes.

5

THE CELL: AN OVERVIEW

CHAPTER HIGHLIGHTS

- Cells come in many shapes and sizes.
- The cells of prokaryotes do not have a nucleus or other membrane-bound organelles, whereas the cells of eukaryotes have a nucleus and other membrane-bound organelles.
- The cells of plants and animals possess many of the same structures/organelles, but they also possess structures that are unique. For example:
- Plant cells possess plastids, such as chloroplasts, but not centrioles, and produce a cell wall that supports and protects it.
- Animal cells possess centrioles but not plastids, and produce an extracellular matrix the organizes the cell exterior.

STUDY STRATEGIES

- The study of cell structure and function is important for understanding other aspects of biology. Therefore, take time now to learn the basic features of cells.
- This chapter contains a lot of material. DO NOT try to go through it in one sitting. Take one section at a time, then work through the companion section(s) in the study guide.
- Draw pictures of cells and their internal structures. Be sure to label the parts and briefly describe their function.

TOPIC MAP

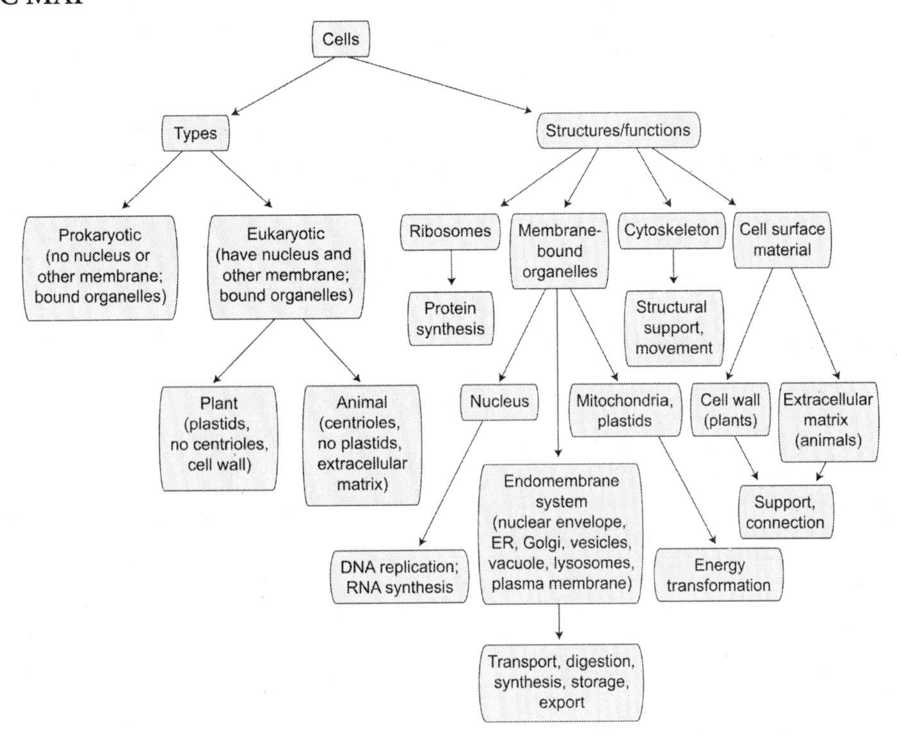

INTERACTIVE EXERCISES

Why It Matters [pp. 91-92]

Section Review

The (1) ____ holds that a) all organisms are composed of cells, b) the cell is the smallest unit that has all of the properties of life, and c) cells only arise from preexisting cells.

5.1 Basic Features of Cell Structure and Function [pp. 92-96]

5.2 Prokaryotic Cells [pp. 96-97]

5.3 Eukaryotic Cells [pp. 97-110]

Section Review

The (2)____ is the part of the cell surrounding the central region, which consists of an aqueous solution, the (3)____, and small organized structures called (4)____. The (5)____ is the outer limit of the cytoplasm. (6)____, making up the Bacteria and Archaea, have a central region called a(n) (7)____ that is not bound by a membrane. The central region of (8)____ contains a membrane-bound (9)____.

The plasma membrane of most prokayotes is surrounded by a rigid (10)____, which in some species is coated with a polysaccharide layer called a(n) (11)____. A(n) (12)____ is a single circular DNA molecule. (13) ____ are small particulate organelles in the cytoplasm specialized for protein synthesis. Many prokaryotes swim by means of a(n) (14)____ that rotates in a socket in the plasma membrane.

A supportive, extracellular (15)____ surrounds the plasma membrane of fungal, plant, and many protistian cells. The nucleus is separated from the cytoplasm by a double membrane (16)____. (17)____ are openings in both membranes of the envelope. The (18)____ is the liquid substance within the nucleus, dissolved in which is (19)____, a combination of DNA and protein. An individual DNA molecule with its associated protein is a(n) (20)____. The nucleus also contains a irregular mass called a(n) (21)____, which forms ribosomal RNA. The (22)____ is a network of membranes sacs and organelles, including the (23)____, (24)____, (25)____, (26)____, and the plasma membrane. The endoplasmic reticulum (ER) consists of an extensive network of (27)____ that surrounds a space called the (28) ____. (29)____ has ribosomes attached to its surface, whereas (30)____ does not have ribosomes attached to it. Proteins secreted from the cell are transported to the plasma membrane in (31)____, which fuse to the membrane and release their contents to the outside by the process of (32)____. Material can be taken into cells by (33)____, which results in the formation of an endocytotic vesicle. The endocytosis of large particles and whole cells is called (34)____. Membrane-bound organelles specialized for respiration are called (35)____, which are enclosed by two membranes, a smooth (36)____ and a(n) (37) ____ that is expanded into numerous folds called (38)____. The inner-most fluid-filled compartment is called the (39)____. (40)____ are small membrane-bound organelles that carry out specialized functions, and include (41)____, which break down hydrogen peroxide. The (42)____ is a network of protein fibers and tubes and are classified as (43)____, (44)____, and (45)____. In animal cells, microtubules radiate out from a site near the nucleus called the (46)____, within which are two barrel-shaped (47)____. (48)____ and (49)____ are two motile structures that connect inside the cells at the (50)____.

Building Vocabulary

Prefixes	Meaning
cyto-	of or relating to a cell
eu-	good, well, "true"
lyso-	loosening, decomposition
micro-	small
pro-	"before"

Suffix	Meaning
-karyo(te)	of or relating to the nucleus
-some	"body"

Prefix	Suffix	Definition
_____	-plasm	51. Contents of cell exclusive of the nucleus
_____	-skeleton	52. Network of protein filaments and tubes within the cell
glyoxy-	____	53. A microbody specialized to convert fats stored in plant seeds to sugars
peroxi-	____	54. A microbody specialized to breakdown hydrogen peroxide
_____	____	55. An organelle containing digestive enzymes
_____	-filaments	56. Solid filaments, 5–7nm in diameter, that make up part of the cytoskeleton of eurkaryotes
_____	-tubules	57. Hollow filaments or tubes, 25nm in diameter, that are part of the cytoskeleton of eukaryotes
_____	-sol	58. Contents of cell exclusive of the nucleus and other organelles
_____	-karyotes	59. Organisms that have cells without nuclei
_____	____	60. Organisms that possess a distinct nucleus surrounded by a nuclear envelope

Choice

For each of the following statements, choose the most appropriate term from the list below.

a. exocytosis b. endocytosis c. phagocytosis

61. ____ The process if releasing material to the outside of the cell
62. ____ The process of bring material into the cell by pinching of small portions of plasma membrane
63. ____ The uptake of large molecules or whole cells into the cell

For each of the following statements, choose the most appropriate structure of the cytoskeleton from the list below.

a. microfilaments b. microtubules c. intermediate filaments

64. ____ Involved in the process of cytoplasmic streaming
65. ____ Major element in animal nails, claws, and hair
66. ____ Involved in moving chromosomes during cell division

For each of the following statements, choose the most appropriate structure of the cytoskeleton from the list below.

a. eukaryotic chromosomes b. bacterial chromosome

67. ____ A single circular DNA molecule (no loose ends)
68. ____ Highly folded single DNA molecule with its associated protein

Matching

Match each of the following structure with its correct definition.

69. ____	organelles	A. Small organized structures that carry out specialized functions inside cell
70. ____	basal body	B. Lipid bilayer that serves as boundary of cell
71. ____	centrosome	C. Membrane-bound central region of eukaryotes
72. ____	ER	D. Nonmembrane-bound organelles that assemble amino acids into proteins
73. ____	ribosome	E. A complex of DNA and protein
74. ____	nucleus	F. The region where rRNA is formed
75. ____	plasma membrane	G. A collection of membrane-bound organelles that includes the nucleus, ER, golgi, and plasma membrane
76. ____	chromatin	H. Location near nucleus from which microtubules radiate outward through cells
77. ____	flagella	I. Network of membranous channels
78. ____	cilia	J. Elongated motile structure usually found signally or in small number on cells
79. ____	nucleolus	K. Motile structure that can occur in large number on cells
80. ____	endomembrane system	L. Innermost end of flagellum or cilium

Distinguish between/among members of the following sets of terms

81. secretory vesicle and endocytotic vesicle ____

____

82. nuclear envelope, nuclear pore, and nucleoplasm ____

____

Labeling

Identify each numbered part of the following illustration.

83. ______________

84. ______________

85. ______________

83

84 85

Identify each numbered part of the following illustration.

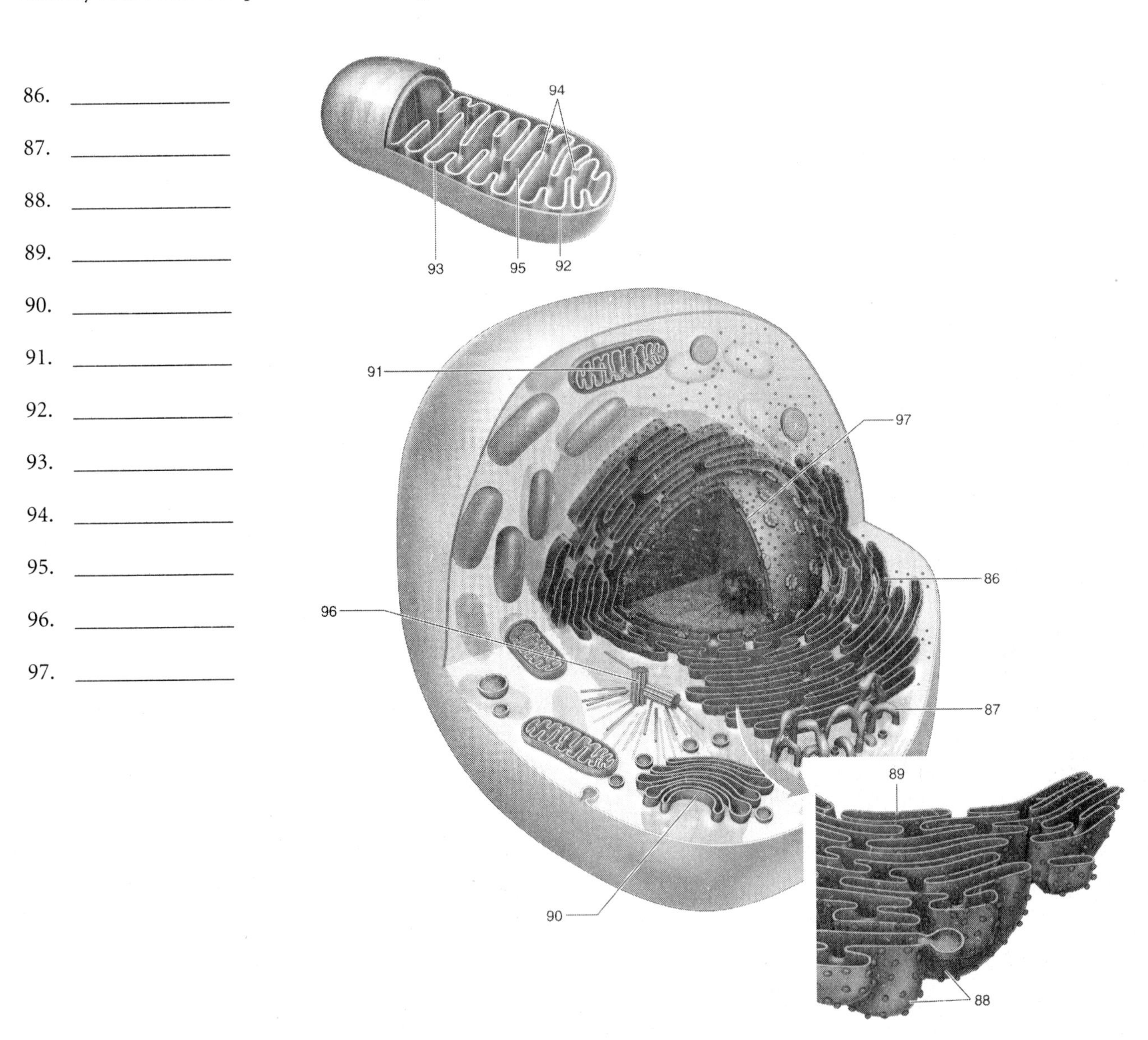

86. ______________

87. ______________

88. ______________

89. ______________

90. ______________

91. ______________

92. ______________

93. ______________

94. ______________

95. ______________

96. ______________

97. ______________

5.4 Specialized Structures of Plant Cells [pp. 110-113]

5.5 The Animal Cell Surface [pp. 113-115]

Section Review

(98)____ are the site of photosynthesis. Their (99)____ is smooth, while their (100)____is highly folded. Within the inner compartment, the (101)____ is a third membrane system comprised of flattened sacs or (102)____; in higher plants, the sacs are stacked to form (103)____. Chloroplasts are a family of plant organelles called (104)____; other members of this family include (105)____ and (106)____. A large (107) ____, surrounded by a membrane called the (108)____, is specialized for storage. Extracellular cell walls are made in sequential layers. The initial flexible layer is the (109)____ and the sequential rigid layer is the (110)____. The walls of adjacent cells are held together by a(n) (111)____. Channels through the cell wall that connect the cytoplasms of adjacent cells are called (112)____.

In animals, the (113)____ supports and protects cells and (114)____ bind cells together. (115)____ seal spaces between cells and provide means of communication between adjacent cells. There are two types of (116)____ that connect neighboring cells: (117)____, which are connected to microfilaments, and (118)____, which are connected to intermediate filaments. (119)____ prevent movement of materials (e.g., ions, water) between cells. (120)____ are direct connections between the cytoplasms of adjacent cells.

Building Vocabulary

Prefixes	Meaning
chloro-	green
chromo-	color

Suffix	Meaning
-plast(id)	formed, molded, "body"

Prefix	Suffix	Definition
____	-phyll	121. A green pigment that absorbs light in photosynthesis
____	____	122. Organelle that containing pigments that give flowers and fruits their color
amylo-	____	123. A nonpigmented organelle that is specialized to store starch and is found primarily in roots and tubers

Choice

For each of the following characteristics, choose the most appropriate structure from the list below that displays it.

a. primary cell wall b. secondary cell wall c. middle lamella

124. ____ Relatively flexible

125. ____ Relatively rigid

126. ____ Contains pectin

Matching

Match each of the following type of junction with its correct definition.

127. ____	Anchoring junctions	A. Glycoproteins embedded in membrane that enable cells to attach to one another
128. ____	Extracellular matrix	B. Proteins and polysaccharides secreted by cells that protect and support cells and tissues
129. ____	Tight junctions	C. Botton-like spots or belts that run around a cell connecting it to adjacent cells
130. ____	Cell adhesion molecules	D. Junction that uses microfilaments to connect to underlying cytoplasm
131. ____	Adherens junctions	E. An anchoring junction that uses intermediate filaments to connect to underlying cytoplasm
132. ____	Desmosomes	F. Junctions that prevent the passage of particles between cells

Distinguish between members of the following set of terms

133. plasmodesmata and gap junction ______________________________

____

Labeling

Identify each numbered part of the following illustration.

134. ______________

135. ______________

136. ______________

137. ______________

138. ______________

139. ______________

140. ______________

141. ______________

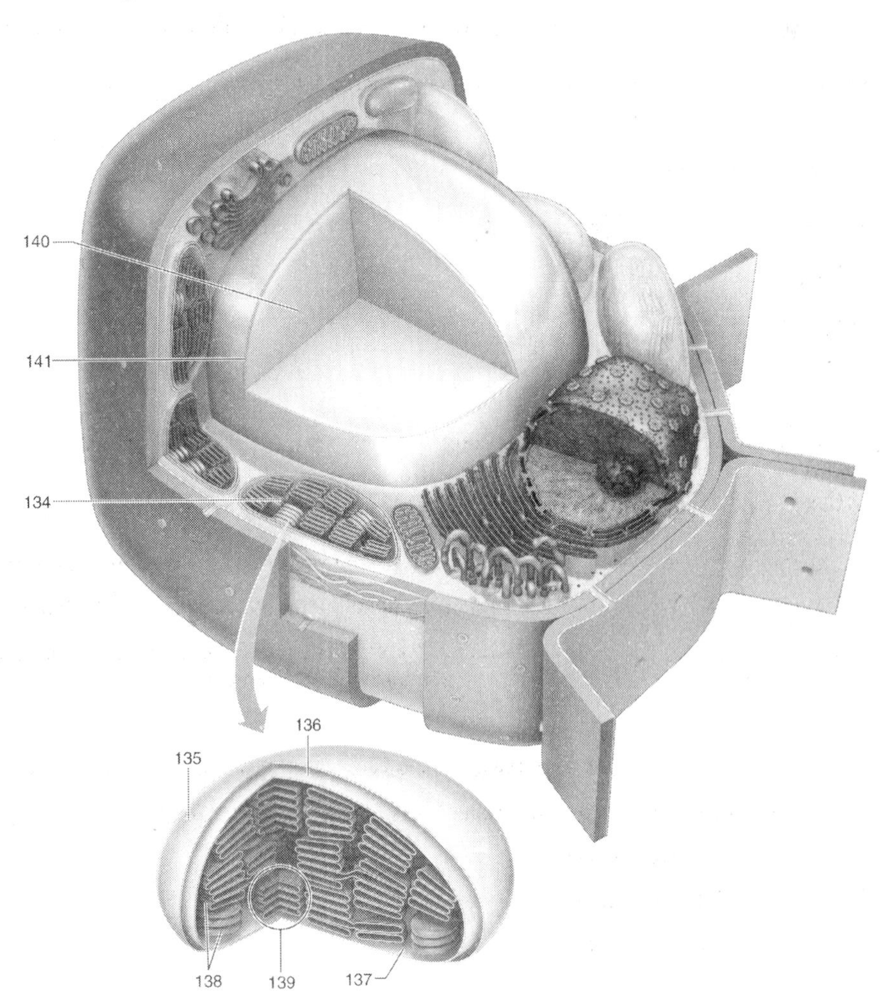

SELF-TEST

1. The cell theory states that _____. [p. 92]

a. new cells arise from preexisting cells

b. cells contain genetic material

c. all cells divide

d. living things are composed of cells

2. Cells are small because, in part, as size increases, the surface area to volume _____. [p. 93]

a. increases

b. doubles

c. decreases

d. reduces adequate nutrient-waste exchange

3. The Golgi complex ________. [p. 103]

a. modifies proteins

b. produces carbohydrates

c. digests organelles

d. assembles proteins using mRNA template

4. The endomembrane system includes ________. [p. 102]

a. endoplasmic reticulum

b. the nuclear envelope

c. the plasma membrane

d. secretory vesicles

e. mitochondria

5. Lysosomes _____. [p. 94]

a. contain digestive enzymes

b. break down complex molecules

c. break down organelles

d. can fuse with endocytotic vesicles

e. produce ATP

6. Mitochondria and chloroplasts both ________. [p. 111]

a. are found in plant cells

b. are found in animal cells

c. contain DNA

d. originated from ancient prokaryotes

7. Cell walls _____. [pp. 112-113]

a. occur in plants

b. occur in fungi

c. occur in algal protistins

d. contain cholesterol

e. contain cellulose

8. The cytoskeleton _____. [pp. 106-107]

a. changes constantly

b. is composed of protein

c. includes microtubules

d. extends into the nucleus

9. Which of the following contain plastids? [p. 111]

a. plants

b. animals

c. algae

d. some eukaryotes

e. some prokaryotes

10. The ATP-generating reactions of the mitochondria occur in/on the ____. [p. 115]

a. cristae of inner mitochondrial membrane

b. outer mitochondrial membrane

c. intermembrane space

d. matrix

11. Extracellular matrix ____. [pp. 115]

a. functions in protection and support

b. is secreted by the cell

c. composed of protein and polysaccharide

d. occurs in plants

e. occurs in animals

12. Tight junctions ____. [p. 114]

a. allow exchange of material directly between adjacent cells

b. are attached to microfilaments that anchor junction to the cytoplasm

c. seal spaces between cells to prevent passage of material

d. are glycoproteins that attach adjacent cells

INTEGRATING AND APPLYING KEY CONCEPTS

1. In cancer, cells transform and can break loose from adjacent cells and migrate to other parts of the body (metastasize). Discuss the possible defects that occur with the cell membrane of cancer-transformed cells.

2. *Insights from the Molecular Revolution:* Explain why scientists don't sequence the genome of all organisms. What criteria should scientists use for selecting species for genomic sequencing?

6

MEMBRANES AND TRANSPORT

CHAPTER HIGHLIGHTS

- The plasma membrane is the boundary of the cell.
- The composition and organization of the membrane underlie its functions.
- The plasma membrane compartmentalizes the contents of the cell from its environment.
- The plasma membrane controls the movement of substances into and out of the cell.

STUDY STRATEGIES

- Review the topic map below to get an overview of the chapter; note that the chapter is divided into two major parts: membrane structure and membrane function.
- Do not try to work through all of this material in one setting. Allocate your study time to concentrate on the membrane structure section first, then take a study break before concentrating on the membrane function sections.
- Re-review topic map to visualize relationships between and among key concepts; be able to clearly distinguish among all of the various forms of transport.

TOPIC MAP

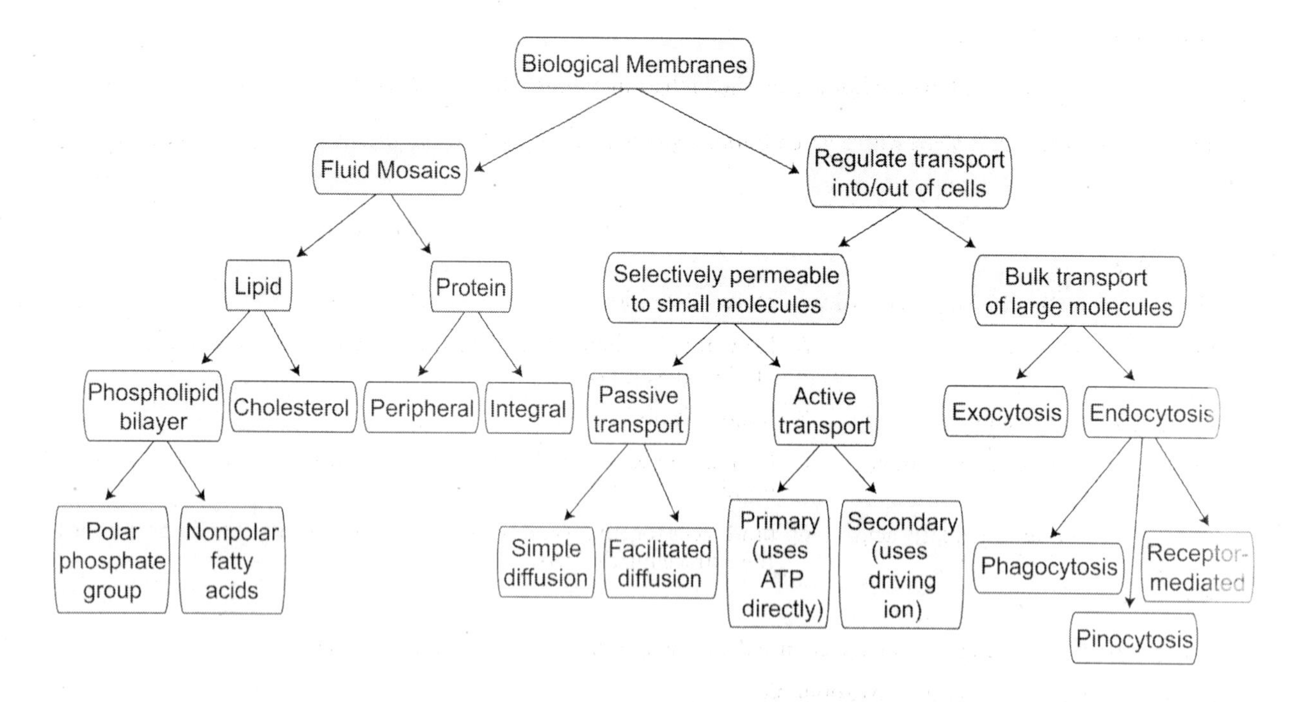

INTERACTIVE EXERCISES

Why It Matters [pp. 123-124]

Section Review (Fill-in-the-Blanks)

The plasma membrane is the primary contact between the cell and its (1) _____.

6.1 Membrane Structure [pp. 124-128]

Section Review (Fill-in-the-Blanks)

The two major types of lipids in membranes are phospholipids and (2) _____. Phospholipids have two ends, one containing nonpolar (3) _____ and the other end containing a (4) _____ linked to alcohols or amino acids, making them polar and capable of being in an aqueous environment. Phospholipid sheets assemble into a (5) _____, in which the polar ends are located on the surfaces _____ where they face the aqueous medium and the nonpolar fatty acids arrange themselves (6) _____in the hydrophobic membrane interior.

The theory of membrane structure known as (7) _____ . A common adaptation that keeps membranes fluid at low temperature is to increase the proportion of (8) _____ fatty acids in the membrane phospholipids. The inside and outside surface of the bilayer contains different mixtures of phospholipids and proteins/orientation of proteins, making the membrane (9) _____ and allow the inside and outside surfaces to have different (10) _____.

Proteins embedded in the phospholipids bilayer are (11) _____, whereas proteins held to the surface of the membrane are (12) _____.

In an experiment where human cells with their membrane proteins labeled with red fluorescent dye were fused with mouse cells labeled with a green fluorescent dye, the colors were completely (13) _____ on fused cells after 40 min.

Matching

Match each of the following terms with its correct definition.

14. ____ cholesterol
15. ____ glycocalyx
16. ____ cell adhesion protein
17. ____ unsaturated fatty acid

A. Network of carbohydrate groups attached to lipid and protein molecules on the exterior face of the membrane
B. Major membrane sterol
C. Contains linear carbon chain with one or more carbon–carbon double bonds
D. binds cells together via receptors of chemical groups on other cells or on the extracellular matrix

Choice

For each of the following statements, choose the most appropriate term from the list below.

a. hydrophilic b. hydrophobic

18. ____ fatty acid segment of phospholipid
19. ____ phosphate-containing segment of phospholipid

Labeling

Identify each numbered part of the following illustration.

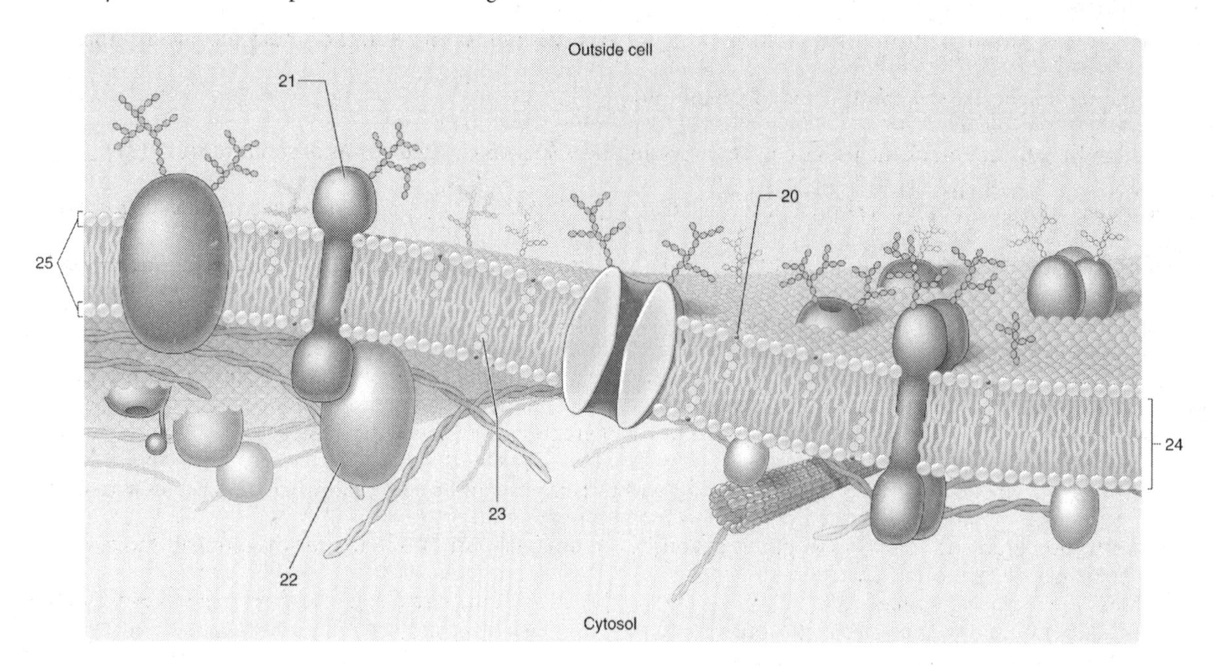

20. ______________

21. ______________

22. ______________

23. ______________

24. ______________

25. ______________

Short Answer

26. List the major functions of proteins associated with the membrane. ______________________________

____

27. Explain how the presence of double bonds in the fatty acids of phospholipids influences membrane fluidity. ___

____

6.2 The Functions of Membranes in Transport: Passive Transport [pp. 128-131]

6.3 Passive Water Transport and Osmosis [pp. 132-135]

6.4 Active Transport [pp. 135-137]

6.5 Exocytosis and Endocytosis [pp. 137-139]

Section Review (Fill-in-the-Blanks)

The primary function of cell membranes is transport. In (28) _____, ions or molecules move across membranes down their concentration gradient, whereas in (29) _____, energy is used to move ions or molecules against their concentration gradient. Passive transport is a form of (30) _____.

Membranes that block the passage of certain substances such as charged atoms or molecules are (31) _____. Small nonpolar substances cross the lipid bilayer by (32) _____. In (33) _____, polar and charged molecules cross the membrane with the help of transport proteins. (34) _____ form hydrophilic channels in the membrane through which substances pass, whereas (35) _____ bind to the solute molecule to transport it across the bilayer.

(36) _____ is the net movement of water across a selectively permeable membrane by passive diffusion, from a solution of lesser solute concentration to a solution of greater solute concentration.

(37) _____ uses energy, directly or indirectly, and carrier proteins to transport ions or molecules across their concentration gradient. The Na^+/K^+ pump establishes an unequal distribution of ions that contributes to a voltage across the membrane called (38) _____. In (39) _____, the cotransported solute and driving ion move in the same direction, whereas in (40) _____, the transported ion or molecule and the driving ion move in opposite directions.

(41) _____ releases molecules from the cell by means of secretory vesicles. In (42) _____, materials outside the cell become enclosed in a segment of plasma membrane, which bulges inward and pinches off to form a vesicle inside the cell.

Building Vocabulary

Prefixes	Meaning
endo-	within
exo-	outside, outer, external
hyper-	over, above
hypo-	under, below
iso-	equal, "same"
phago-	eat, devour
pino-	drink

Suffixes	Meaning
cyto(sis)-	cell or relating to a cell

Prefix	Suffix	Definition
_____	____	43. A process by which materials are encapsulated by a segment of plasma membrane and taken into the cell
_____	____	44. The process by which material is released or secreted from a cell by fusion of vesicle with the plasma membrane
_____	-tonic	45. Having an osmotic pressure or solute concentration greater than a reference solution
_____	-tonic	46. Having an osmotic pressure or solute concentration equal to that of a reference solution
_____	____	47. The ingestion of large molecular aggregates, cell parts, or whole cells by a cell.
_____	____	48. A type of endocytosis in which extracellular water along with small molecules is taken into small vesicles formed at the cell surface

Choice

Choose the most appropriate structure that accomplishes the listed process.

a. gated channel b. channel protein c. carrier protein

49. ____ A membrane protein that forms a hydrophilic channel through which water and ions can pass

50. ____ A transport protein that binds to a single solute particle to move it across the plasma membrane

51. ____ A protein channel that switches between open, closed, or intermediate states

Matching

Match each of the following terms with its correct definition.

52. ____	turgor pressure	A. A network of protein on the cytoplasmic face of the membrane that facilitates the invagination and pinching off of membrane
53. ____	plasmolysis	B. Channel protein that facilitates the diffusion of water
54. ____	clathrin	C. The internal pressure of a plant cell that results from the diffusion of water into the cell
55. ____	aquaporin	D. The shrinkage of a plant cell such that it retracts from its cell wall resulting from the loss of internal osmotic pressure

Complete the Table

Complete the following table by filling in the appropriate description of a listed transport process or by naming the transport process based on the listed description.

Process	*Description/Function*
Simple diffusion	56.
Osmosis	57.
58.	Uses energy to move ions or molecules against their concentration gradient
Facilitated diffusion	59.
60.	Cotransported solute moves in same direction as driving ion
61.	Cotransported solute moves in opposite direction of driving ion
Receptor-mediated endocytosis	62.

SELF-TEST

1. Phospholipids can assemble into a bilayer because of their _____. [p. 124]

a. ability to dissolve in water

b. dual solubility properties

c. inability to associate with other phospholipids

d. lack of fatty acids

2. In a phospholipid bilayer, _____ fatty acids align end-to-end within the bilayer and form a region that excludes water. [p. 125]

a. hydrophobic

b. hygroscopic

c. hydrophilic

d. hypertonic

3. Which of the following is NOT a function of membrane proteins? [pp. 125-126]

a. transport

b. recognition/adhesion

c. receptor

d. none of the above

4. Aquaporins function in transport across the membrane by _____. [p. 130]

a. binding to a molecule of water and carrying it

b. using Na^+ as a driving ion to transport water in the same direction

c. forming a channel to facilitate the rapid diffusion or water

d. inhibiting the activity of the Na^+/K^+ pump

5. Which of the following is LEAST likely to cross a plasma membrane by simple diffusion? [p. 129]

a. O_2

b. H_2O

c. CO_2

d. H^+

6. The passive movement of a substance down its concentration gradient is called _____. [pp. 129-130]

a. active transport

b. diffusion

c. exocytosis

d. symport

7. A plant cell placed in a hypertonic solution will _____.[p. 134]

a. remain unchanged

b. become crenated

c. undergo lysis

d. undergo plasmolysis

8. The energy-requiring transport of solutes against their concentration gradient is called _____. [p. 135]

a. osmosis

b. receptor-mediated endocytosis

c. active transport

d. facilitated diffusion

9. Receptor-mediated endocytosis does NOT involve _____.[p. 137]

a. binding to a specific cell surface receptor

b. fusion of the endocytic vesicle with a lysosome

c. symport with a driver ion

d. formation of a clathrin-coated pit

In a two-compartment system separated by a selectively permeable membrane, Compartment A is filled with a 0.5 M solution of NaCl and Compartment B is filled with a 1M solution of NaCl. Assuming that NaCl dissociates completely in an aqueous solution and that the membrane is freely permeable to Na^+, Cl^-, and H_20, characterize the following aspects of the system in questions 10-12. [pp. 133-135]

10. Compartment B is _____ compared to Compartment A.

a. hypotonic

b. isotonic

c. hypertonic

11. Initially, there is _____.

a. no net movement of solutes (Na^+ and Cl^-)

b. net movement of solutes (Na^+ and Cl^-) from A to B

c. net movement of solutes (Na^+ and Cl^-) from B to A

12. Initially, there is _____.

a. no net movement of water

b. net movement of water from A to B

c. net movement of water from B to A

INTEGRATING AND APPLYING KEY CONCEPTS

1. The blood of striped bass is hypotonic compared to seawater and hypertonic compared to fresh water. Describe the problems that the fish face in terms of ion and water movement when residing in seawater compared to those faced when residing in fresh water. What possible mechanism s could fish use to overcome these problems?

2. Cholesterol is transported in the blood in large complexes known as lipoproteins; a major cholesterol-containing lipoprotein is low density lipoprotein (LDL). Some individuals have a genetic defect that results in chronic elevated levels of cholesterol in their blood. What defect with the plasma membranes of cells would explain this condition?

7

CELL COMMUNICATION

CHAPTER HIGHLIGHTS

- Cells communicate with one another by means of chemical messengers; such communication coordinates the activity of cells.
- The source cell produces the message and the target cell receives the message.
- Receptors for chemical messengers can be located on the plasma membrane (cell surface) or in the interior or the target cell.
- Some cell surface receptors also are enzymes (e.g., tyrosine kinases) while others associate with G-proteins to initiate the formation of second messengers inside the cell (e.g., cAMP, IP_3).
- Internal receptors regulate gene transcription.

STUDY STRATEGIES

- The study of cell communication is important for understanding how the activities of cells are controlled and coordinated. Many of these concepts will be important for understanding whole organism function later (growth, development, reproduction, osmoregulation, etc).
- This chapter contains a lot of material that may be unfamiliar. DO NOT try to go through it in one sitting. Take one section at a time. Start by looking and the figures and legends, then read through the text making note of boldface terms.
- Remember, turning off a pathway is as important as turning it on when considering mechanisms of control. Therefore, carefully review both the activation and inactivation of a particular signaling pathway.
- Draw pictures of cells and their associated messenger systems. Be sure to label the parts of each system. Repeated drawing will assure that you are familiar with all components of each system.

TOPIC MAP

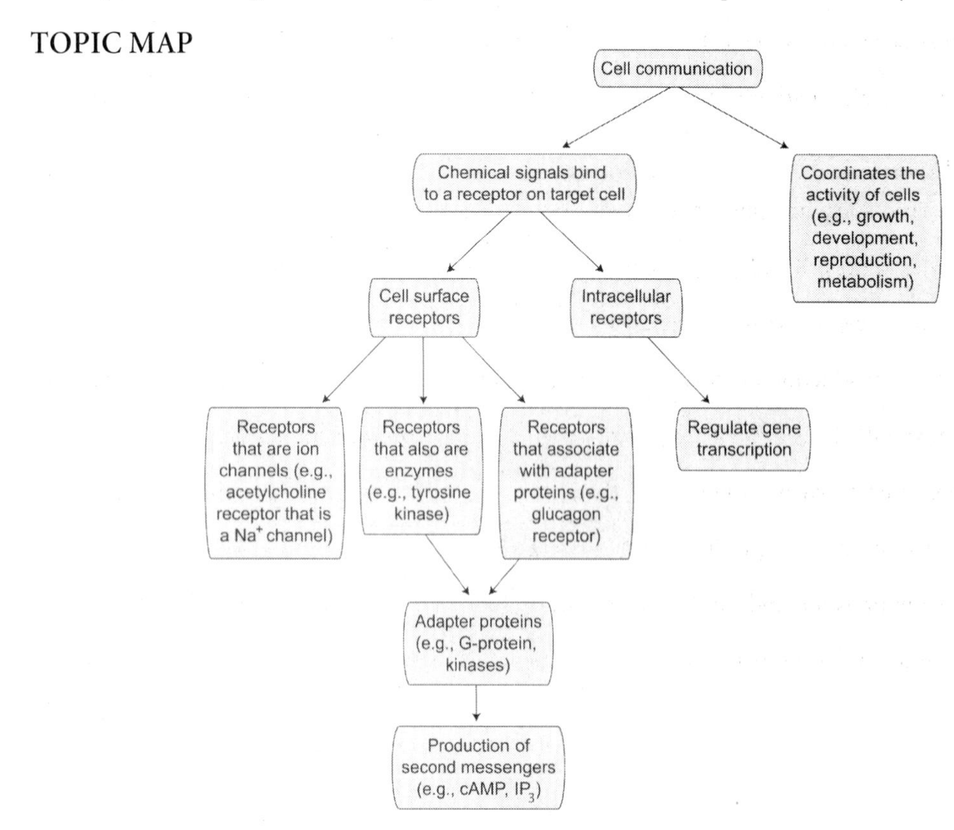

INTERACTIVE EXERCISES

Why It Matters [pp. 143-144]

Section Review

Chemical signals control the (1)____ of individual cells.

7.1 Cell Communication: An Overview [pp. 144–146]

7.2 Characteristics of Cell Communication Systems with Cell Surface Receptors [pp. 147–149]

7.3 Surface Receptors with Built-In Protein Kinase Activity: Receptor Tyrosine Kinases [pp. 149–150]

7.4 G-Protein Coupled Receptors [pp. 150–155]

7.5 Pathways Triggered by Internal Receptors: Steroid Hormone Receptors [pp. 155–156]

7.3 Integration of Cell Communication Pathways [pp. 157–158]

Section Review

In intercellular communication, a controlling (source) cell releases a chemical messenger (signal) that affects the function of target cells. The signal is processes by the target cell in three sequential steps: (2)____, (3)____, and (4)____.

Cell surface receptors recognize two main types of signals, (5)____ and (6)____. Transduction inside of the cell can involve phosphorylation (a form of chemical modification) of cytoplasmic proteins by enzymes called (7)____, which sometimes occurs in a series called a (8)____. Removal of phosphate groups by enzymes called (9)____ can shut off a transduction pathway. The stepwise magnification of events in a transduction pathway is (10)____.

(11)____ are receptors that add phosphate groups to the tyrosine amino acids on the receptor itself as well as on other target molecules in the cell. (12)____ and growth factors such as (13)____, (14)____, and (15)____ are examples of signals that bind receptor tyrosine kinases (RTKs).

(16)____ initiate a transduction process by activating a G-protein. The original chemical signal that binds a G-protein coupled receptor (GPCR) is the (17)____. Signal transduction proceeds from the activated G-protein to a membrane-associated enzyme known as a(n) (18)____ that, in turn, generates internal signal molecules called (19)____. G-protein activation (20)____ produces (21)___ from ATP. G-protein activation of (22)____ produces (23)____ and (24)____ from phospholipids in the membrane.

A major type of internal receptors is (25)____. The interaction of transduction pathways is called (26)____.

Labeling

Identify each numbered part of the following illustration.

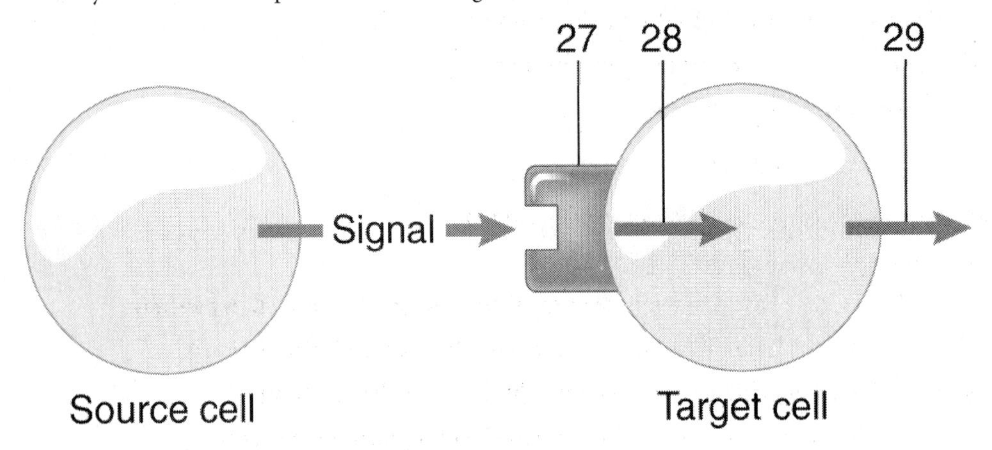

27. ____________

28. ____________

29. ____________

Association/Choice

For each of the following signal elements, choose the most appropriate classification from the list below below.

a. effector b. second messenger

30. ____ adenylyl cyclase

31. ____ inositol triphosphate (IP_3)

32. ____ phospholipase C

33. ____ cAMP

34. ____ diacylglycerol (DAG)

For each of the signal molecules, choose the most appropriate classification from the list below.

a. peptide hormone b. neurotransmitter c. growth factor

35. ____ epidermal growth factor

36. ____ insulin

37. ____ platelet-derived growth factor

38. ____ small molecules released from neurons (e.g., epinephrine)

For each of the following receptors, choose the most appropriate classification from the list below.

a. intracellular receptor b. cell surface receptor

39. ____ steroid hormone receptors

40. ____ receptor tyrosine kinase (RTK)

41. ____ G-protein coupled receptor

For each of the following processes, choose the most appropriate enzyme type from the list below.

a. protein kinase b. phosphodiesterase c. protein phosphatase

42. ____ enzyme that removes phosphate groups from cellular protein

43. ____ enzyme that attaches phosphate groups to cellular protein

44. ____ enzyme that degrades cAMP

Matching

Match each of the following aspect of signaling with its correct description

45. ____ signal transduction

46. ____ integrin

47. ____ G-protein

48. ____ first messenger

49. ____ cross talk

50. ____ amplification

51. ____ phosphorylation cascade

A. extracellular signal molecule

B. increase in magnitude of each step of a signal transduction pathway

C. the series of communication steps, from reception to response

D. interaction between/among different signaling pathways

E. a chain reaction involving stepwise phosphorylation events

F. a molecular switch protein that becomes active when GTP is bound

G. cell surface molecule that triggers a cellular response in another cell

Complete the Table

Messenger	Function
cAMP	52.
53.	stimulates calcium release from ER lumen into cytoplasm
54.	activates protein kinases
Receptor	**Mode of Action**
55.	binds to DNA and activates gene transcription
receptor tyrosine kinase	56.
57.	activates G-proteins which, in turn, stimulates effector enzymes

SELF-TEST

1. Signal transduction involves _____. [pp. 145–146]

a. reception

b. transduction

c. response

d. synthesis of signal molecule

2. _______ serve as signal molecules. [p. 147]

a. Estrogen

b. Insulin

c. Epinephrine

d. Actin

3. ________ activate(s) protein kinases. [p. 148]

a. Steroid hormone receptors

b. IP_3

c. DAG

d. cAMP

4. ________ is/are located on the surface of cells. [pp. 149-150]

a. Steroid hormone receptors

b. G-protein coupled receptors

c. Receptor tyrosine kinase

d. Glycogen phosphorylase

5. _______ breaks down cAMP. [p. 152]

a. Protein kinase

b. Phosphodiesterase

c. Phospholipase C

d. Glycogen synthase

6. G-proteins activate ______. [pp. 149–153]

a. phospholipase C

b. MAP kinase

c. steroid hormone receptors

d. adenylyl cyclase

7. Steroid hormone receptors evoke cellular responses by ______. [p. 156]

a. activating G-proteins

b. binding to DNA

c. promoting Ca^{++} release from ER lumen

d. activating protein kinase

8. Receptor tyrosine kinases [p. 149]

a. bind to DNA.

b. undergo autophosphorylation.

c. activates RAS.

d. phosphorylates other cellular proteins.

e. bind peptide hormones.

9. ______ is/are involved with turning of signal transduction pathways [pp. 148, 152]

a. Phosphodiesterases

b. Protein phosphatases

c. Phospholipase C

d. Phorbal esters

10. Gap junctions between animal cells connect their cytoplasms and allow rapid movement of ____. [pp. 152–153; 157]

a. Ca^{++}

b. cAMP

c. IP_3

d. G-proteins

11. ______ is a cell adhesion molecule that can elicit response in adjacent cells. [pp. 157]

a. Aquaporin

b. Lignin

c. Cytokine

d. Integrin

12. A phosphorylation cascade that results in the breakdown of glycogen to glucose and terminates with the activation of ____. [p. 154]

a. glycogen synthase

b. protein kinase

c. glycogen phosphorylase

d. phosphorylase kinase

INTEGRATING AND APPLYING KEY CONCEPTS

1. *Focus on Research*: Explain why a hormone's action may not be completely blocked when the target cell is cultured in the presence of a calcium ionophore in a calcium-free medium.

2. Individuals with insulin-independent diabetes mellitus can produce insulin, but it is not effective in stimulating glucose uptake. Explain potential defects in insulin signaling that would explain this insulin resistance.

8

HARVESTING CHEMICAL ENERGY: CELLULAR RESPIRATION

CHAPTER HIGHLIGHTS

- All living things metabolize high energy chemicals (glucose, fats, and proteins) to produce energy.
- There are two major pathways: aerobic and anaerobic respiration.
- Each pathway is subdivided into stages.
- Each stage:
  - Has a series of chemical reactions catalyzed by specific enzymes.
  - Has a beginning chemical and end product(s).
  - Produces energy that is stored in chemicals (ATP and NADP) or released as heat.
  - Takes place at a specific location inside the cell.
  - Is regulated by the presence of oxygen or by its end product(s).

STUDY STRATEGIES

- This chapter has extensive chemical pathways. Trying to memorize all of the chemical reactions first is the most common mistake.
- Remember the goal of this chapter is to understand how living things produce energy.
- Concentrate on the key concepts.
- Focus on what each pathway is about.
- Once the key concepts are understood, you can then learn the steps within each pathway.

TOPIC MAP

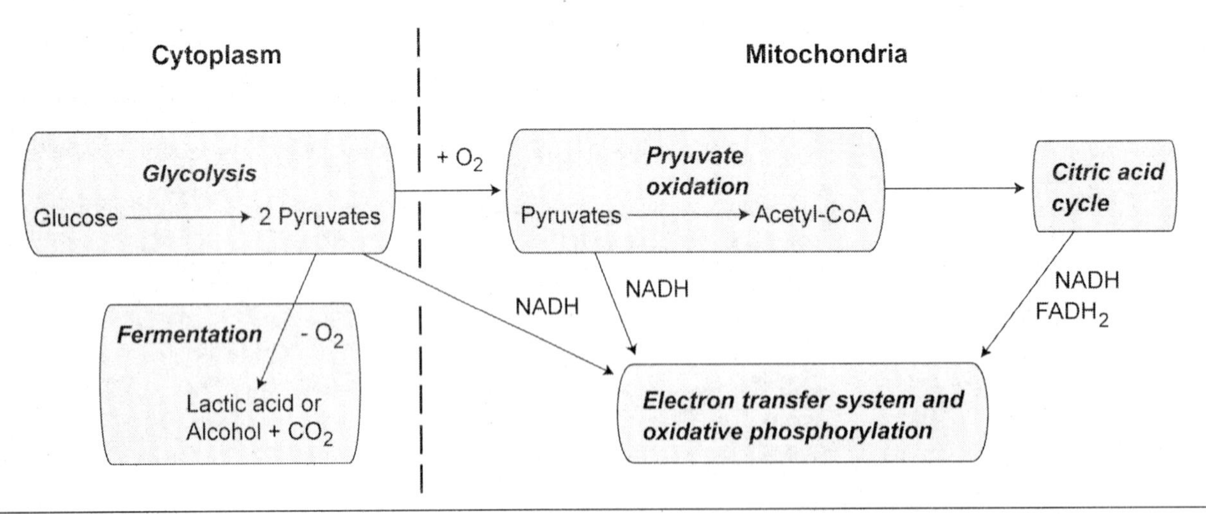

INTERACTIVE EXERCISES

Why It Matters [pp. 157–158]

Section Review (Fill-in-the-Blanks)

Cellular respiration refers to the breakdown of food molecules to produce (1) _____ _____, (2) _____ , and energy. Part of this energy is stored in (3) _____ molecules, and the rest is released as (4) _____. Stored energy is used by the eukaryotes and prokaryotes to support the (5) _____ reactions taking place inside their cells. In eukaryotes, most of this process takes place in the (6) _____ of the cells. Defects in structure and function of this organelle are now known to be the cause of many diseases, such as (7) _____, and contribute to many (8) _____ - _____ problems.

Photosynthesis, a process that occurs in most (9) _____, some (10) _____, and (11) _____, is quite opposite of cellular respiration. It uses (12) _____ energy, (13) _____, and (14) _____ _____ to make carbohydrates and (15) _____. It provides the fuel, (16) _____ and (17) _____ , to cellular respiration. Together, photosynthesis and cellular respiration cycle carbon chemicals by using or releasing (18) _____.

8.1 Overview of Cellular Energy Metabolism [pp. 158–162]

Section Review (Fill-in-the-Blanks)

Removal of electrons from a chemical is called (19) _____, while the chemical from which electrons have been removed is being (20) _____. During this process of removal of electrons, (21) _____ are often removed and (22) _____ must be added. Addition of electrons to a chemical is called (23) _____, while the chemical to which electrons are being added is (24) _____. This process of adding electrons is frequently accompanied by the addition of (25) _____ to the chemical. Chemical reactions where one chemical loses electrons and another chemical receives them are referred to as (26) _____ reactions.

Photosynthesis provides chemicals that have high energy (27) _____. During this process, (28) _____ splits to release its electrons that get energized by absorption of (29) _____ energy. High-energy electrons, H^+, and carbon dioxide are then combined to make (30) _____ molecules. Cellular respiration passes high-energy electrons from (31) _____ through (32) _____ reactions to release energy. The released energy is stored in (33) _____ molecules that will be used by the cells to perform several functions. Living things are supported by this flow of (34) _____ between two major processes: (35) _____ and (36) _____ _____.

Matching

Match each of the following terms with its correct definition.

37. ____ Oxidation
38. ____ Oxidized
39. ____ Reduction
40. ____ Reduced
41. ____ Redox reactions
42. ____ Oxidative phosphorylation
43. ____ Glycolysis
44. ____ Pyruvate oxidation
45. ____ Citric acid cycle
46. ____ Electron transfer system
47. ____ ATP synthase

A. A cycle of reactions that generates ATP, NADH, and $FADH_2$
B. Process where pyruvates are converted to acetyl-CoA
C. A series of carriers that transfer electrons
D. Process of adding electrons to a chemical
E. Process where glucose is partially oxidized in the cytoplasm to form pyruvates
F. A chemical from which electrons have been removed
G. Process of removing electrons from a chemical
H. An enzyme that uses a H^+ gradient to generate ATP
I. Process where a H^+ gradient is used to add phosphate to ADP and make ATP
J. A chemical to which electrons have been added
K. Coupled reactions where electrons released from one chemical are added to another chemical

Complete the Table

48. Electron flow connects two processes: photosynthesis and cellular respiration. Complete the following table by selecting your answer from the parenthesis, giving specific information about each process.

Events	*Photosynthesis*	*Cellular Respiration*
A. Sugars (makes/breaks)		
B. O_2 (uses/releases)		
C. CO_2 (uses/releases)		
D. Net energy (stores/releases)		

8.2 Glycolysis [pp. 162–165]

8.3 Pyruvate Oxidation and the Citric Acid Cycle [pp. 165–168]

8.4 The Electron Transfer System and Oxidative Phosphorylation [pp. 168–172]

Section Review (Fill-in-the-Blanks)

Glycolysis is a series of ten (49) _____ reactions that take place in the (50) _____ of the cell. The pathway involves a cascade of chemical reactions that break down each molecule of (51) _____, a six carbon sugar, into two molecules of (52) _____. The initial reactions of glycolysis require (53) _____ molecules of ATP but subsequently produce (54) _____ molecules of ATP. Therefore, there is a net gain of (55) _____ ATP molecules during this pathway. Electrons and protons removed during some of the oxidation reactions are accepted by a nucleotide-based electron carrier called (56) _____; as a result, (57) _____ is formed. Glycolysis is regulated by the concentration of (58) _____ and (59) _____ in the cytoplasm. These energy storing chemicals inhibit the enzyme (60) _____, which catalyzes the (61) _____ step of glycolysis.

Two molecules of pyruvate produced during glycolysis are actively transported into the (62) _____ of mitochondria. After removing two molecules of (63) _____, a pair of (64) _____ and two protons, the (65) _____ groups from the pyruvates are combined with (66) _____ to form acetyl-CoA. Electrons and protons removed during this reaction are accepted by NAD^+ to form two molecules of (67) _____.

The citric acid cycle is a cycle of (68) _____ reactions that take place in the (69) _____ of the mitochondria. During each cycle, a molecule of (70) _____ enters the cycle; (71) _____ molecules of CO_2 are released; and (72) _____ molecules of NAD^+ and (73) _____ molecule of another electron acceptor, FAD^+, are reduced to (74) _____ and (75) _____. One molecule of ATP is formed by (76) _____ - _____ phosphorylation. During glycolysis, (77) _____ molecules of pyruvate were formed from each molecule of (78) _____, and each pyruvate was converted to an acetyl group. Therefore, the number of products of the citric acid cycle must be doubled. The citric acid cycle is regulated by the concentration of (79) _____ that inhibits the first enzyme of the cycle: (80) _____ _____.

Complex carbohydrates, such as starch, must be broken down first into (81) _____, which then enter glycolysis. Besides carbohydrates, proteins and lipids can also act as energy sources. Proteins must be broken down to form (82) _____ _____; and after removal of the (83) _____ group, the remaining molecule enters cellular respiration as pyruvates or in the citric acid cycle. Fats are broken down first into (84) _____ and (85) _____ _____ and then into smaller chemicals before they can be used for energy.

When one molecule of glucose is broken down through glycolysis and the citric acid cycle, a total of only (86) _____ ATP molecules are formed. The remaining energy is stored in the high energy (87) _____ of NADH and $FADH_2$. The electrons flow through a series of (88) _____ _____ proteins located in the inner membrane of the (89) _____. (90) _____ released by the oxidation of NADH and $FADH_2$ are passed from one protein to another and ultimately accepted by (91) _____ as the final (92) _____ acceptor. Oxygen then combines with two protons and forms (93) _____. Energy released during the flow of (94) _____ through (95) _____ _____ proteins is used to transfer (96) _____ from the matrix to the intermembrane compartment of the (97) _____. This creates a (98) _____ gradient across the inner membrane of the (99) _____. The flow of (100) _____ back into the matrix through an enzyme (101) _____ _____, located in the inner membrane, generates ATP by (102) _____ phosphorylation. Approximately (103) _____ ATP are generated from NADH and $FADH_2$ through glycolysis and citric acid cycles.

From one molecule of glucose, a total of (104) _____ molecules of ATP are generated: (105) _____ from glycolysis, (106) _____ from the citric acid cycle, and (107) _____ from the electron transfer step. Only (108) _____ percent of the energy in glucose is cashed by the cell and stored in (109) _____, while the remaining energy is released as (110) _____.

Matching

Match each of the following terms with its correct definition.

111. ____	Substrate-level phosphorylation	A. Proteins that contain iron to accept and donate electrons
112. ____	Cytochromes	B. H^+ gradient that powers ATP synthesis
113. ____	Mitochondrial electron transfer system	C. Transfer of phosphate from a substrate to ADP to make ATP
114. ____	Proton-motive force	D. A proton and voltage gradient that generates energy
115. ____	Chemiosmotic hypothesis	E. A series of carriers in the inner mitochondrial membrane that transfer electrons

Building Vocabulary

Prefixes	Suffixes	Meaning
glykys-	____	Sweet
____	-lysis	To break down
____	-ate	The ionized form of an organic acid

Short Answer

116. Differentiate between oxidative phosphorylation and substrate-level phosphorylation. ____________________

____

____

117. Give a short description of NADH and $FADH_2$ and their role in cellular respiration. ____________________

____

____

118. Explain how glycolysis and the citric acid cycles are regulated. ____________________

____

____

119. Explain the role of electron transfer proteins in generating ATP. ____________________

____

____

120. Explain the location and the role of ATP synthase. ____________________

____

____

121. Compare the number of ATP molecules produced during the steps of cellular respiration. ____________________

____

____

Sequence

122. Arrange the following steps of cellular respiration in the correct sequence.

A. Citric acid cycle

B. Electron transfer system

C. Glycolysis

D. Pyruvate oxidation

_____ _____ _____ _____

Complete the Table

123. There are ten steps in the glycolysis pathway. Complete the following table with specific information about these steps.

Events of glycolysis	*Step number(s) in which the event takes place*
A. ATP is used	
B. ATP is produced	
C. NADH and H^+ are produced	
D. Key step for regulation	

124. There are eight steps in the citric acid cycle. Complete the following table with specific information about these steps.

Events of citric acid cycle	*Step number(s) in which the event takes place*
A. ATP is produced	
B. NADH and H^+ are produced	
C. $FADH_2$ is produced	
D. Key step for regulation	

125. Complete the following table with specific information on each step of cellular respiration.

Steps of cellular respiration	*Location in the cell*	*# of ATP produced*	*# of NADH and $FADH_2$ produced*	*O_2 used*	*# of CO_2 produced*
A. Glycolysis					
B. Pyruvate oxidation					
C. Citric acid cycle					
D. Electron transfer system					

Matching

Match each of the following scientists with their contribution to cellular respiration.

126. ____ Peter Mitchell
127. ____ Paul Boyer and John Walker
128. ____ Hans Krebs
129. ____ Gustav Embden and Otto Meyerhof
130. ______ Rolf Luft

A. Determined the sequence of reaction in glycolysis
B. Proposed the chemiosmotic hypothesis for generation of ATP
C. Explained the mechanism by which ATP synthase makes ATP
D. Worked out most of the reactions of the citric acid cycle
E. Correlated defective mitochondria with a muscle disease

Match each of the following chemicals with its correct definition.

131. ____ Glucose
132. ____ Phosphofructokinase
133. ____ Pyruvates
134. ____ ATP
135. ____ NADH
136. ____ ATP synthase
137. ____ Citrate synthase
138. ____ Electron transport proteins
139. ____ Phosphorylation

A. End product of glycolysis
B. Chemical that provides energy for most of the activities of the cell
C. Beginning chemical of glycolysis
D. Reduced form of an electron carrier
E. Enzyme of glycolysis that is regulated
F. Enzyme of the citric acid cycle that is regulated
G. Transfer of phosphate to a molecule
H. Enzyme that makes maximum ATP during cellular respiration
I. A series of electron carriers that use energy to transport protons across the mitochondrial membrane

Labeling

140. Identify each part of the mitochondrion.

A. ______________________

B. ______________________

C. ______________________

D. ______________________

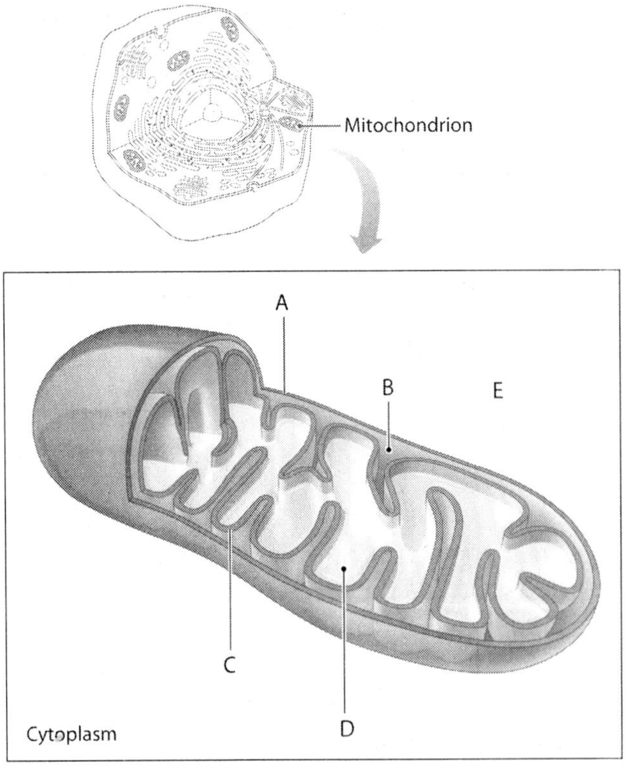

Locate the following structures or processes in the cell:

141. ____ Glycolysis

142. ____ Pyruvate oxidation

143. ____ Citric acid cycle

144. ____ Electron transport proteins

145. ____ ATP synthase

True/False

Mark if the statement is true or false. If the statement is false, justify your answer in the line below each statement.

146. ____________ In cellular respiration, all of the ATP is produced in the mitochondria.

____

147. ____________ All of the energy released by the breakdown of sugars is stored in ATP.

____

148. ____________ Carbohydrates are used by the cell to produce energy, while proteins and fats are used to store energy in the cell.

____

149. ____________ The electron transfer system and chemiosmosis produce the maximum number of ATP molecules.

____

150. ____________ Cellular respiration takes place only in eukaryotic cells.

____

8.5 Fermentation [pp. 172–175]

Section Review (Fill-in-the-Blanks)

Glycolysis is a series of ten (151) _____ reactions that take place in the (152) _____ of the cell. Glucose is broken down to form two molecules of (153) _____. During this process, the released energy is stored in four molecules of (154) _____ and two molecules of (155) _____. When (156) _____ is available, pyruvates are actively taken into the (157) _____ for further oxidation. However, when (158) _____ is absent or scarce, pyruvates remain in the (159) _____ and go through (160) _____. This step does not produce additional ATP. Therefore, in the absence of oxygen, the only step that produces ATP is the (161) _____ step.

In animal muscles and certain bacteria, pyruvates are converted to lactate; this process is called (162) _____ _____. In yeast, pyruvates are converted to alcohol and (163) _____ _____ this process is called (164) _____ _____.

Some bacteria and fungi do not have oxidative phosphorylation enzymes. They make all of the ATP through fermentation. They are referred to as (165) _____ _____.

Yeast and some bacteria can produce ATP through fermentation or cellular respiration, depending upon the availability of (166) _____. They are called (167) _____ _____.

Some bacteria and most eukaryotic cells must have (168) _____ to produce ATP and are known as (169) _____ _____.

Matching

Match each of the following terms with its correct definition.

170. ____ Fermentation
171. ____ Lactate fermentation
172. ____ Alcohol fermentation
173. ____ Strict anaerobes
174. ____ Facultative anaerobes
175. ____ Strict aerobes

A. Organisms that must have oxygen to survive
B. Organisms that require an oxygen-free environment to survive
C. Process that converts pyruvates into lactate or alcohol
D. Process that produces alcohol and carbon dioxide
E. Process that produces lactate in muscles
F. Organisms that can do aerobic or anaerobic respiration

Complete the Table

176. Complete the following table with specific information on each step of anaerobic respiration.

Steps of anaerobic respiration	*Location in the cell*	*# of ATP produced*	*# of NADH and $FADH_2$ produced*	*O_2 used*	*CO_2 produced*
A. Glycolysis					
B. Fermentation					

True/False

Mark if the statement is true or false. If the statement is false, justify your answer in the line below each statement.

177. __________ In anaerobic respiration, the fermentation step produces four ATP molecules.

____

178. __________ The fermentation step takes place in the mitochondria.

____

179. __________ In anaerobic respiration, glycolysis does not require oxygen, but the fermentation step requires oxygen.

____

180. __________ All bacteria undergo fermentation.

____

SELF-TEST

1. How many ATP molecules are invested in glycolysis? [p. 162]
a. 2
b. 4
c. 32
d. 38

2. Which of the following is the end product of glycolysis? [p. 162]
a. glucose
b. pyruvate
c. citric acid
d. acetyl-CoA

3. Where in the cell does glycolysis take place? [p. 162]
a. cytoplasm
b. mitochondrial matrix
c. intermembrane compartment
d. inner membrane of mitochondria

4. Before processing fat through cellular respiration, it must be broken down into ________. [p. 168]
a. amino acids
b. glycerol and fatty acids
c. nucleotides
d. monosaccharides

5. All of the following steps of cellular respiration produce ATP except ________. [p. 171]
a. the citric acid cycle
b. pyruvate oxidation
c. glycolysis
d. the electron transfer system

6. Which of the following steps of cellular respiration generates the maximum number of ATP molecules? [p. 171]
a. citric acid cycle
b. pyruvate oxidation
c. glycolysis
d. electron transfer system

7. How many ATP molecules are produced by the citric acid cycle? [p. 171]
a. 1
b. 2
c. 4
d. 32

8. Only about 30 percent of the energy from glucose is stored in ATP. What happens to the remaining energy? [p. 172]
a. It is stored in NADH.
b. It is released as heat.
c. It is converted to fat.
d. It is wasted as light energy.

9. During exercising, when oxygen is limited, additional ATP molecules are produced in the muscles by ________. [p. 172]
a. aerobic respiration
b. anaerobic respiration
c. the citric acid cycle
d. the electron transfer system

10. Where in the cell does the fermentation step take place? [p. 172]
a. cytoplasm
b. mitochondrial matrix
c. intermembrane compartment
d. inner membrane of mitochondria

11. Yogurt is a bacterial culture. As this culture gets older, it becomes increasingly sour due to the accumulation of lactic acid. Which of the following steps in bacteria is responsible for producing this acid? [p. 174]
    a. glycolysis
    b. citric acid cycle
    c. electron transfer system
    d. fermentation

12. While making the dough for bread, addition of yeast causes the dough to rise. Which of the following chemical(s) released by yeast makes this happen? [p. 174]
    a. lactic acid
    b. alcohol and carbon dioxide
    c. ATP and NADH
    d. pyruvate and acetyl-CoA

13. What is the net production of ATP molecules during anaerobic respiration of one glucose molecule? [pp. 172–174]
    a. 2
    b. 4
    c. 32
    d. 38

14. Per glucose molecule, which of the following organisms makes more ATP molecules? [p. 175]
    a. strict aerobes
    b. strict anaerobes

15. The chemical whose presence or absence decides if pyruvates will go in the mitochondria for oxidation or remain in the cytoplasm for fermentation is ________. [p. 174]
    a. carbon dioxide
    b. oxygen
    c. ATP
    d. NADH
    e. water

INTEGRATING AND APPLYING KEY CONCEPTS

1. In different tissues and organs of the human body, glucose is broken down by aerobic and/or anaerobic respiration. Discuss how glucose requirements, its storage, and the process of energy production differ in some of these tissues/organs.

2. Compare the process of glucose breakdown in bacteria, protists, fungi, plants, and animals.

3. Some migrating birds store fat instead of carbohydrates or proteins, have greater numbers of mitochondria in their muscles, and have a special lung structure. Explain how these factors help them in their migratory flight.

9

PHOTOSYNTHESIS

CHAPTER HIGHLIGHTS

- Autotrophs use chemical or light energy to convert simple inorganic chemical into complex organic chemicals.
- Some living things use light energy to make organic chemicals, a process called photosynthesis.
- Photosynthesis has two steps:
    - Light-dependent reaction—uses light energy to make ATP and NADPH.
    - Light-independent reaction (Calvin cycle)—Uses carbon dioxide, ATP, and NADPH to make glucose.
- Each step:
    - Consists of a series of events or chemical reactions.
    - Produces or uses energy chemicals (ATP and NADPH).
    - Takes place at a specific location inside the chloroplast.

STUDY STRATEGIES

- This chapter has extensive pathways. Do not get lost in the details of each pathway.
- Remember the goal of this chapter is to understand how autotrophs use light energy to make organic chemicals such as glucose.
- Concentrate on the key concepts and focus on how the pathways fit with each other.
- Once the model is clear, you can then learn the steps within each pathway.

TOPIC MAP

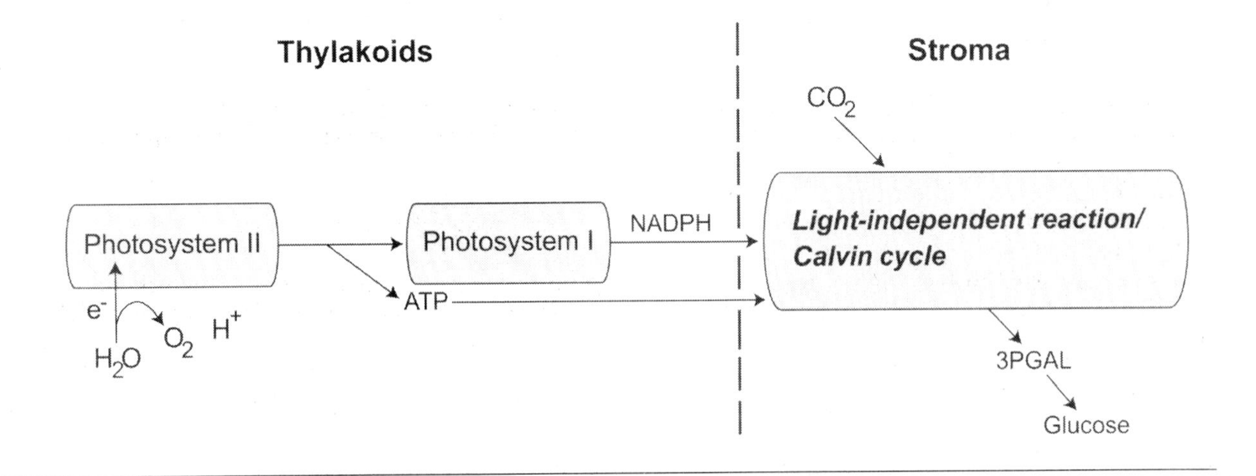

INTERACTIVE EXERCISES

Why It Matters [pp. 177–178]

Section Review (Fill-in-the-Blanks)

Photosynthesis refers to using (1) ____________ energy and simple (2) ____________ chemicals to make complex organic chemicals and release (3) ____________ in the atmosphere. A German scientist, (4) ____________ ____________, used a green algae, *Spirogyra,* and (5) ____________ requiring bacteria to prove that (6) ____________, ____________, and ____________ color light are most effectively used for photosynthesis. Through his experiments, he constructed a curve called the (7) ____________ ____________ which correlated the color of light with their effectiveness on photosynthesis.

9.1 Photosynthesis: An Overview [pp. 178–180]

Section Review (Fill-in-the-Blanks)

Plants, some bacteria and archaeans, and some protists are able to make their own organic chemicals. They are referred to as (8) ____________. They not only make these chemicals for themselves but are the major sources of other organisms. Therefore, they are considered as the primary (9) ____________ of this planet. Animals eat other organisms to get organic chemicals. They are called (10) ____________ that consume the primary (11) ____________. Therefore, they are referred to as the (12) ____________ of this planet. Fungi, some bacteria and archaeans, and some protists decompose plants and animals to get organic chemicals and called the (13) ____________.

Photosynthetic organisms use (14) ____________ energy and (15) ____________ chemicals to make organic chemicals. (16) ____________ and (17) ____________ use these organic chemicals to release energy. Part of this energy supports their (18) ____________ ____________ and the remaining energy is returned to the atmosphere as (19) ____________.

Photosynthesis has two phases: Phase 1 is called (20) ____________ ____________ reaction. Phase 2 is called (21) ____________ ____________ reaction, also called the (22) ____________ cycle.

During phase 1, plants absorb (23) ____________ energy from the sun, convert into (24) ____________ energy, and stores in the energy molecule (25) ____________ and electrons of (26) ____________.

During phase 2, these high energy chemicals, (27) ____________ and ____________, are used to convert CO_2 into (28) ____________, a process that is referred to as CO_2 (29) ____________. This conversion also requires the addition of protons and high energy (30) ____________ that come from splitting of (31) ____________ molecules. Splitting of (32) ____________ molecules by photosynthetic organisms also releases (33) ____________ into the atmosphere, a major source of this gas to support aerobic respiration. The evidence for this came from the experiments using heavy isotope of (34) ____________ incorporated into water and CO_2.

In eukaryotic cells, photosynthesis takes place in the (35) _______________. The organelle is surrounded by (36) ___________ and (37) ___________ membranes. Inside, there are membrane sacs called (38) _________________, suspended in a fluid referred to as the (39) ______________. The stack of membrane sacs is called (40) ____________. The photosynthetic pigments, electron transfer carriers, and ATP synthase enzymes are located in the membrane of (41) _________________.

Phase 1 of photosynthesis takes place in the (42) _________________, whereas phase 2 takes place in the (43) ____________.

CO_2 that is needed for photosynthesis, and (44) ____________ that is released during this process, are able to get into or out of the plants through very tiny opening on the surface of the leaves called (45) _____________. Water is absorbed from the ground by the (46) __________. (47) ________________ made during photosynthesis are transported to all parts of the plant.

Matching

Match each of the following terms with its correct definition.

48.	____	Photosynthesis	A. Organisms that cannot make their own organic chemicals from simple inorganic chemicals
49.	____	Action spectrum	B. Organelles in eukaryotic cells where photosynthesis takes place
50.	____	Primary producers	C. Process by which certain organisms use light energy to convert simple organic chemicals into organic chemicals
51.	____	Consumers	D. Organisms that break down plants and animals to obtain organic chemicals
52.	____	Decomposers	E. Openings on the under surface of the leaves to allow diffusion of gases in and out of the leaves
53.	____	Autotrophs	F. Membrane sacs inside a chloroplast that has the photosynthetic pigments
54.	____	Heterotrophs	G. Plants and other photosynthetic organisms that produce organic chemicals
55.	____	CO_2 fixation	H. A curve that expresses the effectiveness of different color light on photosynthesis
56.	____	Chloroplasts	I. A stack of membrane sacs inside a chloroplast
57.	____	Thylakoids	J. Fluid inside the chloroplast
58.	____	Stroma	K. Conversion of CO_2 to carbohydrates during phase 2 of photosynthesis
59.	____	Granum	L. Organisms that live by eating plants and animals
60.	____	Stomata	M. Organisms that use light energy to convert simple organic chemicals into organic chemicals

Complete the Table

61. Electron flow connects two processes: photosynthesis and cellular respiration. Complete the following table by selecting your answer from the parentheses, giving specific information about each process.

Events	*Photosynthesis*	*Cellular Respiration*
A. Sugars (makes/breaks)		
B. O_2 (uses/releases)		
C. CO_2 (uses/releases)		
D. Net energy (stores/releases)		

9.2 The Light-Dependent Reaction Of Photosynthesis [pp. 180–189]

Section Review (Fill-in-the-Blanks)

The main purpose of light-dependent reaction is to convert (62) __________ energy into (63) __________ energy. This energy form is stored in (64) __________ and (65) __________.

The electromagnetic spectrum ranges from (66) __________ waves to (67) __________ rays. (68) __________ light is a small part of this spectrum and consists of a mixture of rainbow colors. It ranges from (69) ______nm red light to (70) ______nm blue light. Each color has a specific (71) ______-length. A unit of light energy is referred to as a(n) (72) __________; the amount of energy contained in each unit is inversely proportional to the (73) __________ of the visible light. Therefore, red light with a longer (74) __________ has (75) __________ energy; blue light with a shorter (76) __________ has (77) __________ energy.

In plants, light energy is absorbed by the pigments in the (78) __________ membranes of the chloroplasts. There are two major groups of pigments: the green colored pigments are called chlorophylls and the orange-yellow pigments are called the (79) __________. When visible light falls on a pigment, it absorbs certain colors and it reflects or transmits the colors that are not absorbed. The color (80) __________ back is the one we see. Therefore, (81) __________ appear green because they absorb red and blue colors while transmitting the (82) ______ color light.

On the other hand, the colors that are (83) __________ by the pigment are used for photosynthesis. The (84) __________ of the pigment molecule absorb the energy of the light photon, moving from their (85) __________ state to a(n) (86) __________ state. The (87) __________ return from this unstable, temporary state by losing their energy as heat and light as in (88) __________, or transfer their high energy (89) __________ to another molecule called the primary (90) __________ molecule.

Just as chlorophylls serve as major photosynthetic pigments in plants and algae, (91) __________ are the pigments in bacteria.

Chlorophylls in plants are made of a carbon ring structure with (92) _______________ atom in the center and a long (93) __________________ side chain. There are two forms of chlorophylls: (94) _____ and _____, where chlorophyll (95) ____ is the most important form. (96) ___________________ are yellow-orange pigments that are also involved in photosynthesis and commonly found in yellow fruits and vegetables. Even though all of these pigments absorb light energy, all high energy electrons are eventually passed on to (97) _________________ ___.

The amount of light energy absorbed at different wavelengths of light is referred to as the (98) _________________ spectrum. The effectiveness of each wavelength in driving photosynthesis is called the (99) ________________ spectrum, which is determined by the amount of (100) _____________ released during this process.

Pigments are organized in groups called (101) _______________ that are embedded in the (102) __________________ membranes. There are two types of groups: (103) _______________ I and II. Each group or (104) ______________ consists of an aggregate of chlorophyll and carotenoids molecules called the (105) _____________ complex. Chlorophyll a serves as the (106) ___________ center molecule; P700 in photosystem (107) ___ and P680 in photosystem (108) ___.

In light dependent reaction, (109) _________ splits to provide electrons. These electrons first flow through photosystem (110) ___. They absorb (111) ________ energy, become excited, and are passed on to chlorophyll a—P 680. This (112) ____________ center molecule transfers electrons to a(n) (113) ______________ molecule and then through a chain of electron transfer proteins, similar to those found in the (114) ________________. The energy from these electrons is used to create a(n) (115) _____ gradient which triggers (116) _______ synthesis by a process called (117) _________________. After losing the energy, the electrons are received by the pigments of photosystem (118) ___. They once again absorb additional (119) _____________ energy to become excited and are passed to chlorophyll a—P700. This (120) ____________ center molecule transfers electrons to a(n) (121) ______________ molecule which then triggers (122) __________ synthesis with the help of the enzyme NADP+ (123) _______________. Due to one-way flow of (124) _____________, from water to NADPH, this type of light dependent reaction is called (125) ______________ electron flow.

Photosystem I can work independently by a process called (126) ___________ electron flow. This flow makes additional (127) ________ but no (128) ____________. As the pigments of photosystem absorb (129) __________ energy, the electrons are excited and are passed on to chlorophyll al— (130) _______. This (131) ____________ center molecule transfers these electrons to the (132) ____________ molecule and then to electron transfer proteins instead of directing them to making NADP+ (133) _____________ for making (134) __________. As the electrons pass through the carrier proteins, the energy is used to create a(n) (135) _____ gradient which triggers (136) _______ synthesis.

Overall, light dependent reaction makes more (137) __________ than (138) _________, as is required by the second step of photosynthesis, (139) _________-________________ reaction.

Matching

Match each of the following terms with its correct definition.

140. _____ Electromagnetic spectrum
141. _____ Chlorophylls
142. _____ Carotenoids
143. _____ Ground state electrons
144. _____ Excited state electrons
145. _____ Fluorescence
146. _____ Primary acceptor
147. _____ Absorption spectrum
148. _____ Action spectrum
149. _____ Photosystem
150. _____ Antenna complex
151. _____ Reaction center
152. _____ Electron transfer proteins
153. _____ ATP synthase
154. _____ Photophosphorylation
155. _____ Noncyclic electron flow
156. _____ Cyclic electron flow
157. _____ NADP+ reductase

A. Electrons with relatively low energy level
B. Light energy released by electrons returning from excited state to ground state
C. All forms of radiant energy ranging from radio waves to gamma rays
D. Synthesis of ATP by electrons that have absorbed light energy
E. Enzyme that generates ATP
F. Green photosynthetic pigments present in plants and algae
G. Yellow-orange photosynthetic pigments present in plants and algae
H. A one-way flow of electrons in light dependent reaction of photosynthesis
I. The amount of light energy absorbed at different wavelengths of light
J. Flow of electrons that only involves photosystem I and generates additional ATP for light-independent reaction
K. Enzyme that reduces NADP+ to NADPH during noncyclic light dependent reaction
L. A chain of proteins that use energy from electrons to create a H+ gradient in chloroplast and mitochondria
M. Chemical that accepts electrons from reaction center molecule
N. Chlorophyll a (P700 and P680) molecules that receive energized electrons from pigments of photosystems
O. The effectiveness of each wavelength in driving photosynthesis
P. Electrons that have absorbed light energy and have relatively higher energy level
Q. An aggregate of chlorophyll and carotenoids molecules in a photosystem
R. Cluster of light absorbing pigments located in thylakoid membranes

Matching

Match each of the following scientists with their contribution to cellular respiration.

158. ____	Peter Mitchell	A. Determined the chain arrangement of the electron carriers
159. ____	Andre Jagendorf and Earnest Uribe	B. Demonstrated the use of different colors of light in photosynthesis
160. ____	Engelmann	C. Explained the mechanism by which ATP synthase makes ATP
161. ____	Robert Hill and Fay Bendall	D. Elucidated the light-independent reactions of photosynthesis
162. ____	Melvin Calvin	E. Proved that H+ gradient in chloroplast generates ATP

Sequence

163. Arrange the following steps of electron flow in light dependent reaction in the correct sequence.

A. Photosystem I

B. Photosystem II

C. Water

D. NADPH

E. Electron transfer proteins

____ ____ ____ ____ ____

Short Answer

164. [p. 188] Differentiate between cyclic and noncyclic electron flow. ______________________________

____

____

165. [p. 178] Differentiate between absorption spectrum and action spectrum. ______________________

____

____

166. [p. 179] Explain the role of water in light dependent reaction of photosynthesis. ______________

____

____

167. [p. 179] Explain the major role of light dependent reaction in photosynthesis. ________________

____

____

168. [p. 184] Explain the difference between antenna complex and reaction center molecule. ____________

____

____

169. [p. 183] Explain why plants have several types of photosynthetic pigments. ____________________

____

____

Complete the Table

170. Complete the following table with specific products generated during each step.

Events	*Product/s generated*
A. Splitting of water	
B. Electron transfer proteins and ATP synthase	
C. Photosystem I and NADH+ reductase	

9.3 The Light-Independent Reactions Of Photosynthesis [pp. 189–194]

Section Review (Fill-in-the-Blanks)

The main purpose of light-dependent reaction is to convert (171) __________ energy into (172) __________ energy, stored in (173) _______ and (174) _________. These chemicals are used by the light-independent reaction to fix (175) ________ and form (176) ____________. The cycle of reaction is also called (177) _________ cycle. While light dependent reactions take place in (178) _____________, light-independent reactions take place in the (179) ____________ of chloroplast.

CO_2 combines with (180) ___________ _____________, a five-carbon sugar, to form a six-carbon chemical. This reaction is commonly referred to as the (181) __________-____________ step and uses an enzyme called (182) _____. This chemical splits into two three-carbon molecules of (183) _________________, which then use energy of phosphate bond from three molecules of (184) _________ and high energy electrons from two molecules of (185) ____________, both made by (186) ________-____________ reactions. As a result, two molecules of (187) ______________________ are formed. Through (188) __________ cycles of chemical reactions, (189) ________ molecules of this chemical are formed; out of which most are (190) _______ molecules are used to regenerate (191) __________ _________________ while (192) ______ molecule is used to form (193) __________. However, the cycle must go around (194) ______ times before (195) _______ molecules of 3PGAL can be spared to make one glucose molecule.

Once glucose is formed in plants, it is transported to other cells as (196) _____________. Some of it is used by the plants for making (197) _______ by cellular respiration, while the excess is stored as (198) _____________ or (199) ____________.

Matching

Match each of the following terms with its correct definition.

200. ____ Rubisco

201. ____ CO_2 fixation

A. Enzyme helps CO_2 combine with ribulose biphosphate

B. Combining of CO_2 with ribulose biphosphate

Complete the Table

202. Complete the following table with the specific role of each chemical.

Chemicals	*Its role in light-independent reaction*
A. CO_2	
B. 3PGAL	
C. Glucose	
D. Sucrose	
E. Starch	

203. Complete the following table with specific information on each step of photosynthesis.

Steps of Photosynthesis	Location in chloroplast	ATP used/produced	NADPH used/produced	O_2 produced	CO_2 used	Glucose produced
A. Light-dependent reaction						
B. Light-independent reaction/ Calvin cycle						

Short Answer

204. Discuss the role of rubisco in light-independent reaction. ______________________________

____

____

205. Discuss the number of Calvin cycles that must take place before one glucose molecule is produced.

____

____

Labeling

206. Identify each part of the chloroplast.

A. ______________________

B. ______________________

C. ______________________

D. ______________________

Locate the following reactions in the chloroplast:

207. Light-dependent reactions _____

208. Light-independent reactions/ Calvin cycle _____

A

B

C

D

True/False

Mark if the statement is true or false. If the statement is false, justify your answer in the line below each statement.

209. ____________ Calvin cycle produces one molecule of glucose per cycle.

____

210. ____________ All of the glucose produced during the Calvin cycle is stored by plants as starch.

____

211. ____________ ATP and NADPH used by Calvin cycle comes from mitochondria as a result of cellular respiration.

____

212. ____________ Photosynthesis takes place only in plants.

____

9.4 Photorespiration And The C4 Cycle [pp. 194–198]

Section Review (Fill-in-the-Blanks)

Rubisco, the enzyme needed for (213) _____ ____________, binds to (214) ______ which eventually results in production of 3PGAL and glucose. However, in plants that are known as C3 plants, when (215) ______________ levels are high, it can bind to rubisco to produce (216) ______________ which gets converted to a toxic chemical— (217) ____________. This not only prevents glucose production, but it eventually results in release of (218) __________. Since this process uses (219) __________ and releases (220) ___________, it is often referred to as (221) ________________. On a hot summer day, since the temperatures go up, stomata (222) ____________. As a result, (223) __________ builds up and (224) __________ drops, the rate of (225) __________________ increases.

To prevent this depletion of energy, some plants have developed another (226) ______________ enzyme that is unaffected by high oxygen levels. In these plants, CO_2 combines with phosphoenolpyruvate (PEP) to form a four-carbon chemical called (227) ______________, a reaction that is catalyzed by (228) ______ ________________. Plants that can make this four-carbon chemical are referred to as the (229) ______ plants, such as corn and sugar cane. They reduce this four-carbon chemical to eventually release (230) _________, which can be used for Calvin cycle. In these plants, oxaloacetate is formed in the cells closer to the surface, the (231) _____________ cells, whereas release of CO_2 and Calvin cycle take place in deeper cells, the (232) _____________ _______________ cells.

In cacti, called the (233) _______ plants, the two processes occur at different times, formation of the four-carbon chemical at night and Calvin cycle during the day.

Matching

Match each of the following terms with its correct definition.

234. ____	Photorespiration	A. Plants that fix CO_2 to make 3PGAL and glucose
235. ____	Oxaloacetate	B. Light stimulated use of oxygen and release of CO_2 in chloroplasts
236. ____	C3 plants	C. Plants that make oxaloacetate at night and release CO_2 for Calvin cycle in the day
237. ____	C4 plants	D. Plants that simultaneously make oxaloacetate and release CO_2 for Calvin cycle
238. ____	CAM plants	E. Four carbon chemical produced by some plants as storage of CO_2

SELF-TEST

1. All of the following groups include organisms that photosynthesize except [p. 178]

a. Eubacteria.

b. Archaea.

c. Protista.

d. Plantae.

e. Fungi.

2. Which of the following are produced by light dependent reactions? [p. 178]

a. glucose

b. ATP and NADPH

c. CO2

d. lactic acid and alcohol

3. Which of the following acts as the reaction center molecule? [p. 183]

a. Carotenoids

b. Chlorophyll a

c. Chlorophyll b

d. All of the above

4. Which of the following are present in photosystems? [p. 183]

a. Carotenoids

b. Chlorophyll a

c. Chlorophyll b

d. All of the above

5. Which of the following provides electrons to light dependent reactions? [p. 185]

a. water

b. chlorophyll

c. ATP

d. NADPH

6. Where are photosynthetic pigments present in plant cells? [p. 181]

a. cytoplasm

b. stroma of chloroplast

c. double membranes surrounding chloroplast

d. thylakoid membranes

7. Which of the following produces ATP and NADPH? [p. 185]

a. Calvin cycle

b. cyclic electron flow

c. noncyclic electron flow

8. All of the following are part of Calvin cycle except [p. 191]

a. chlorophyll.

b. rubisco.

c. RUBP.

d. 3PGAL.

9. How many times Calvin cycle must go around before one molecule of glucose is made? [p. 191]

a. 1

b. 2

c. 3

d. 6

10. How many molecules of 3PGAL are made during each Calvin cycle? [p. 191]

a. 1

b. 2

c. 3

d. 6

11. Most of 3PGAL formed during Calvin cycle are used to regenerate [p. 191]

a. rubisco.

b. glucose.

c. RUBP.

d. CO_2.

12. Plants and other photosynthetic organisms that produce organic chemicals are referred to as [p. 178]

a. autotrophs.

b. heterotrophs.

c. consumers.

d. decomposers.

13. Structures through which plants exchange oxygen and CO2 are called [p. 180]

a. mesophyll cells.

b. bundle sheath cells.

c. granum.

d. stomata.

14. Which of the following types of plants undergo photorespiration during hot days? [p. 194]

a. C3 plants

b. C4 plants

c. CAM plants

15. C4 and CAM store CO_2 as [p. 196]

a. rubisco.

b. RUBP.

c. 3PGAL.

d. oxaloacetate.

INTEGRATING AND APPLYING KEY CONCEPTS

1. Living things adapt to different environments. Discuss how plants have modified photosynthesis for different temperature conditions.

2. Briefly discuss how the thylakoids membranes are organized to allow smooth processing of light dependent reactions.

3. The Calvin cycle requires 3 molecules of ATP and 2 molecules of NADPH. Show how a light dependent reaction is able to produce more ATP than NADPH.

10

CELL DIVISION AND MITOSIS

CHAPTER HIGHLIGHTS

- Overview of the cycle of cell growth and division is presented.
- Mitotic cell cycle is defined and each component of the cycle is explained.
- Formation and action of the mitotic spindle are presented in context with the cell cycle.
- Regulatory points of the cell cycle is explained and associated with possible points of error leading to diseased states.
- Cell division in prokaryotes is described.

STUDY STRATEGIES

First, look at the big picture of cell division, focusing on the parent cell and daughter cells.
Evaluate the stages of cell cycle with respect to DNA amount.
Next, focus on the individual stages of mitosis, with respect to chromosome activity.
Understand the components and role of the mitotic spindle.
Review the overall process, associating the events of the cell cycle with the starting and end products of mitosis.

TOPIC MAP

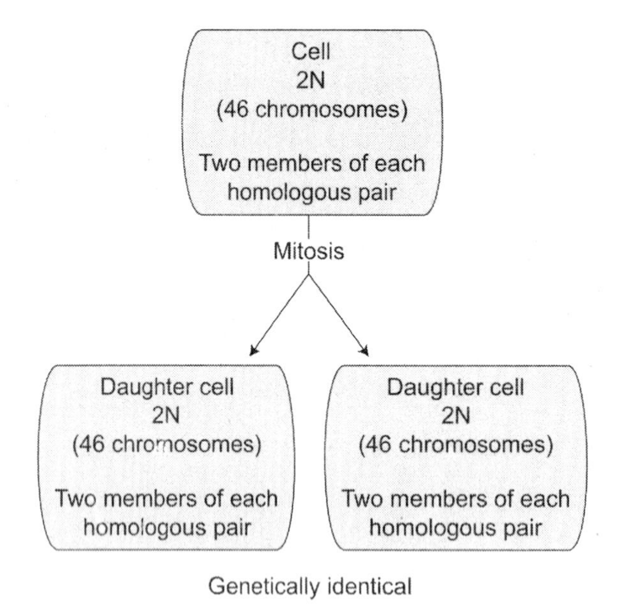

INTERACTIVE EXERCISES

Why It Matters [pp. 201-202]

Section Review (Fill-in-the-Blanks)

Red blood cells, which carry oxygen to your tissues, only live for about 120 days. The process of (1) _____ _____ replaces the old, dying red blood cells. Each cell will divide into (2) _____ daughter cells. The accuracy of this process is amazing, with very few errors. If cell division were stopped, the ability to carry (3) _____ to our tissues would be greatly (4) _____. Death would result within several weeks.

10.1 The Cycle of Cell Growth and Division: An Overview [pp. 202-203]

Section Review (Fill-in-the-Blanks)

Eukaryotic cell division is a series of complex events. There are (5) _____separate but interrelated systems. These systems are DNA (6) _____, mitotic spindle production which separates the duplicated (7) _____ between (8) _____ daughter cells, and a series of molecular (9) _____, which regulate the entire cell division process. The (10) _____ cycle in a eukaryotic cell consists of three phases or periods, (11) _____, (12) _____, and (13) _____. With respect to the parent cell, nuclear division occurs by (14) _____, where daughter cells are genetically (15) _____ or by (16) _____, where daughter cells are genetically (17) _____ _____. Hereditary information is located in the nucleus in the form of linear (18) _____ and associated proteins, in the form of (19) _____. In mitosis, each chromosome undergoes exact (20) _____, forming (21) _____copies, these copies are called (22) _____ _____. Each daughter cell nucleus receives the same (23) _____ and (24) _____ of (25) _____ as the parent cell. Distribution of the daughter chromosomes to the daughter cells is chromosome (26) _____. In summary, mitosis starts with (27) _____ parent cell and ending up with (28) _____ genetically (29) _____ daughter cells.

Building Vocabulary

Prefixes	Meaning	Suffixes	Meaning
mitos	thread	-kinesis	movement
cyto-	cell	-some	body
chroma-	color		

Prefix	Suffix	Definition
_____	-some	30. Linear DNA with associated proteins, which stains with dyes
_____		31. Division of nuclear material (threadlike structures)
Cyto_	_____	32. Division of the cytoplasm

Choice

For each of the following statements, choose the most appropriate structure of the cytoskeleton from the list below.

b. haploid c. diploid d. sister chromatids

Duplicated chromosomes

Chromosome sets

Two copies of each chromosome type, two sets

36. ____ One copy of each chromosome type, one set

10.2 The Mitotic Cell Cycle [pp. 203-209]

Section Review (Fill-in-the-Blanks)

The cell cycle accomplishes one of two tasks, (37) ____ or (38) ____. There are two major phases of the cell cycle, each with specific "subphases." These are (39) ____ and (40) ____. (41) ____ is the major phase of cell (42) ____ and preparation for nuclear division, (43) ____. By definition, interphase is the time in the cell cycle from the (44) ____ of mitosis to the (45) ____ of the next (46) ____. There are (47) ____ major phases within interphase, (48) ____, (49) ____, and (50) ____. The (51) ____ phase is variable in length and may enter a state of division arrest, the (52) ____ phase. Mitosis is a (53) ____ process with 5 phases, (54) ____, prometaphase, (55) ____, (56) ____, and (57) ____. Some scientists do not classify prometaphase as a separate phase. In that case there would be (58) ____ phases of (59) ____. During the last phase of mitosis, (60) ____, the cytoplasm of the parent cell divides. This is known as (61) ____.

Matching

Match each of the following phases of the cell cycle with its correct definition.

62. ____ G1 phase
63. ____ S phase
64. ____ G2 phase
65. ____ G0 phase
66. ____ prophase
67. ____ prometaphase
68. ____ metaphase
69. ____ anaphase
70. ____ telophase
71. ____ cytokinesis
72. ____ interphase
73. ____ mitosis

A. State of cellular arrest, growth occurs, but no additional nuclear division
B. Sister chromatids are apparent for the first time; spindle begins to develop
C. Nuclear division, consists of 4–5 phases
D. DNA replicates
E. Cell has twice the amount of DNA as G1 phase
F. Phase of cell growth after cytokinesis
G. After nuclear division cytoplasm divides and two daughter cells result
H. Sister chromatids are moving towards the middle of the developing spindle
I. Phase of cell growth and DNA replication
K. Centromere separates and sister chromatids move to opposite poles
L. Spindle disassembles, nuclear envelope reappears, chromosomes become less condensed
M. Sister chromatids are aligned at the midpoint of the spindle

Short Answer

74. Explain the difference between chromosomes and sister chromatids. ______________________________

____

Identification

75. For each image, identify the phase (top line) and the amount of DNA using ploidy number (bottom line). The cell A and B is in the G1 phase with 2N DNA.

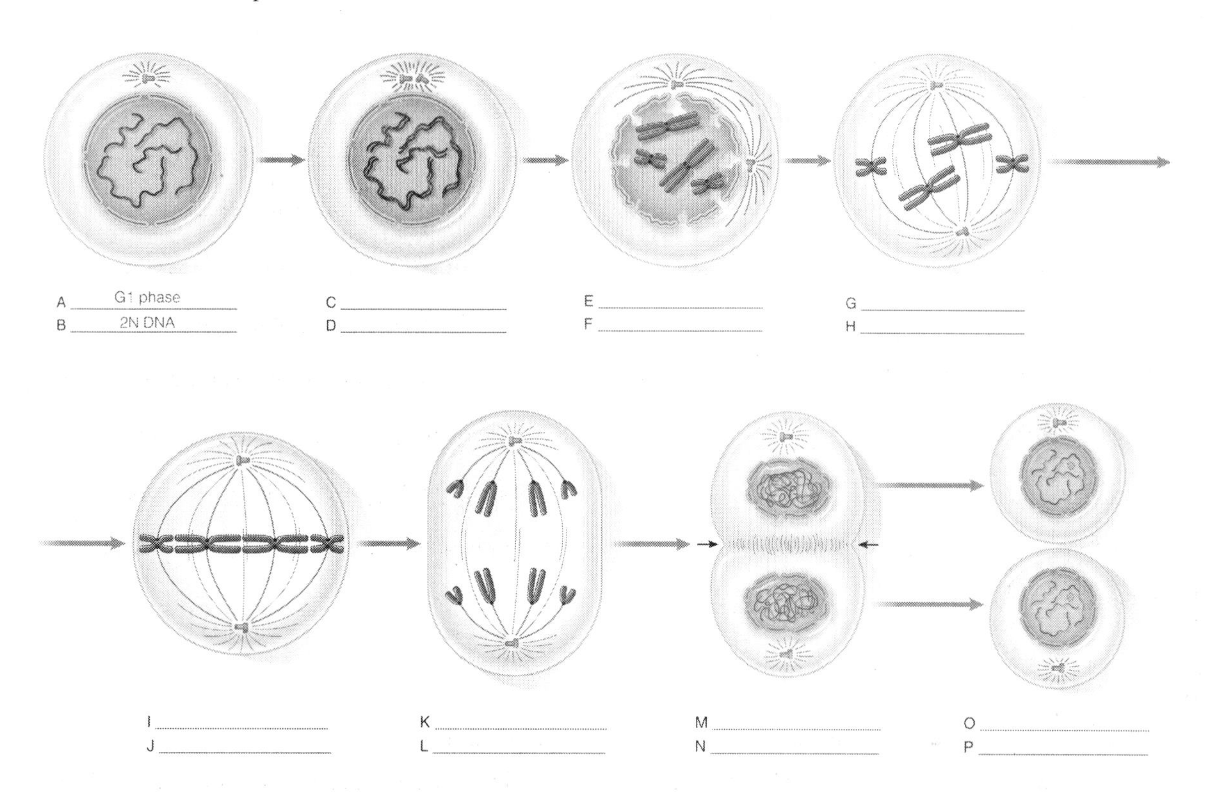

Short Answer

76. Compare and contrast animal and plant cell cytokinesis. ______________________________

____

10.3 Formation and Action of the Mitotic Spindle [pp. 209-212]

Section Review (Fill-in-the-Blanks)

Mitosis and cytokinesis are dependent on the formation of the (77) _____ _____. Spindle organization develops by (78) _____ different pathways, depending on the presence of the (79) _____. (80) _____ cells and most protists have a(n) (81) _____, most (82) _____ cells don't have a(n) (83) _____. The centrosome can be considered as the cell's primary (84) _____ _____ _____. The MTOC consists of a(n) (85) _____ of (86) _____. (87) _____ surround the centrioles, from which the (88) _____ is (89) _____. During (90) _____, the centrioles (91) _____. Each pair has one (92) _____ centriole and one (93) _____ centriole. As the (94) _____ (95) _____ move to opposite poles, (96) _____ form between, so that each daughter cell will receive a pair of (97) _____. In cells (98) _____ a centrosome, the MTOC surrounds the entire (99) _____. The (100) _____ forms from (101) _____ in all directions. Each sister chromatid has a (102) _____, a group of proteins, at the centromere. (103) _____microtubules which (104) _____ to the kinetochore of each sister chromatid, while (105) _____ microtubules extend between opposite poles without connections to chromosomes. Chromosomes move by (106) _____ proteins in the (107) _____, so chromosomes (108) "_____" to opposite poles.

10.4 Cell Cycle Regulation [pp. 212-216]

Section Review (Fill-in-the-Blanks)

The cell cycle has internal (109) _____, which regulate the division process. External regulation primarily occurs by (110) _____ and (111) _____ _____. A key player in regulation is (112) _____, a protein complex, and an enzyme (113) _____-_____ _____. The (114) _____ enzymes only function when in combination with (115) _____. The function of CDK is adding (116) _____ groups to target proteins, directly affecting the cell cycle. The role of (117) _____ is to activate or inactivate (118) _____. There are (119) _____ major checkpoints in the cell cycle, prior to the cell entering (120) _____ and leaving the G1 phase and entering the (121) _____ phase. Cyclin (122) _____ and CDK (123) _____ combination determines if (124) _____ occurs and Cyclin (125) _____ and CDK (126) _____ combination determines if the cell will leave the (127) _____ phase or cellular arrest.

Matching

Match each of the following terms/concepts with its correct definition.

128. _____ Contact inhibition A. Mass of cell as a result of unregulated cell division
129. _____ Growth factors B. Cell stays in G1 phase due to presence of adjacent cells
130. _____ Oncogene C. CDK activity primary regulation
131. _____ Tumor D. External control of cyclin—CDK activity
132. _____ Cyclin-dependent E. Gene which produces uncontrolled cell division

10.5 Cell Cycle Division in Prokaryotes [pp. 216-218]

True/False

If the statement is true, write a "T" in the blank. If the statement is false, make it correct by changing the underlined word(s) and writing the correct word(s) in the answer blank.

133. ____________ DNA replication is a <u>small</u> part of the cell cycle in prokaryotes.
134. ____________ <u>Binary fission</u> is the replication of DNA and dividing the cytoplasm into two portions.
135. ____________ In prokaryotes, DNA replication starts or proceeds from the same region, the origin of replication which is located <u>at the poles</u> of the cell.

SELF-TEST

1. Assume that no errors occur in DNA replication. Mitosis produces two daughter cells that are genetically _________. [pp. 202-203]

a. unique to each other

b. identical to the parental cell

c. accurate but unique from the parental cell

d. unique from the parental cell

2. Which of the following phases is the most variable with respect to ________ length? [p. 204]

a. G1 of interphase

b. G2 of interphase

c. S phase of interphase

d. mitosis

3. According to Russell, et al., when does the nuclear envelope breakdown? [p. 205]

a. prophase

b. G2 of interphase

c. telophase

d. prometaphase

4. Mature red blood cells lack a nucleus. Which phase of the cell cycle would you expect to find this cell? [pp. 203-204]

a. G1 of interphase

b. G2 of interphase

c. G0 of interphase

d. S phase of interphase

5. Which of the following is FALSE with respect to pair of sister chromatids? [pp.204-206]

a. They are genetically identical.

b. They will end up in different daughter cells.

c. They are held together by a single kinetochore.

d. Once separated are called daughter chromosomes.

6. Plants differ from animal cells in that plants ________. [pp. 209-210]

a. lack a karyotype

b. lack microtubules

c. lack centrioles

d. lack a microtubule organizing center

7. If you wanted to block the cell from going from G2 to the M phase, which of the following would accomplish that goal? [pp. 212-214]

a. increase the amount of cyclin B

b. increase the amount of phosphate

c. increase the amount of CDK1

d. decrease phosphatase

8. If contact inhibition were blocked, predict a likely response. [pp. 214-215]

a. A tumor might develop.

b. Wounds would never heal.

c. The cell would go into G0 phase.

d. Both B and C are possible.

9. A primary difference between mitosis in eukaryotic cells and binary fission is ___________. [pp. 216-217]

a. binary fission lacks cytokinesis

b. binary fission has few to no checkpoints

c. mitosis requires the presence of centrioles

d. DNA replication only occurs in mitosis

10. Which of the following statement about cells that undergo mitosis is FALSE? [pp. 212-214]

a. DNA replication starts at a specific region, the origin of replication.

b. G2 phase has twice as much DNA as G1 phase.

c. Chromatids move to opposite poles of the cell.

d. Checkpoints ensure that one phase is complete before the next phase is initiated.

INTEGRATING AND APPLYING KEY CONCEPTS

1. With respect to p53 as anticancer therapy, which would prove the best type of therapy, prevention of the abnormalities in p53 or finding another way to produce the same action of p53?

2. The *Ameoba*, a protist, reproduces by asexual reproduction. Assuming the amoeba is not destroyed, discuss the statement, "an *amoeba* never dies."

3. Explain why prokaryotes can successfully divide every 20 minutes and eukaryotic division is significantly longer.

11

MEIOSIS: THE CELLULAR BASIS OF SEXUAL REPRODUCTION

CHAPTER HIGHLIGHTS

- Overview of meiotic cell cycle, a primary source of genetic variation.
- Homologous chromosomes are defined.
- Mechanisms of genetic variation—recombination and chromosome segregation are explained.
- General patterns of genetic variation between organismal groups are presented.

STUDY STRATEGIES

- First, look at the big picture of meiosis, focusing on the parent cell and daughter cells.
- Evaluate the stages of meiosis I and II with respect to DNA amount and homologous pairs.
- Explain when genetic recombination occurs and the ultimate outcome of this process.
- Next, focus on when homologous pairs separate and when sister chromatids separate.
- Review the overall process and make comparisons with cell division or mitosis.

TOPIC MAP

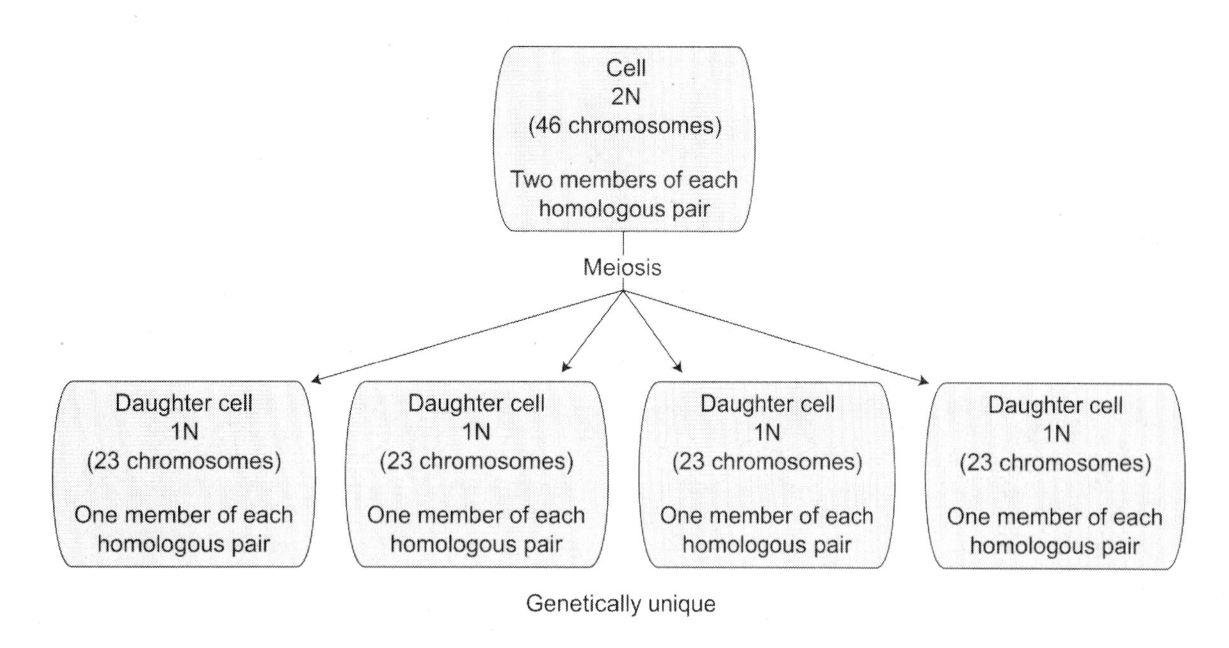

INTERACTIVE EXERCISES

Why It Matters [pp. 221-222]

Section Review (Fill-in-the-Blanks)

Sexual reproduction involves the production of (1) _____ which are produced by a specialized type of cell division, (2) _____. There is a great deal of variation in meiosis between different organisms. With respect to animals, both the male and female of a species produce gametes with (3) _____ the number of chromosomes, the (4) _____ number. If (5) _____ occurs, the male gamete unites with the female gamete to form a(n) (6) _____. The chromosome number of the newly formed (7) _____ is (8) _____ to the chromosome number of both the male and female of the species, or the (9) _____ number. The genetic material in the zygote is (10) _____ due to genetic (11) _____ which occurred during meiosis.

11.1 The Mechanisms of Meiosis [pp. 222-227]

Section Review (Fill-in-the-Blanks)

The (12) _____ number represents the total number of (13) _____ in a species. Each chromosome type comes in a(n) (14) _____, made of one (15) _____ chromosome from the (16) _____ parent and one (17) _____ chromosome from the (18) _____ parent. When (19) _____ are formed during (20) _____, each gamete is (21) _____ or has (22) _____ member of each (23) _____ pair. The chromosomes of the homologous pair have the same (24) _____ in the same (25) _____. For each gene of the homologous pair, the gene version may be the (26) _____ or (27) _____. The different versions are called (28) _____. Some genes have 1 or 2 different versions, other genes have multiple versions or (29) _____. During the process of (30) _____, the (31) _____ pairs line up next to each other and (32) _____ alleles, this is known as genetic (33) _____. Then two nuclear divisions produce (34) _____ daughter cells, each with (35) _____ member of each (36) _____ _____. In addition, to genetic recombination, the segregation of the homologous pairs is (37) _____. The daughter cells, or (38) _____ are (39) _____ and (40) _____ unique.

Boldfaced, Terms or Concepts

41. Allele ____

____

42. Interkinesis ____

____

43. Synapsis & Tetrad ____

____

Choice

For each of the following events, choose the most appropriate division of meiosis.

A. Meiosis I B. Meiosis II

44. ____ DNA replication of meiosis occurs during the S phase prior to this division
45. ____ Recombination of alleles
46. ____ 4 daughter cells are produced
47. ____ Homologous pairs line up and undergo synapses
48. ____ Sister chromatids separate
49. ____ Homologous pairs separate

Labeling

50. For each of the labeled sections, respond to the following:

A. Number of homologous chromosome pairs: _____. This cell is _____. (haploid or diploid)

B. Assume this is prophase I, DNA replication has _____. (occurred or not occurred)

C. Identify the event that occurs during this phase that increases genetic variability: _____.

D. During this division, the _____ _____ separate. This is done in a _____ fashion. Does DNA replication occur again? _____ (Yes or No)

E. During this division, the _____ _____ separate. These cells are _____ (haploid or diploid). Each cell contains _____ member(s) of each _____ pair.

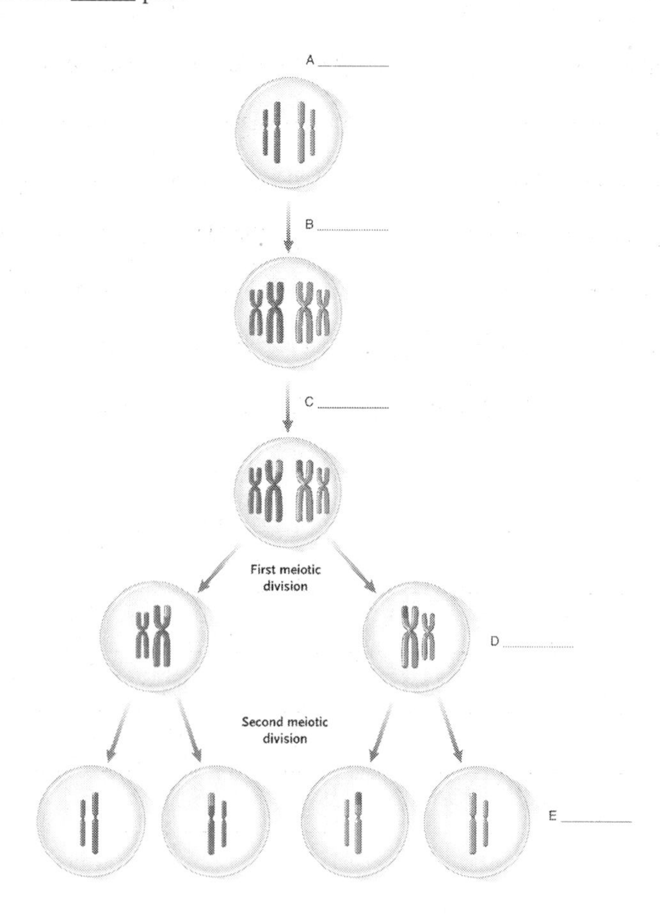

11.2 Mechanisms That Generate Genetic Variability [pp. 227-230]

Section Review (Fill-in-the-Blanks)

Genetic variability comes from (51) _____ sources. These are (52) _____, (53) _____ combinations of (54) _____ and (55) _____ chromosomes moving to (56) _____ poles of the forming daughter cells, and (57) _____ chance that any formed (58) _____ will be involved in (59) _____.

Matching

Match each of the following terms or concepts with its correct definition.

60. _____ chiasmata
61. _____ crossing over
62. _____ paternal chromosome
63. _____ maternal chromosome
64. _____ homologous pair

A. Chromosomes of the same size, shape, and order of genes
B. Member of homologous pairs from the male
C. Exchange of genetic material between sister chromatids of a tetrad
D. Site of cross over
E. Member of homologous pairs from the female

Sequence

Arrange the following events of meiosis in the correct hierarchical order. If an event takes place in a phase, list the phase first, followed by the event.

65. _____
66. _____
67. _____
68. _____
69. _____
70. _____
71. _____
72. _____
73. _____
74. _____
75. _____
76. _____
77. _____
78. _____

A. Homologous pairs undergo synapsis
B. Telophase I
C. Telophase II
D. Crossing over
E. Anaphase I
F. Metaphase I
G. Prophase/Prometaphase II
H. Haploid daughter cells result
I. Sister chromatids separate
J. Homologous pairs separate
K. Anaphase II
L. Metaphase II
M. DNA replication
N. Prophase/Prometaphase I

Matching

In the diagram, match the labeled regions with the appropriate term or concept

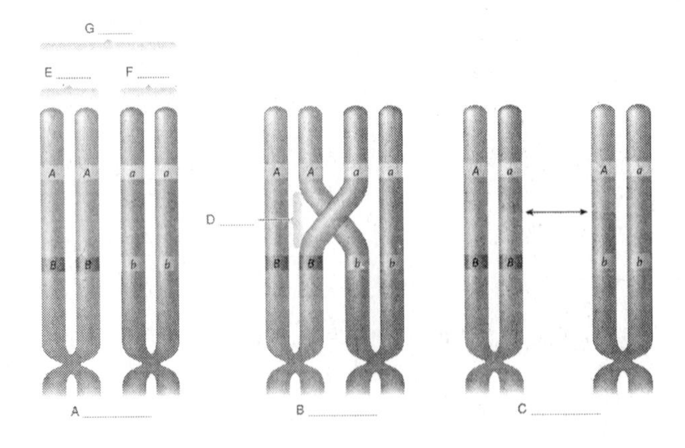

79. Sister chromatids of paternal/maternal source _____

80. Crossing over site _____

81. Chromatids exchange alleles _____

82. Occurs in Anaphase I _____

83. Homologous pair ____

84. Occurs in Prophase I _____

11.3 The Time and Place of Meiosis in Organismal Life Cycles [pp. 230-233]

Section Review (Fill-in-the-Blanks)

In eukaryotes, there are (85) _____ major patterns in the life cycle. In animals, the (86) _____ phase is dominant, with the process of (87) _____ producing (88) _____ which are (89) _____ and genetically(90) _____. In most plants and fungi, (91) _____ alternate between (92) _____ and (93) _____. The last pattern occurs in a few of the fungi and algae, where the (94) _____ phase is (95) _____, and the (96) _____ are produced by (97) _____. The gametes in this pattern are genetically (98) _____.

SELF-TEST

1. If the 2N number is 16, this many chromosomes are maternal: _____; this many chromosomes are paternal: _____. [p. 222]

a. 32, 32

b. 16, 16

c. 8, 8

d. 4, 4

2. Alleles are different versions of the same gene. On a pair of homologous chromosomes, the alleles ______. [pp. 222-223]

a. are always the same

b. are always different

c. could be the same or different

d. none of these

3. In the process of meiosis, how many times does DNA replication occur? [p. 223]

a. once

b. twice

c. prior to prophase I and II

d. both b and c are correct

4. Homologous pairs separate during ________. [pp. 223-227]

a. anaphase I

b. anaphase II

c. telophase I

d. prophase II

5. If the 2N number is 6 and one homologous pair underwent nondisjunction, how many chromosomes would be present in the daughter cells of that division? [pp. 226-227]
   a. Both daughter cells would have 3.
   b. One would have 2 and the other 3.
   c. One would have 1 and the other 5.
   d. One would have 2 and the other 4.

6. With respect to the sex chromosomes in humans, females and males will produce gametes with which possible sex chromosomes? [p. 227]
   a. females only X; males only Y
   b. females only X; males both X and Y
   c. females both X and Y; males only Y
   d. both females and males can produce both types of gametes—X and Y

7. The gametes that united to form you and your other siblings have a combination of alleles from both your parents. Predict with who of the following that you would have the most alleles in common. [pp. 227-230]
   a. your mother
   b. your father
   c. your siblings
   d. any of these; recombination, random separation of homologous pairs, and random gamete selection all play a role in common alleles.

8. Assume the DNA content of a cell about to undergo mitosis is X. Immediately after crossing over has occurred, the amount of DNA present is ______. [pp. 229-230]
   a. 0.5 X
   b. 0.75 X
   c. X
   d. 2 X

9. Which of the following life cycle patterns would you predict to have the potential to produce the most genetic variability? [pp. 230-233]
   a. diploid dominant
   b. haploid dominant
   c. diploid – haploid alternate
   d. difficult to determine the most potential of variability—all of these life cycle patterns produce genetic variability

10. Regardless of the life cycle pattern, gametes that are produced by mitosis are genetically ________. [pp. 230-233]
   a. unique
   b. identical
   c. the same as the previous generation
   d. none of these

INTEGRATING AND APPLYING KEY CONCEPTS

1. Predict the effects if synapsis were prevented.
2. Discuss advantages and disadvantages of gametes which are genetically identical to gametes which are genetically unique.
3. Predict the effects, if any, chromosomes could undergo crossing over, not just homologous pairs.

12

MENDEL, GENES, AND INHERITANCE

CHAPTER HIGHLIGHTS

- Overview of the foundational components of genetics and inheritance patterns.
- Principles of segregation and independent assortment are associated with allele recombination and chromosome movement in meiosis.
- Using simple genetic examples and interactions, the concepts of gene frequency and probably of gene expression are explained.
- Concept of multiple alleles, polygenic inheritance, and gene interactions are presented.

STUDY STRATEGIES

- First, focus on terminology and Mendel's experiments and careful evaluation.
- Understand how the scientific method is the basic framework of understanding complex cellular and molecular interactions.
- Next, using test crosses and Punnett square analysis, predict the probability of gene expression in offspring.
- Be able to relate allele and chromosome activities in meiosis to gene expression.
- Finally, focus on the complexity of multiple alleles, gene interactions, and the enormous potential for variation.

TOPIC MAP

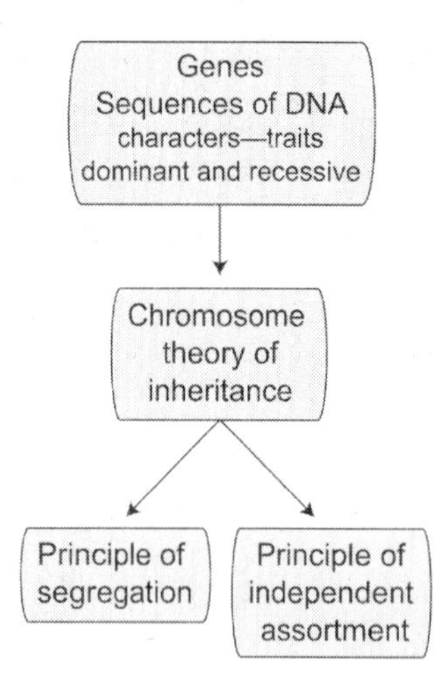

INTERACTIVE EXERCISES

Why It Matters [pp. 235-236]

Section Review (Fill-in-the-Blanks)

All living organisms have observable characteristics in common and others that are unique to the individual. These characteristics, or (1) _____, are (2) _____, which can be passed from parent to (3) _____. Red blood cells that have a (4) _____ shape is one example. By using the scientific method, careful observations and meticulous documentation, (5) _____ _____ in the mid-1800s was able to discover some of the fundamental rules of an area of science now known as (6) _____.

12.1 The Beginnings of Genetics: Mendel's Garden Peas [pp. 236-245]

Section Review (Fill-in-the-Blanks)

Gregor Mendel's experiments with (7) _____ _____ are not only famous, but foundational to understanding (8) _____. Given the level of cellular and molecular understanding of the time, Mendel was able to make conclusions about (9) _____ (variations in genes) and (10) _____ movement during the process of (11) _____. Mendel's term for gene was (12) _____, and (13) _____ for allele. In hind-sight, the selection of the garden pea as an experimental tool was fortuitous; these plants are (14) _____, or true-breeding, and can be manually (15) _____. This allowed Mendel to evaluate effects of mating pea plants with varying (16) _____ or traits. In addition, Mendel's work used large (17) _____ and (18) _____ experiments. The initial cross of the true-breeding plants were the (19) _____ or (20) _____ generation. The first generation offspring were the (21) _____ generation. The (22) _____ generation was the result of (23) _____ of the (24) _____ generation.

Matching

Match each of the following terms and concepts with its correct definition.

25.	_____	Dominance	A. Test cross to evaluate one gene (character)
26.	_____	Dominant trait	B. Both alleles (traits) are the same
27.	_____	Recessive trait	C. Test cross to evaluate two genes (characters)
28.	_____	Homozygote	D. Trait that is masked, unless in the homozygous state
29.	_____	Heterozygote	E. Trait that is expressed, when in the heterozygous state
30.	_____	Monohybrid	F. Masking of a allele (trait)
31.	_____	Dihybrid	G. Alleles (traits) for a given gene (character) are different

Choice

For each of the following statements, choose the most appropriate genetic term.

A. Genotype B. Phenotype

32. _____ Expressed alleles
33. _____ When you look in the mirror
34. _____ Purple flowers
35. _____ Plant which is heterozygous for flower color
36. _____ Alleles that are present

Complete the Table

Identify the number and type of all possible gametes (even if gametes are identical)—see example below:

	Number of potential gametes	*Types of all possible gametes*			
Tt	2	*A*		*a*	
TT	2	*A*		*A*	
Rr	37. _____	38. _____		39._____	
TTRr	4	40. _____	41. _____	42. _____	43. _____
TtRr	44. _____	45. _____	46. _____	47. _____	48. _____

Note: Individuals that are homozygous for an allele, will still produce the same number of gametes as an individual that is heterozygous. The difference is the homozygous individual only produces one type.

Section Summary (Fill-in-the-Blanks)

Given the table above, "T" (Tall) is dominant over "t" (short) for plant height, and "R" (red) is dominant over "r" (white) for flower color. The Rr individual is a (49) _____ for genotype, while the flower color is (50) _____. Rr identifies the (51) _____, while red identifies the (52) _____. The phenotype of the TTRr individual is (53) _____ in height and (54) _____ in color. Another way to indicate the genotype of the TTRr individual is (55) _____ _____ for height and (56) _____ for flower color. The TtRr individual is tall and red, which is the (57) _____. The genotype for this individual is (58) _____ for (59) _____ traits. With respect to these examples, the only way to have a white flower (60) _____ is to have a genotype of (61) _____ _____ or (62) _____. The only way to have a short plant (63) _____ is to have a genotype of (64) _____ _____ or (65) _____.

Short Answer

66. Explain the product rule and sum rule of probability. ______________________________

____

67. Define the principle of segregation. ______________________________

____

68. Define the principle of independent assortment. ______________________________

____

Problems

69. In garden pea plants, tall (*T*) is dominant over dwarf (*t*). In the cross *Tt* × *tt*, the *Tt* parent would produce a gamete carrying *T* (tall) and a gamete carrying *t* (dwarf) through segregation; the *tt* parent could only produce gametes carrying the *t* (dwarf) gene. Use the Punnett-square method to determine the following of a *Tt* x *tt* cross:

a. The genotypes of the possible gametes from both parents: ______________________________

b. The genotype probabilities of the offspring: ______________________________

c. The phenotype probabilities of the offspring: ______

70. The parental generation is a cross between two true-breeding plants. One plant is homozygous dominant for both plant height (Tall) and stem strength (Strong), and the other plant is homozygous recessive for both traits. Use the Punnett-square method to determine the following, assuming the production of F1 and F2 generations are as presented in the text:

a. The genotype of the possible gametes for each P generation plant ______

b. The genotype probabilities of the F1 offspring ______

c. The phenotype probabilities of the F1 offspring ______

d. The genotype probabilities of the F2 offspring ______

e. The phenotype probabilities of the F2 offspring ______

Short Answer

71. Describe the chromosome theory of inheritance. ______

______

12.2 Later Modifications and Additions to Mendel's Principles [pp. 245-251]

Section Review (Fill-in-the-Blanks)

Some of the garden pea plant characters (genes) that Mendel studied did not show (72) _____ dominance. Instead, there was a "blending" of the traits, or (73) _____ _____, and in some characters, both traits were expressed, (74) _____. As we stated earlier, Mendel was fortunate to look at characters (genes) that were determined by only (75) _____ alleles. In a given individual organism, there are only (76) _____ alleles, one from the (77) _____ source and one from the (78) _____ source, however, many genes have (79) _____ _____. In addition, genes can (80) _____ with each other. In the example of (81) _____, one or more genes will (82) _____ or mask the (83) _____ of another gene. It is now known that many characteristics such as height or skin color are determined by (84) _____ _____.

True/False

If the statement is true, write a "T" in the blank. If the statement is false, make it correct by changing the underlined word(s) and writing the correct word(s) in the answer blank.

85. ____________ Labrador retriever coat color is determined by <u>pleiotropy</u> effects on a gene that determines pigment distribution.

86. ____________ In <u>epistasis</u>, the expression of one gene will have affects on several different characteristics.

SELF-TEST

1. You have a good understanding of genetics. You are asked by your classmates, how do you know if an allele is recessive or dominant? Your response is it ________. [pp. 238-239]

a. depends on if the allele is on an autosomal or sex chromosome

b. depends on the associated linkage group

c. depends on whether it is on the maternal or paternal chromosome

d. depends if that allele or another allele determines the phenotype

2. Given the following genotypic patterns, which adds the most variation? [pp. 238-239]

a. homozygous dominant

b. homozygous recessive

c. heterozygous

d. sex-linked

3. Bob has blood type A and his children all have blood type O. Bob is very upset with his wife, who is also blood type A. You come to the rescue and explain to Bob that he is getting the _____ confused with the ________. [pp. 238-239]

a. phenotype, genotype

b. dominance, recessive

c. monohybrid, dihybrid

d. homozygous, heterozygous

4. An individual genotype is AaBB. What is/are the type(s) and number of different gamete combinations? [pp. 241-242]

a. Aa and BB, two

b. AaB, one

c. AB and aB, two

d. A and a and B, three

5. Following Mendel's experiment patterns for a monohydrid cross, if the F1 generation was a different phenotype than either of the parental individuals, which would be the simplest explanation? [pp. 246-247]

a. codominance or incomplete dominance

b. multiple alleles

c. pleiotropy

d. sex linkage

6. Blue flower color is dominant over white. Predict the phenotypic ratio if a heterozygous blue flower is crossed with a white flower. [pp. 241-243]

a. 100% blue flowers

b. 75% blue flowers and 25% white (3:1)

c. 50% blue flowers and 50% white (1:1)

d. 25% blue flowers and 75% white (1:3)

7. Refer to question #6. Predict the genotypic ratio. [pp. 241-243].

a. all Bb

b. 75% Bb and 25% bb

c. 50% Bb and 50% bb

d. 25% BB and 75% bb

8. Albert has type A blood, and Barbara has type B blood. Not knowing their genotypes, identify all the possible blood types of their children. [pp. 241-243]

a. type A and B

b. type A, B, AB or O

c. type A, B, or AB

d. not able to determine

9. You have a new puppy, a yellow lab. The parents of your new puppy were both black labs. The only way this could have occurred is that both parents must have been ______. [pp. 249-250]

a. heterozygous for black and the Epistasis gene

b. heterozygous for black and the pleiotropic gene

c. homozygous recessive for black

d. heterozygous for black

10. Harry knew that he was adopted. He was 6 ft 2 in, and both parents were under 5 ft 6 in. In fact, Harry checked his pedigree for 3 generations and no one was over 5 ft 8 in. How would you counsel Harry? [pp. 251-252]

a. Harry is correct, he must be adopted.

b. Height is determined by polygenic inheritance.

c. Height is a result of the product rule.

d. Height is due to multiple alleles.

INTEGRATING AND APPLYING KEY CONCEPTS

1. Often recessive alleles are defective or produce an unwanted phenotype. Design an experiment that would eliminate one recessive allele. Assume you have everything you need for your experimental crosses.

2. Suzie has sex-linked hemophilia. Please explain to her and her parents how this occurred. Include what must be true about the parents.

3. Joe is accused of fathering a baby with type AB blood. Joe is type 0 blood and the mother is type B blood. Does Joe have a case? Defend your response.

13

GENES, CHROMOSOMES, AND HUMAN GENETICS

CHAPTER HIGHLIGHTS

- Overview genetic linkage groups and recombination of alleles is presented.
- Sex-linked genes, present on the sex chromosomes, are described.
- Abnormalities in meiosis are associated with chromosomal alterations with genetic consequences.
- Examples of common dominant, recessive, and sex-linked genetic disorders are identified and explained.
- Other factors which influence inheritance are discussed.

STUDY STRATEGIES

- First, focus on the concepts of linkage groups and genetic recombination events.
- Be able to relate sex-linked disorders to sex linked genes.
- Apply abnormalities with the meiotic process to chromosomal abnormalities.
- Given either genotypes or phenotypes of parents with genetic disorders and predict probabilities of these genetic disorders in offspring.
- With the process of meiosis in mind, look at the big picture of chromosomes, genes, alleles, and expression of the alleles in a given genotype.

TOPIC MAP

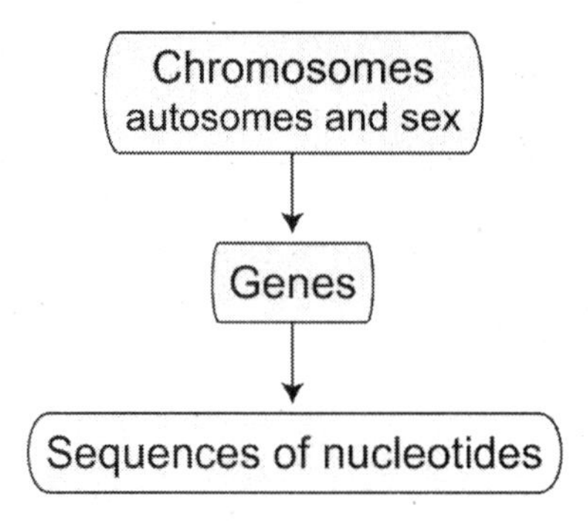

INTERACTIVE EXERCISES

Why It Matters [pp. 255-256]

Section Review (Fill-in-the-Blanks)

The old cliché "what you see is what you get" is true in many cases; however, when it comes to (1) _____, that is not always the case. All organisms have a (2) _____, the actual genes that are present, however, the (3) _____ of these genes, or the (4) _____, is really "what we get." Humans, like all organisms, are products of our inherited (5) _____, as well as multiple gene (6) _____ with other genes and the environment.

13.1 Genetic Linkage and Recombination [pp. 256-261]

Section Review (Fill-in-the-Blanks)

Once again, it is important to note that Gregor Mendel's choice of characters was very fortuitous. He evaluated (7) _____different characters (genes), and each was controlled (8) _____ of each other. During (9) _____ formation, these genes (10) _____ and (11) _____ independently of each other. Further investigations have now shown that genes located on (12) _____ have a specific (13) _____. Genes located on (14) _____ chromosomes will separate into gametes (15) _____, but if the genes are on the (16) _____ chromosomes, independent assortment is (17) _____ _____ the case. When (18) _____ on the (19) _____ chromosome that are inherited as a(n) (20) _____, these genes are considered to be (21) _____, and the concept is referred to as (22) _____. Through studies using the (23) _____ _____, (24) _____, we now understand that if two genes on the (25) _____ chromosome are (26) _____ in proximity, there is a greatly likelihood they will undergo (27) _____ or (28) _____ _____as a (29) _____ _____. If the location of two genes are (30) _____ _____, they are likely not to be (31)_____ and will separate independently of each other. When chromosomes undergo (32) _____ _____, or recombination, the physical constraints of the chromosome has a large influence on recombination location as well as (33) _____ _____. One advantage to linkage groups and the (34) _____ _____ has been their use in producing chromosome (35) _____.

Short Answer

36. Define linkage group. ____

____

37. Provide one type of evidence for linkage groups. ____

____

13.2 Sex-Linked Genes [pp. 261-266]

Section Review (Fill-in-the-Blanks)

In many organisms a pair or pairs of (38) _____, the (39) _____ chromosomes, are unique or different in (40) _____ and (41) _____. (42) _____ on these chromosomes are referred to as (43) _____ genes. The remaining chromosomes that are the same in (44) _____ and (45) _____ are referred to as (46) _____. In humans, as well as many other species, females have a (47) _____ _____ of sex chromosomes or (48) _____ (49) _____ chromosomes. Males are considered to be hemizygous, having (50) _____ (51) _____ and (52) _____ (53) _____ chromosome. With this pattern in mind, females have (54) _____ alleles for all genes on the (55) _____ chromosomes, while males have (56) _____ allele for the same genes. In addition, males have (57) _____ allele for genes on the (58) _____ chromosome. In males, with respect to allele expression on the sex chromosomes, both (59) _____ and (60) _____ alleles will be (61) _____. In females, an equalizing mechanism exists that results in (62) _____ of the X chromosome becoming (63) _____ forming a (64) _____ _____. However, this mechanism is (65) _____ with respect to whether it is the

(66) _____ or (67) _____ X chromosome. Females are therefore a mosaic or composite of cells, some with the (68) _____ X chromosomes and some cells with (69) _____ X chromosome expressed.

Short Answer

70. Explain how a female can have a sex-lined genetic disorder. _____

_____

71. An XY individual has a defective SRY gene, explain the phenotypic sex of the individual. _____

_____

Problem

72. Suzie's dad has a homozygous recessive sex-linked disorder, SQ. Sam, Suzie's husband, has no known genetic disorders in his family. Suzie and Sam want to have children but decide to seek genetic counseling prior to starting a family. You are that genetic counselor. With respect to the genetic disorder SQ please explain the following questions to the couple. Hint, the Punnett-square method will help.

a. The genotype and phenotype of Suzie _____

b. The probability of this couple having a child with genotype and phenotype of SQ _____

c. The possible phenotypes and genotypes of this couple's male children _____

d. The possible phenotypes and genotypes of this couple's female children _____

13.3 Chromosomal Alterations That Affect Inheritance [pp. 266-269]

Matching

Match each of the following terms/concepts with its correct definition.

73.	_____	Duplication	A. Failure of homologous pairs to separate during meiosis I
74.	_____	Euploids	B. Individual with a 2N that is either -1 or +1 chromosome
75.	_____	Translocation	C. Segment of a chromosome is missing
76.	_____	Nondisjunction	D. Individual is 2N with no abnormalities
77.	_____	Aneuploids	E. Extra set of chromosomes
78.	_____	Deletion	F. Segment of a chromosome is present more than once
79.	_____	Polyploids	G. Segment of a chromosome that is present, but in the reverse order
80.	_____	Inversion	H. Segment of a chromosome that is attached to a nonhomologous chromosome

13.4 Human Genetics and Genetic Counseling [pp. 269-272]

Section Review (Fill-in-the-Blanks)

Genetic counseling allows couples with known genetic disorders in the family (81) _____ to make (82) _____ decisions about the probably that their children may (83) _____ the disorder. Not all genetic conditions can be predicted, however, there are several major autosomal disorders that follow predictable patterns. In some conditions such as cystic fibrosis, offspring can have a defective phenotype, yet both parents can have a normal phenotype. In this case, the genotype of the parents is (84) _____, while both alleles of the offspring are (85) _____, the condition being (86) _____ _____. With other conditions such as polydactyly, both parents can have the condition, and yet both the (87) _____ and (88) _____ of a child are normal. In this case, the parents are (89) _____ and the child has (90) _____ _____ alleles, the condition being (91) _____ _____. In addition to counseling prior to conception of a child, (92) _____ _____ can provide valuable information after conception as well as (93) _____ _____ after the child is born.

13.5 Nontraditional Patterns of Inheritance [pp. 272-274]

Short Answer

94. Define cytoplasmic inheritance. ____

____

95. Explain the statement, "You have your mothers mitochondria." ______________________

____

SELF-TEST

1. All individuals of a group of cats have rounded ears and short whiskers. The simplest explanation of this pattern of inheritance is ________. [pp. 256-259]

a. sex-linked inheritance
b. linkage
c. pleiotropy
d. independent assortment

2. Two moths with short wings have offspring with the following wing length: 100 short-winged males, 97 long-winged males, and 205 short-winged females. The gene that determines wing length is _______. [pp. 261-265]

a. sex-linked and recessive
b. sex-linked and dominant
c. autosomal recessive
d. autosomal dominant

3. Betty Sue and Barney just had a baby. The doctor comes in and tells the happy couple the child is a carrier for red-green color blindness. The sex of the child is ________. [pp. 261-265]

a. male
b. female
c. impossible to tell
d. not one child but two males

4. You are analyzing karyotypes of two patients, a male and a female. The phone rings and you are distracted for a few minutes. When you return you know you are looking at the female karyotype because you clearly see ______. [p. 265]

a. mitochondria
b. centrioles
c. a barr body
d. only one Y chromosome

5. A normal chromosome sequence is ABCDEFG. You discover the sequence in your patient to be ABFEDCGEFG. Which of the following genetic alteration(s) most likely occurred? [pp. 266-267]

a. Deletion and dupliction
b. Duplication and translocation
c. Inversion
d. Inversion and duplication

6. After six months of experimental crosses with a plant species that has 50 chromosomes, you discover offspring with 150 chromosomes. What must have happened? [p. 267]

a. nondisjunction
b. euploidy
c. tetraploidy
d. triploidy

7. You need to determine if a patient has trisomy 21 or Down's syndrome. The best way to make an accurate diagnosis is to evaluate a ________. [pp. 267-268]

a. pedigree
b. karyotype
c. maternal inheritance
d. mitochondria

8. Refer to question #7, this condition most likely occurred because of ________. [pp. 267-268]

a. nondisjunction
b. polyploidy
c. translocations
d. inversions

9. If 75% of the offspring have a genetic condition, it can be concluded the condition is ____________. [pp. 269-272]

a. autosomal recessive
b. sex-linked autosomal recessive
c. autosomal dominant
d. sex-linked autosomal dominant

10. Assume you are expecting a child. Your doctor want to test for PKU. You are really nervous because the pregnancy is only in the 5th month. Your genetic counselor is explaining the procedure, which is a(n) ______________. [pp. 271-272]

a. chorionic villus sampling
b. screening test done after birth
c. amniocentesis
d. karyotype

INTEGRATING AND APPLYING KEY CONCEPTS

1. Predict the possible advantages and disadvantages of polyploidy.

2. Given the information in your text, predict which is better tolerated, aneuploidy in autosomes or sex chromosomes. Defend your response.

3. Explain why aneuploidies or genetic abnormalities, such as deletion, duplications, translocations, or inversions are typically viewed as "bad." Could these events every be considered as "good"? How would scientists know they occurred? Defend your responses.

4. You are in a debate about genetic screening. Pick one side of the issue (either for or against) and defend your position.

14

DNA STRUCTURE, REPLICATION, AND ORGANIZATION

CHAPTER HIGHLIGHTS

- DNA is identified as the genetic material by evaluation of historical scientific data.
- Nucleotides are the major structural component of DNA.
- DNA of daughter cells is identical to parent cell DNA as a result of replication mechanisms.
- Chromosomes have structure and organization.

STUDY STRATEGIES

- First, focus on the structure of DNA—the three basic components of a nucleotide, including the two types of nitrogenous bases, and the underlying structural reasons for complementary base pairing.
- If given a series of DNA nucleotides, be able to provide the complementary bases.
- DNA replication has multiple steps; focus on one step at a time.

TOPIC MAPS

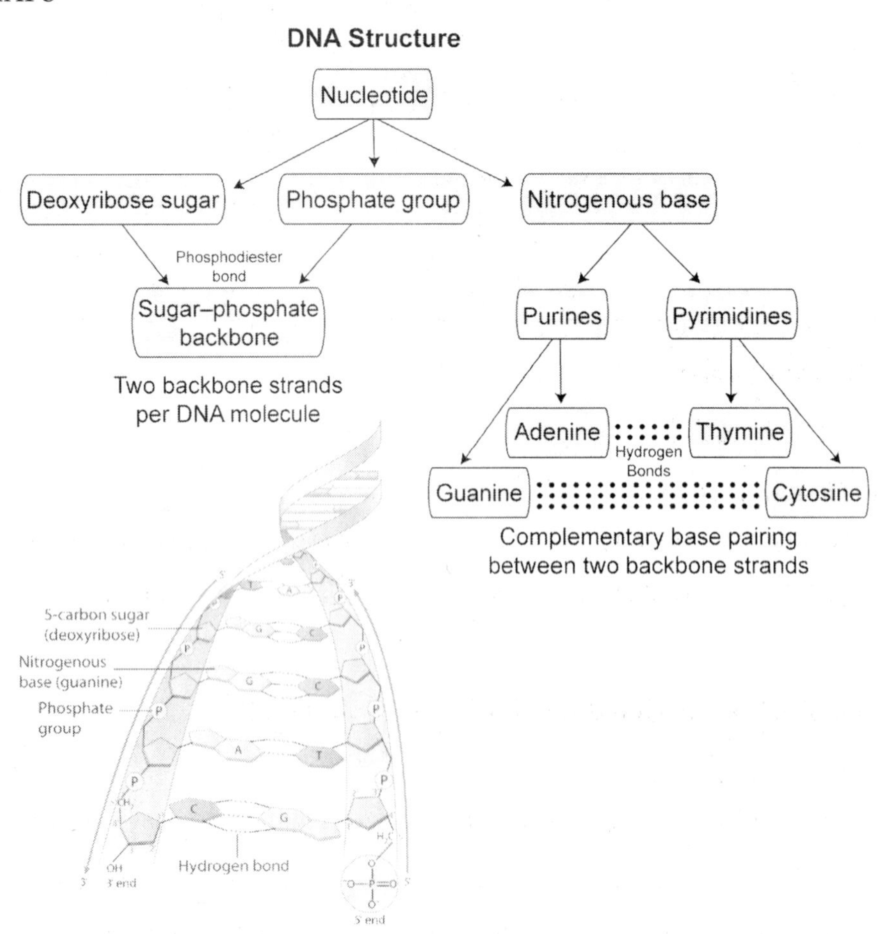

DNA Replication

Starting material: Parent DNA—double-stranded (chromatin fiber) DNA, histone, + non-histone proteins

↓

Preparation to replicate:
1. DNA Helicase—unwinds chromatin into two single strands of DNA
2. Single-stranded binding protein—stabilizes each single strand
3. Topoisomerase—prevents overtwisting and excessive strain of single strand

↓

Replication:
1. Start—multiple sites on each single strand of DNA; RNA primer and primase
2. Elongation—DNA polymerases and DNA nucleotides
3. Connection of DNA pieces—DNA ligase
4. Proofreading and repair

↓

Product: Daughter DNA: one parental strand and one antiparallel new strand

Semiconservative Replication

KEY
Parental DNA
Replicated DNA

1st replication

INTERACTIVE EXERCISES

Why It Matters [pp. 277-278]

Section Review (Fill-in-the-Blanks)

The material that Miescher extracted from pus was (1) _____ with a high (2) _____ content, which was primarily from the (3) _____ of white blood cells. Formally known as (4) _____, it is now known as (5) _____ _____ or (6) _____.

14.1 Establishing DNA as the Hereditary Molecule [pp. 278-281]

Section Review (Fill-in-the-Blanks)

Scientists knew that the hereditary material would need to contain a vast amount of information. Given the basic biological molecules, (7)_____, (8) _____, (9) _____ _____, and (10) _____, it seemed likely that (11) _____ provided the necessary variety due to (12) _____ different (13) _____ _____. Griffith's experiments with pathogenic

(*S*-strain) and nonpathogenic (*R*-strain) bacteria showed that one type of bacteria could be altered due to the process of (14) _____. Something in the *S*-strain was passed to the *R*-stain. Avery's experiments showed that by (15) _____ _____ the *S*-strain bacteria the proteins were (16) _____, but the transforming factor still caused the *R*-stain bacteria to become pathogenic. Then Hershey and Chase, using a virus that infected bacteria, a(n) (17) _____, showed that (18) _____ were not the (19) _____ _____, but rather (20) _____ was the genetic material.

Once (21) _____ was identified as the (22) _____ _____, Watson, Crick, Franklin, and others were able to determine the structure of DNA as a (23) _____ _____. The basic structural unit of DNA is the (24) _____.

Matching

Match each of the following scientists with their experiment or experimental conclusions.

25. _____ Watson and Crick

26. _____ Griffith

27. _____ Hershey and Chase

28. _____ Wilkins and Franklin

29. _____ Avery and coworkers

A. Found that material in heat-killed bacteria could still transform nonpathogenic bacteria into pathogenic bacteria

B. Using mice, found that nonpathogenic bacteria could be transformed into pathogenic bacteria

C. Using conclusions from numerous experiments, identified the molecular structure of DNA, complementary base pairing, and the means of replication

D. Using bacteriophages and radioactive isotopes of sulfur and phosphate, concluded that DNA is the genetic material rather than proteins

E. Analyzed X-ray diffraction patterns of DNA, identifying a helical structure

14.2 DNA Structure [pp. 281-284]

Section Review (Fill-in-the-Blanks)

Each nucleotide is composed of a sugar, (30) _____, a (31) _____ group, and a (32) _____ base. The sugar and phosphate groups alternate forming a (33) _____ - _____ _____ chain. The phosphate group acts as a (34) _____ between the two sugars, binding with the (35) _____ of one sugar and the (36) _____ of the next sugar. The bond between the sugar and the phosphate is a (37) _____ _____. There are (38) _____ sugar-phosphate chains that run (39) _____ to each other, and two (40) _____ _____ are located between the two chains, one associated with each sugar-phosphate chain. The nitrogenous bases are composed of either (41) _____ or _____. (42) _____ and _____ are the purines, and (43) _____ and _____ are the pyrimidines. (44) _____ base pairing binds one (45) _____ with one (46) _____. The complementary (47) _____ _____ are (48) _____ with _____ and (49) _____ with _____.

With respect to reading and replication, the (50) _____ strands of DNA are (51) _____. The (52) _____ end represents the 3' (53) _____ of the sugar, while the (54) _____ end represents the 5' (55) _____ of the sugar.

Labeling

56. Given the backbone of DNA, label the directional ends of the backbone and number the carbons of deoxyribose.

A. ____________

B. ____________

C. ____________

D. ____________

E. ____________

F. ____________

G. ____________

H. ____________

I. ____________

J. ____________

K. ____________

L. ____________

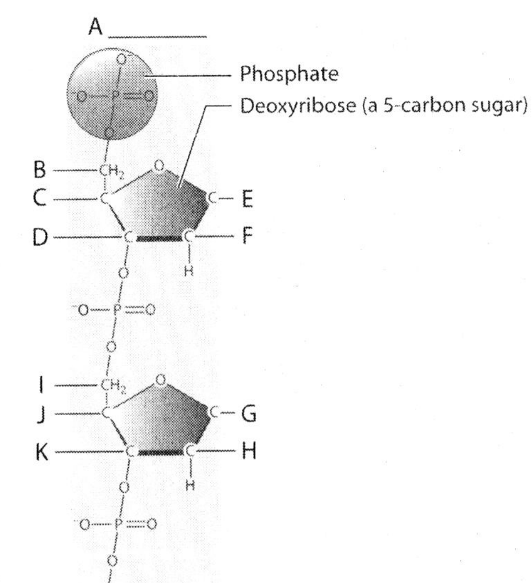

57. Given the double strand of DNA, label the nitrogenous bases of three sets of complementary base pairs.

A. ____________

B. ____________

C. ____________

D. ____________

E. ____________

F. ____________

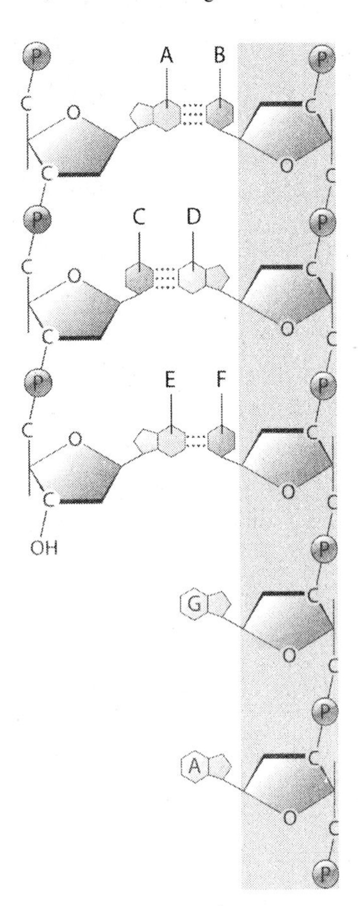

14.3 DNA Replication [pp. 284-292]

Section Review (Fill-in-the-Blanks)

Experimental findings have identified that each strand of DNA acts as a template during (58) _____ replication. The two strands of DNA are (59) _____ to each other. The enzyme DNA (60) _____ unwinds the DNA strands, producing a site called a (61) _____ _____. Short pieces of a(n) (62) _____ _____, about (63) _____ nucleotides in length are produced by an enzyme, (64) _____. The template is read in the (65) _____ → _____ direction and assembled in the (66) _____ → _____ direction. (67) _____ _____ then adds nucleotides to the (68) _____ _____. During assembly of a DNA strand, the next nucleotide that is added to the growing chain has a(n) (69) _____ group exposed. Since DNA polymerase can only add a nucleotide at the (70) _____ end of an existing nucleotide chain, the direction of chain assembly is always in the (71) _____ → _____ direction.

The DNA strand that is synthesized (72) _____ in the direction of DNA unwinding is called the (73)_____ _____. The second DNA strand is synthesized (74) _____ and is called the (75) _____ _____. Replication of the (76) _____ strand produces (77) _____ lengths of nucleotides called (78) _____ _____. These short lengths as well as any gaps in the nucleotide strands are connected by (79) _____ _____.

DNA replication is partially influenced by the length of (80) _____, segments at the (81) _____ of chromosomes. Each time DNA (82) _____, (83) _____ become (84) _____ in length. An enzyme, (85) _____ affects the length of the (86) _____ by adding (87) _____ _____ to the chromosome end. The (88) _____ process and (89) _____ development may be associated with (90) _____ length and (91) _____ activity.

Choice

For each of the following statements, choose the most appropriate enzyme that catalyzes the reaction in DNA replication.

a. DNA polymerase b. DNA helicase c. DNA ligase d. telomerase e. primase

92. _____ Catalyzes the unwinding of the DNA double helix
93. _____ Produces an RNA strand that acts as a starting point for DNA replication
94. _____ Connects Okazaki fragments in the lagging strand
95. _____ Adds repeating units to the end of chromosomes
96. _____ May be responsible for cancer cell development
97. _____ Adds nucleotides only at the 3' end of an existing nucleotide chain

Short Answer

98. Given the following DNA molecule, identify which strand is the leading strand and which is the lagging strand:

3' AATCCGTACGGT 5'

5' TTAGGCATGCCA 3'

____

____

99. Refer to the previous question and explain which strand will have continuous replication and which will have discontinuous replication. Include explanations for why this occurs. ______________________

____

____

14.4 Mechanisms That Correct Replication Errors [pp. 292-294]

14.5 DNA Organization in Eukaryotes and Prokaryotes [pp. 295-298]

Section Review (Fill-in-the-Blanks)

DNA polymerase has a built-in enzyme, (100) _____, which can remove (101) _____ nucleotides and insert the correct base pair. If a base pair mistake remains, additional (102) _____ _____ _____ function to ensure accuracy in replication. (103) _____ _____ will not form between mismatched base pairs. DNA proofreading and repair mechanisms ensure a high level of (104) _____ during replication.

In the nucleus of eukaryotes, (105) _____ is composed of DNA and two major types of proteins, (106) _____ and _____. DNA, which is (107) _____ charged, binds with the (108) _____ charged (109) _____ proteins. Histone proteins and DNA form an organizational structure, the (110) _____. (111) _____ proteins, as well as histone proteins, are thought to play a role in (112) _____ _____. Chromatin fibers have loosely packed regions, (113) _____, and densely packed regions, (114) _____. It is thought that most gene expression occurs in the (115) _____.

Short Answer

116. Explain the effect of inhibition of DNA repair mechanisms. ______________________

____

____

117. Compare and contrast nucleoid and plasmid. ______________________

____

____

Labeling

118. Label the various levels of organization of eukaryotic chromatin and chromosomes.

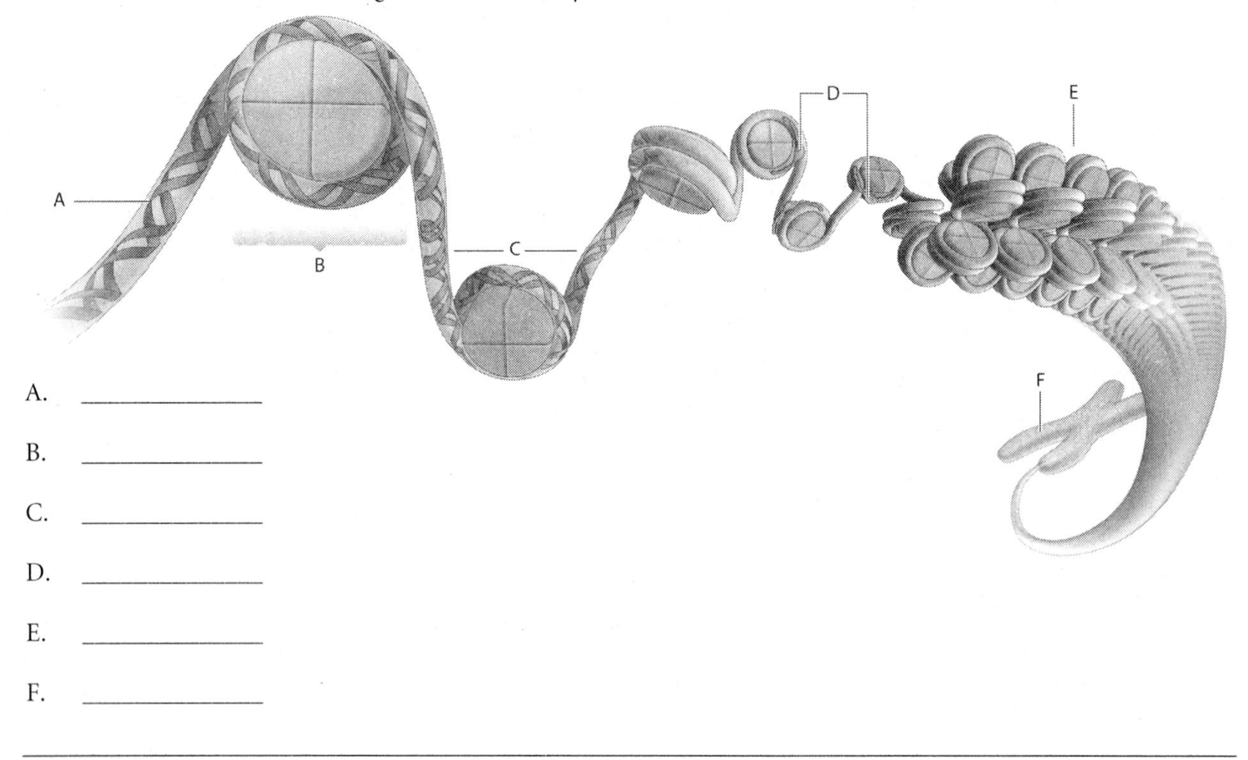

A. ______________

B. ______________

C. ______________

D. ______________

E. ______________

F. ______________

SELF-TEST

1. In the early 1900s, the hereditary material was thought to be ________. [p. 278]
   a. steroids
   b. nucleic acids
   c. carbohydrates
   d. proteins

2. Griffith's experiments with bacteria identified an agent responsible for the ________ of the *R* strain (nonpathogenic) to the *S* strain (pathogenic). [pp. 278-279]
   a. Chargaff conversion
   b. transformation
   c. diffraction
   d. complementary exchange

3. Hershey and Chase used radioactive labeling experiments to identify the hereditary material. Radioactive isotopes of both sulfur and phosphate were used. Sulfur is found in ______ but not in ________. [pp. 280-281]
   a. DNA; proteins
   b. proteins; DNA
   c. carbohydrates; DNA
   d. DNA; lipids

4. Nucleotides consist of deoxyribose, a phosphate group, and ________. [p. 281]
   a. a purine
   b. a pyrimidine
   c. either a purine or pyrimidine
   d. a phosphodiester bond

5. Which of the following techniques led to the conclusion that the DNA molecule was a double helix? [p. 282]
   a. complementary base pairing
   b. transformation
   c. X-ray diffraction
   d. replication

6. Experiments of DNA replication identified that the daughter DNA consisted of one strand of parent DNA and one strand of new DNA. This type of replication is ________. [pp. 284-285]
   a. semiconservative
   b. conservative
   c. dispersive
   d. semidisruptive

7. When DNA is replicating, a nucleotide can only be added to the ________ end of an existing nucleotide chain. [p. 287]
   a. 3'
   b. 5'
   c. either 3' or 5'
   d. phosphodiester end

8. Activity in which enzyme would indicate that DNA was about to replicate? [p. 288]
   a. topoisomerase
   b. DNA polymerase
   c. primase
   d. DNA helicase

9. Due to the directionality of DNA polymerase, the lagging strand of DNA is replicated in a ________ fashion. [p. 288]
   a. discontinuous
   b. continuous
   c. Okazaki
   d. antiparallel

10. If the replication fork were blocked, the activity of ________ would be affected first. [p. 288]
   a. helicase
   b. DNA polymerase
   c. primase
   d. DNA ligase

11. If DNA proofreading fails, ________ may occur. [p. 293]
   a. transformations
   b. mutations
   c. telomerization
   d. mismatch base pairing

12. If you wanted to study DNA replication in a eukaryote, the most likely portion of the chromosome to look for changes in activity would be the ________. [pp. 294-296]
   a. heterochromatin
   b. nucleoid
   c. nucleosome
   d. euchromatin

INTEGRATING AND APPLYING KEY CONCEPTS

1. For a chromosome with 1000 nucleotide base pairs, predict the mutation rate for 1000 replications, given that one mistake in base pairing occurs for every 10 million nucleotides assembled.

2. Predict the effects of mutations in histone and nonhistone proteins.

3. *Unanswered Questions:* If telomerase activity could be conclusively linked to cancer cell development, predict some of the possible types of treatment or agents to inhibit or stop cancer growth and spread to other tissues.

15

FROM DNA TO PROTEIN

CHAPTER HIGHLIGHTS

- All living things use their genetic program to make proteins.
- Specific regions on DNA represent a gene or transcription unit.
- There are two major steps of protein synthesis: transcription and translation.
- Transcription:
    - Base sequence on DNA is read.
    - RNA polymerase is used to make a corresponding mRNA sequence.
- Translation:
    - Triplet base sequence on mRNA represents a coding system.
    - Ribosome reads the codes on mRNA.
    - tRNA brings the coded amino acid.
    - Amino acids are joined to form a polypeptide.
- Although the basic process of making proteins is similar in prokaryotic and eukaryotic cells, there are a few detail differences.

STUDY STRATEGIES

- Remember the goal of this chapter is to understand how cells make proteins.
- There are step-by-step processes that must be understood and followed.
- Concentrate on the key concepts of transcription and translation.
- Once the key ideas are understood, you can then learn the steps within each pathway.

TOPIC MAP

Prokaryotic cells

DNA → mRNA → Protein

Eukaryotic cells

DNA → Pre-mRNA → mRNA → Protein

INTERACTIVE EXERCISES

Why It Matters [p. 301–302]

Section Review (Fill-in-the-Blanks)

Mytilus, a mussel, produces fibers made of (1) ________________ that allow the animal to attach to the substrate. In order to produce large amounts of this strong adhesive, the (2) ____________ from the mussel is transferred to yeast cells. Yeast cells are then able to use this (3) ____________ to make fibers.

This chapter is about reading the code in (4) __________ to make a corresponding copy of messenger (5) ___________. This messenger is then read by the (6) _______________ to make (7) _________________.

15.1 The Connection Between DNA, RNA, and Protein [pp. 302–307]

Section Review (Fill-in-the-Blanks)

In 1896, English physicians (8) ______________ ____________ and (9) _____________ ______________ studied human genetic disease, *alkaptonuria,* to connect inheritance of (10) _____________ defect with change in genetic information.

It was the work of (11) _______________ _____________ and (12) ______________ ___________ with *Neurospora* which proved that the genetic information codes for specific (13) ________________ required to catalyze metabolic reactions. They found that the wild variety of *Neurospora* can grow on (14) ______________ medium containing salts, sugar, and vitamins. After exposing the fungus to X rays, it needed additional (15) _______________ added to the medium. These new variants of the fungus are referred to as (16) ___________________ or nutritional (17) ________________. The scientists hypothesized that X-ray caused changes in the (18) ___________, as a result *Neurospora* could not make specific (19) ________________ and therefore blocked specific (20) ______________ pathways. They proposed one (21) ___________ – one (22) ________________ hypothesis and received a Nobel Prize for their work.

Cells have many other types of proteins in addition to enzymes, and proteins are made of one or more chains of (23) ____________ __________ called (24) ________________. The hypothesis was then changed to one (25) ___________ – one (26) ________________ hypothesis.

Protein synthesis has two steps: Step I is called (27) ________________________, where the specific sequence in one of the two strands of (28) _________, called a(n) (29) ____________, is copied to make a complementary messenger (30) _________. The enzyme involved in this process is called (31) _______ ___________________. Step II is called (32) ________________________, where the message code in messenger (33) ________ is read by the organelle called (34) _______________, and a specific sequence of (35) _______________ ___________ is assembled to make a polypeptide. Francis Crick named this flow of information from DNA to (36) ________ to proteins as the (37) ____________ _________.

Although the two steps of protein synthesis are similar in (38) ________________ and eukaryotic cells, one major difference does exist. Eukaryotic cells make a larger (39) ________-mRNA in the nucleus, which is processed to form functional (40) __________ before it is translated in the (41) ______________ of the cell.

The four nucleotide bases found in DNA are (42) ______________, ______________, ______________, and ______________. RNA, on the other hand, has (43) ______________, ______________, ______________, and ______________. While making mRNA, for every adenine in DNA, mRNA would have (44) ______________; for thymine in DNA, mRNA would have (45) ______________; for guanine in DNA, mRNA would have (46) ______________; for cytosine in DNA, mRNA would have (47) ______________.

Once the program is transferred from DNA to mRNA, the code in mRNA is read to put a sequence of (48) ___________ ___________ together to make a(n) (49) ______________. This code in mRNA is called the (50) ______________ code. In 1968, two scientists, (51) ______________ ______________ and (52) ______________ ______________, received a Nobel Prize for cracking this code. They achieved this by making artificial (53) ____________, with known nucleotide sequence, and then determining the amino acids used by the ribosomes to make a(n) (54) ______________. With (55) _______ types of amino acids available, they proposed that a set of (56) _____ nucleotides in mRNA coded for an amino acid. Using (57) _______ types of nucleotides that make up mRNA, there are (58) _______ possible combinations, more than what is needed to code (59) _______ amino acids. Obviously, some amino acids are coded by more than one (60) ____________. This multiple coding of an amino acid is referred to as the (61) ______________ or (62) ______________. Out of (63) _______ possible combinations of nucleotides, (64) _______ code for amino acids, and these are called the (65) _________ codons. One of these codons, AUG, not only codes for the amino acid (66) ______________, but is also a(n) (67) __________ codon. The remaining three codons, UAA, UAG, and UGA, are called (68) ___________ codons. They do not code for any amino acid, instead they represent the end of (69) ___________ sequence.

Because the coding system is similar in all living things, eukaryotic or prokaryotic, it is referred to as the (70) ______________ coding system.

Matching

Match each of the following terms with its correct definition.

71.	____	Auxotroph	A. Organelles that read mRNA to assemble a polypeptide
72.	____	Polypeptide	B. AUG—that codes for methionine and starts translation process
73.	____	Transcription	C. Mutants that require additional nutrients added to the minimal medium
74.	____	Translation	D. UAA, UGA, UAG—that code for the termination of translation process
75.	____	Central dogma	E. A chain of amino acids
76.	____	RNA polymerase	F. A longer mRNA made by eukaryotic cells
77.	____	Ribosomes	G. The process of copying DNA sequence to make mRNA
78.	____	Codons	H. The flow of information from DNA to RNA to proteins
79.	____	Pre-mRNA	I. Refers to genetic code being same in all living things
80.	____	Start codon	J. The process of reading codes on mRNA to assemble amino acids
81.	____	Stop codon	K. The enzyme that helps in making mRNA
82.	____	Universal coding	L. A set of three nucleotides on mRNA that codes for a specific amino acid

Matching

Match each of the following scientists with their contribution to cellular respiration.

83.	____	Archibald Garrod and William Bateson	A. Proposed one gene–one enzyme hypothesis
84.	____	George Beadle and Edward Tatum	B. Named the flow of information from DNA to RNA to protein as the central dogma
85.	____	Francis Crick	C. Received Nobel prize for cracking the genetic code.
86.	____	Marshall Nirenberg and Hargobind Khorana	D. Correlated a change in genetic information with development of certain diseases

Labeling

87. Label the major steps of protein synthesis in eukaryotic cells:

A. ______________________

B. ______________________

C. ______________________

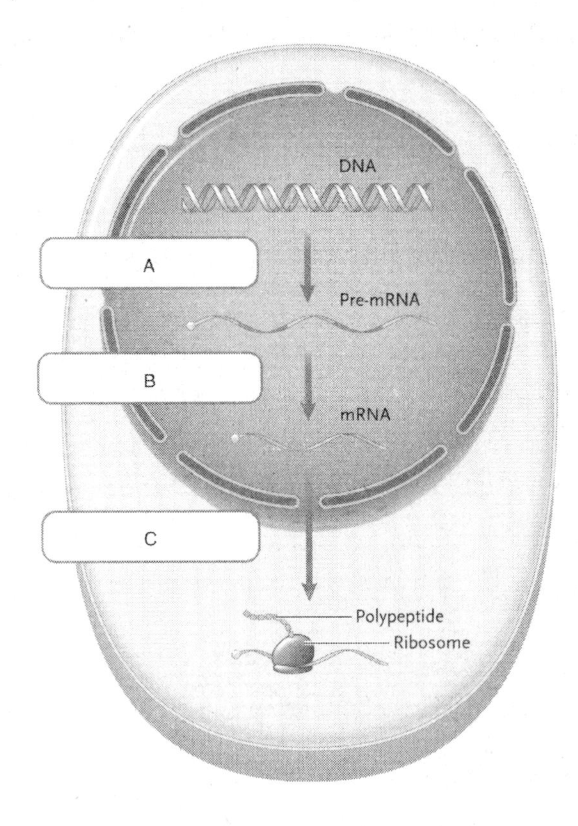

Complete the Table

88. Complete the following table for the nitrogenous bases found in the two types of nucleic acids.

Nucleic Acids	*Types of Nitrogenous Bases*
A. DNA	
B. RNA	

15.2 Transcription: DNA-directed RNA Synthesis [pp. 307–309]

Section Review (Fill-in-the-Blanks)

Copying DNA sequence to make a mRNA is called (89) ____________________. This process is similar to making copies of DNA, known as DNA (90) ____________________.

The differences between the two process are:

A. In making DNA, adenine pairs with (91) ______________, and cytosine pairs with (92) ______________. Whereas in making mRNA, everything is the same except that adenine in DNA is copied as (93) __________ in RNA.

B. In making DNA, both the strands of (94) _____________ are copied. In making RNA, only one strand of (95) __________ is copied.

C. In making DNA, a(n) (96) ____________-stranded DNA is formed. In making mRNA, a(n) (97) ______________-stranded RNA is formed.

D. DNA polymerase is used to make new (98) ________ molecules. In making (99) ________, RNA polymerase is used.

E. DNA polymerase requires a short RNA called the (100) ___________ before it starts making a complimentary strand. RNA polymerase does not require any (101) ____________.

The part of DNA that has the program for making a polypeptide is called the (102) _______________ unit or the (103) ________. According to the human genome project, there are approximately 20,000–30,000 (104) __________ needed to code an entire human. Out of these, most are transcribed and translated to make proteins called the (105) ____________-______________ genes. The remaining genes code for RNA that are not translated, called the (106) ________-____________-______________ genes such as the ribosomal RNA and transfer RNA.

The enzyme (107) ________ ________________ binds to a region before the transcription unit called the (108) ________________ region. The enzyme unwinds the two strands of the transcription unit on (109) _______ molecule and moves from (110) _____ end → (111) _____ end of the unit to start reading the base sequence. Complementary (112) _________________ that match the sequence on DNA are added in (113) _____ end → (114) _____ end. The new mRNA temporarily binds to the (115) ________ template. The enzyme reaches the end of the transcription unit and at that point the enzyme and (116) __________ come of the (117) ________ template.

As one RNA polymerase proceeds with transcription of the (118) _______ template, another enzyme molecule can begin making another (119) ________. In a short period of time, multiple copies of (120) _________ are formed.

Even though the major steps of transcription are similar for prokaryotic and eukaryotic cells, there some significant differences:

A. In prokaryotic cells, there is only one enzyme, RNA polymerase, that is responsible for transcribing (121) ____ types of genes. In eukaryotes, RNA polymerase (122) _____ that transcribes protein-coding genes, whereas RNA polymerases (123) ____ and (124) ______ transcribe non–protein-coding genes, such as genes that code for (125) __________ and (126) ___________.

B. In prokaryotes, RNA polymerase binds directly to the (127) ______________ region. In eukaryotes, the (128) ______________ region of the protein-coding genes has a(n) (129) ____________ box that is not recognized by RNA polymerase (130) ____. Proteins called the (131) ______________ factors bind to the (132) __________ box which then allows RNA polymerase (133) ____.

C. In prokaryotic cells, there is (134) ______________ signal at the end of the transcription unit, where a(n) (135) ____________ binds to trigger release of RNA polymerase and mRNA. There is no such signal in eukaryotic cells, instead the 3' end of (136) ____________ itself determines the end of transcription.

Matching

Match each of the following terms with its correct definition.

137. ____ DNA polymerase
138. ____ RNA polymerase
139. ____ RNA polymerase II
140. ____ RNA polymerases I and II
141. ____ Transcription unit
142. ____ Protein-coding genes
143. ____ Non–protein-coding genes
144. ____ Promoter region
145. ____ RNA Primer
146. ____ TATA box
147. ____ Transcription factors
148. ____ Terminator region

A. Transcription units that code for proteins
B. Enzyme that transcribes protein-coding genes in eukaryotic cells
C. Enzyme for DNA replication
D. Sequence located before transcription unit, where RNA polymerase binds
E. Sequence located at the end of the transcription unit in prokaryotes that signals the end of transcription
F. Enzyme responsible for transcription of protein-coding and non–protein-coding genes in prokaryotes
G. Region of DNA that has the program for specific mRNA and protein
H. Enzyme that transcribes protein-coding genes in eukaryotic cells
I. A sequence in the promoter region of eukaryotic cells that is recognized by transcription factors before RNA polymerase binds
J. Genes that code for RNA that do not get translated into proteins
K. A short RNA sequence that is needed before enzymes can replicate DNA
L. Proteins that must bind to the promoter region before RNA polymerase binds in eukaryotic cells

Sequence

149. Arrange the following steps of transcription in the correct sequence.

A. The new mRNA temporarily binds to DNA template.

B. RNA polymerase binds to the promoter region.

C. The enzyme reaches the end of the transcription unit.

D. The enzyme moves along transcription unit.

E. The enzyme unwinds the two strands of DNA.

F. mRNA and the enzyme are released from DNA.

G. Corresponding nucleotides are added.

____ ____ ____ ____ ____ ____ ____

Complete the Table

150. Complete the following table for the differences between DNA replication and transcription.

Points of Comparison	*DNA Replication*	*Transcription*
Base pairing		
Number of DNA strands copied		
Number of new strands formed		
Enzyme involved		
Primer formed		

151. Complete the following table for differences in transcription of prokaryotic and eukaryotic cells.

Points of Comparison	*Prokaryotic Cells*	*Eukaryotic Cells*
Transcription enzyme/s		
Binding of RNA polymerase to promoter region		
Termination of transcription		

Short Answer

152. Compare the types of RNA polymerase/s found in prokaryotic and eukaryotic cells and how they differ in their functions.

____

____

153. What is a TATA box and what is its role in eukaryotic transcription? _______________

____

____

154. What is the difference between protein-coding and non–protein-coding genes? _______________

____

____

True/False

Mark if the statement is true or false. If the statement is false, justify your answer in the line below each statement.

155. ______________ Base pairing in DNA is the same as the one used by RNA polymerase for making mRNA.

____

156. ______________ Enzymes used for transcription are same in prokaryotic and eukaryotic cells.

____

157. ______________ Transcription takes place in the cytoplasm of prokaryotic cells and in the nucleus of eukaryotic cells.

____

15.3 Production of mRNAs in Eukaryotes [pp. 309–313]

Section Review (Fill-in-the-Blanks)

Prokaryotic mRNA contains coding region that is exactly the size needed to code for a(n) (158) ____________. It also has the non-coding regions, the 3' and 5' (159) ________________ regions, or UTR, that play important roles in prokaryotic protein synthesis.

The transcript in eukaryotic cells is quite different from that made by the prokaryotic cells.

A. The protein-coding gene is transcribed in the (160) ____________ to form a longer (161) _____-___________. It is edited to form a(n) (162) ___________ before it reaches the cytoplasm for (163) ____________________.

B. The 5' end of pre-mRNA gets a(n) (164) ____________ triphosphate as the 5'cap that serves as the binding site for the (165) ______________ during translation.

C. Eukaryotic cells do not have a(n) (166) ________________ sequence at the end of the protein-coding genes. A sequence at the end of the gene allows pre-mRNA to be released, to which is then added a chain of (167) _______________, called the (168) __________ ____ tail. This addition to the (169) ______ end of pre-mRNA is to protect the transcript from digestive enzymes while it is moving from the (170) ____________ to the (171) _______________.

D. Two scientists, (172) ______________ ________________ and (173) _____________ _______________, received a Nobel Prize for reporting that (174) ______________, the non–protein-coding sequence in eukaryotic pre-mRNA, are interspersed in (175) ______________, the protein coding sequence.

E. Before pre-mRNA exits the nucleus, a process called (176) _______ __________________ takes place where (177) _____________ are removed and (178) ______________ are joined together. A complex of (179) ____________ _________________ RNA and several proteins, called (180) __________ ____________________ _________________, also referred to as (181) _______________, binds to pre-mRNA. This complex is called the (182) _______________, pre-mRNA is edited, and translatable (183) ___________ is formed.

Advantages of RNA processing:

1. It allows eukaryotes to make (184) ______________ splicing and create multiple types of mRNA and therefore proteins using the same gene.

2. It may allow evolution of new proteins by (185) __________ ______________ by selecting different regions of the gene and combining them to make a variety of proteins.

Matching

Match each of the following terms with its correct definition.

186. ____	Untranslated regions	A. Sequence of adenines added to pre-mRNA to protect it from degradative enzymes
187. ____	5' GTP cap	B. Non–protein-coding sequence in eukaryotic pre-mRNA
188. ____	3' poly A tail	C. Sequence in prokaryotic mRNA that regulate protein synthesis
189. ____	Pre-mRNA	D. Process where introns are removed and exons are joined together
190. ____	Introns	E. Small ribonucleoprotein particles used for splicing
191. ____	Exons	F. RNA found in the nucleus that combines with proteins to form snRNPs
192. ____	mRNA splicing	G. 5' end of pre-mRNA that serves as the binding site for the ribosomes
193. ____	Spliceosome	H. Multiple ways to edit pre-mRNA to form different types of mRNA
194. ____	snRNPs	I. A process where different regions of a gene are combined to form different types of mRNA and proteins
195. ____	Small nuclear RNA	J. Protein-coding sequence in eukaryotic pre-mRNA
196. ____	Alternative splicing	K. Larger form of mRNA made by eukaryotes
197. ____	Exon shuffling	L. A complex formed by ribonucleoproteins attached to pre-mRNA during editing process

Complete the Table

198. Complete the following table with specific information on eukaryotic transcript.

Eukaryotic Transcription	Role in Transcription
A. Pre-mRNA	
B. 5' GTP cap	
C. 3' poly A Tail	
D. Introns and Exons	
E. mRNA Splicing	

15.4 Translation: mRNA-Directed Polypeptide Synthesis [pp. 313–326]

Section Review (Fill-in-the-Blanks)

Translation is about reading the base sequence on (199) ____________, using its 3 nucleotide coding system to put (200) _____________ ____________ together, and forming a(n) (201) __________________. In both prokaryotes and eukaryotes, this process takes place in the (202) ________________ of the cell. The difference is that in prokaryotes, as mRNA is being made, (203) ____________________ can begin simultaneously; whereas in eukaryotes, (204) _____-____________ must be edited and exit the (205) ________________ of the cell before translation can begin.

Translation requires (206) ______________ that is made during transcription, (207) __________________ the organelles that facilitate this process, (208) ______________ RNA or tRNA that carry (209) ______________ ______________, and (210) ________________ that join amino acids together to form a(n) (211) ____________________.

tRNA is made of only (212) _____-_____ bases and folded to form a shape of (213) ________ leaf. At the tip of tRNA is a set of 3 (214) ________________, the (215) ______________, whose sequence matches specific codon sequence on (216) ______________. At the other end of tRNA, there is a specific (217) __________ __________ attached which is coded by the (218) ______________ on mRNA and carried by tRNA whose (219) ______________ matches the codon.

Amino acids are attached to tRNA by a process called (220) __________________ or charging by using enzymes called (221) ________________-________ ________________, one for each of 20 amino acids. This process is active, that is, it requires (222) _________.

Ribosomes are made of two subunits: (223) ____________ and (224) ____________ subunits. Each subunit is made of (225) ________________ RNA and (226) ________________ proteins. Prokaryotic ribosomes are (227) ________________ than eukaryotic ribosomes. This difference allows physicians to treat bacterial infections by prescribing antibiotics such as streptomycin and erythromycin, which inhibit (228) ________________ ribosomes.

Ribosomes have special binding sites:

(229) _____ site: aminoacyl site, where aminoacyl-tRNA (carrying specific amino acid) binds.

(230) _____ site: peptidyl site, where tRNA shifts after its amino acid has joined the growing peptide.

(231) _____ site: exit site, where tRNA leaves the ribosomes after its peptide chain has joined to the new tRNA.

Translation has 3 major steps:

1. (232) ______________: A(n) (233) ____________ and a(n) (234) ____________ subunit of the ribosome bind to the start codon (235) __________ on (236) ____ end of mRNA. The first tRNA, the initiator tRNA with anticodon (237) __________ and amino acid (238) ______________ binds to the (239) ____ site of the ribosome. GTP and (240) ______________ factors are involved in this process.

2. (241) ______________: A second (242) __________ with its specific anticodon and amino acid attaches to the (243) ____ site of the ribosome. Enzyme (244) __________ ________________ catalyzes the attachment of

methionine, the first amino acid, to the specific (245) __________ __________ carried by the second tRNA. This enzyme is a ribozyme and it is unusual because of it is a(n) (246) ________ instead of being a protein. Ribosome uses GTP to move to the next codon, that makes the first tRNA (without an enzyme), to now shift to the (247) ___ site and leave the ribosome; the second tRNA (with two amino acids) shifts to the (248) ___ site; and the (249) ___ site is now available for the next tRNA. This process continues, ribosome continues to move from 5' end towards the (250) ___ end, and the chain amino acids called the (251) ____________ continues to grow.

3. (252) ____________: Ribosome finally reaches the last codon called the (253) _______ codon. (254) ______________/________________ factor binds to the (255) ___ site, the growing (256) ____________________ chain is released, and the ribosomal (257) ____________ separate.

Multiple ribosomes can translate a(n) (258) ________ simultaneously. The complex is referred to as a(n) (259) ______________. In prokaryotic cells, absence of a nuclear membrane allows both the steps of protein synthesis, (260) _________________ and (261) _________________, to take place simultaneously.

Modification of polypeptide:

1. For some proteins, the chain of (262) __________ __________ must be edited.

2. Some proteins need (263) _____________ to help them fold the chain to achieve a specific 3-dimensional shape.

3. Other proteins, such as (264) ______________, are made as large inactive proteins that must be enzymatically digested by to form the shorter, active protein.

Proteins that must remain in the cytoplasm are made by (265) ________ ribosomes. Those proteins that are transported to other locations of the cell or out of the cell are made by (266) ______________ ribosomes. These proteins have additional amino acids, (267) __________ sequence that help in attaching to the (268) _______ receptors on rough ER for further processing.

Mutation refers to a change in (269) _____________ sequence. There are different types of mutations:

1. Base substitution mutation: where one (270) ______________ is replaced by another type. This may or may not change the (271) ___________ on mRNA and may not change the (272) ______________ ____________ of the protein. This type of mutation is called the (273) __________ mutation. If the amino acid sequence changes to alter the protein and its function, this type of mutation is called the (274) ____________ mutation. If due to mutation, the chain abruptly terminates, it is referred to as the (275) ____________ mutation.

2. Frame shift mutation: where a(n) (276) ______________ is deleted or added in the sequence. This would cause all codons on (277) ___________ to change and usually result in major changes in (278) ______________ __________ sequence of the protein making the protein nonfunctional.

Matching

Match each of the following terms with its correct definition.

279. ____	Translation	A. RNA that carries specific amino acid to ribosomes
280. ____	Ribosomes	B. A set of 3 nucleotides on tRNA that matches a specific codon on mRNA
281. ____	tRNA	C. Process of reading the codes on mRNA to put amino acids together and forming a polypeptide
282. ____	Codon	D. Set of 20 enzymes that join a specific amino acid to a tRNA
283. ____	Anticodon	E. Exit site on the ribosome where tRNA binds after its peptide transfers to the new tRNA
284. ____	Aminoacylation/charging	F. An enzyme that is made of RNA instead of a protein
285. ____	Aminoacyl-tRNA synthetase	G. Organelles that assist translation
286. ____	rRNA	H. Aminoacyl site where tRNA binds, bringing its specific amino acid
287. ____	E site	I. Peptidyl site where tRNA shifts carrying the growing peptide
288. ____	Ribozyme	J. Process of attaching specific amino acid to a tRNA
289. ____	A site	K. A set of 3 nucleotides on mRNA that codes for a specific amino acid
290. ____	P site	L. Type of RNA that makes up ribosomes
291. ____	Polyribosome	M. A series of ribosomes bound to a mRNA
292. ____	Pepsinogen	N. Proteins that help fold certain polypeptides to achieve a specific 3-D shape.
293. ____	Chaperones	O. A larger, inactive protein that must be cleaved to make it shorter and active

Matching

Match the following terms related to mutation with their correct definition.

294. ____	Base substitution mutation	A. Mutation that results in a change in the amino acid sequence to alter the functions of the protein
295. ____	Missense mutation	B. Mutation where a nucleotide is deleted or added that changes all the codons on mRNA.
296. ____	Nonsense mutation	C. Mutation where a nucleotide is replaced by another type of nucleotide.
297. ____	Frame shift mutation	D. Mutation where the change in DNA sequence does not change the amino acid sequence or the functions of a protein.
298. ____	Silent mutation	E. Mutation that results in abrupt termination of the polypeptide, leaving the protein totally dysfunctional.

Short Answer

299. Prokaryotic cells can transcribe and translate simultaneously. Explain why that is not possible in eukaryotic cells.

____

____

300. Substitution mutation may not make a change in protein sequence. Explain. ____________________

____

____

301. Ribosomes have 3 binding sites. What are they called and what binds to each site? ____________________

____

____

302. Methionine is always the first amino acid in a protein. Give an explanation. ____________________

____

____

Problem solving:

303. A gene has a sequence of: TACTTCGCAAATCCCGCAGTCACGTTGATC

Give the mRNA sequence: ____

Copy the mRNA sequence and mark codons:

____

Give amino acid sequence coded in the above mRNA:

____

SELF-TEST

1. In which step of protein synthesis is RNA polymerase involved? [p. 307]
   a. Transcription
   b. Translation
   c. In both the steps
   d. In neither of the two steps

2. Codons are read in [p. 305]
   a. Transcription
   b. Translation
   c. Both the steps
   d. Neither of the two steps

3. Which of the following is the start codon? [p. 306]
   a. TAC
   b. UAA
   c. AUG
   d. UAC

4. Which of the following is the amino acid coded by the start codon? [p. 306]
   a. glutamine
   b. methionine
   c. lysine
   d. phenylalanine

5. How many codons are possible using the four types of nitrogenous bases? [p. 305]
   a. 20
   b. 40
   c. 46
   d. 64

6. Where does protein synthesis take place in a prokaryotic cell? [p. 305]
   a. In the cytoplasm
   b. In the nucleus
   c. Part in nucleus and part in cytoplasm
   d. In mitochondria

7. Which of the following is formed when a gene is transcribed in a eukaryotic cell? [p. 309]
   a. protein
   b. mRNA
   c. pre-mRNA
   d. pepsinogen

8. In eukaryotic transcript, all of the following are present except [p. 309]
   a. introns and exons.
   b. 5' GTP cap.
   c. methionine.
   d. 3' poly A tail.

9. Which of the following helps in splicing pre-mRNA in eukaryotic cells? [p. 311]

a. snRNP

b. ribosome

c. RNA polymerase

d. tRNA

10. Which of the following has the anticodon? [p. 313]

a. mRNA

b. DNA

c. tRNA

d. rRNA

11. Which of the following site in the ribosome allows aminoacyl-tRNA to bind? [p. 315]

a. A

b. P

c. E

d. D

12. At which site in ribosome does the polypeptide grow? [p. 315]

a. A

b. P

c. E

d. D

13. The enzyme peptidyl transferase is: [p. 317]

a. a protein.

b. RNA.

c. DNA.

d. a carbohydrate

14. Which of the following type of mutation is usually considered as most devastating? [p. 326]

a. base substitution

b. silent

c. frame shift

d missense

15. Only one ribosome can translate an mRNA at one time. [p. 322]

a. True

b. False

INTEGRATING AND APPLYING KEY CONCEPTS

1. The genetic code has not changed through years of evolution. How have we used this fact to benefit us?

2. Prokaryotic and eukaryotic cells have major differences in their structure. How does that impact protein synthesis?

16

CONTROL OF GENE EXPRESSION

CHAPTER HIGHLIGHTS

- Prokaryotic organisms:
  - Are simpler, single-cell organisms.
  - Adapt quickly to changes in their environment.
  - Mostly regulate their gene expression at the transcriptional level.
  - Can make short-term, rapid, and reversible alterations in their metabolic pathways.
- Eukaryotic organisms:
  - Are more complex unicellular or multicellular organisms.
  - Undergo long-term changes in gene expression during differentiation.
  - Undergo short-term changes in gene expression to adapt to changes in their environment.
  - Regulate genes at transcriptional, posttranscriptional, translational, and posttranslational levels.

STUDY STRATEGIES

- This chapter has extensive gene control pathways. Trying to memorize all of the steps is the most common mistake.
- Remember the goal of this chapter is to understand how genes are controlled.
- Concentrate on the key steps for protein synthesis: transcription and translation.
- Focus on what each step is about and where changes can be made.
- Once the key steps are understood, you can then learn the regulatory pathways.

TOPIC MAP

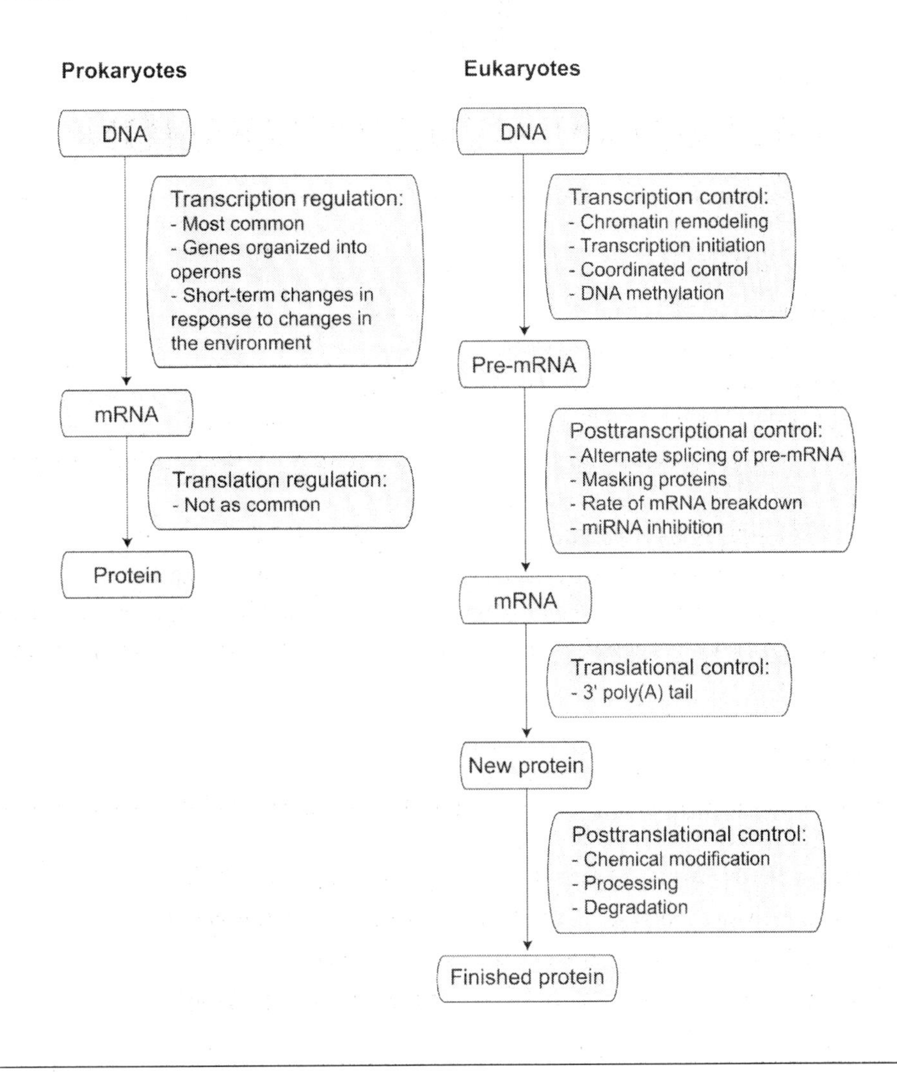

INTERACTIVE EXERCISES

Why It Matters [pp. 329–330]

Section Review (Fill-in-the-Blanks)

At the time of release, a human egg is metabolically (1) ______________. As soon as the (2) ______________ fertilizes the egg, the egg cell becomes (3) ___________ and starts dividing. Initially all the embryonic cells are structurally and functionally similar, but they soon start to (4) ____________________. That is, even though they have the same set of (5) ___________, certain genes are turned on and others are turned (6) ______.

Prokaryotic organisms make rapid, (7) __________-term changes in their transcriptional step in order to respond to changes in their environment. Eukaryotic organisms make (8) ________-term changes to respond to changes in their environment but (9) __________-term changes for differentiation of their cells.

16.1 Regulation of Gene Expression in Prokaryotes [pp. 330–335]

Section Review (Fill-in-the-Blanks)

Prokaryotic cells have very rapid changes in their gene expression to adapt to the changes in their (10) ________________.

Escherichia coli can metabolize a wide range of (11) ____________ for producing energy. Even though glucose is a preferred sugar, when lactose is present in the medium, the bacterium makes the three (12) ______________ needed to metabolize lactose. However, if lactose is absent, the bacteria does not make these (13) ______________.

The two scientists (14) ________________ ____________ and (15) ______________ ____________ proposed the lac (16) ______________ model to explain the gene expression for lactose metabolism in *E. coli* where an operon refers to a cluster of genes and associated (17) _______ sequences in (18) ________________ organisms.

Negative gene regulation of lac operon: The cluster of genes, (19) ________, ________, and ________, for lactose metabolism code for 3 enzymes: (20) ______________ for the breakdown of disaccharide lactose into glucose and galactose, (21) ________________ for the transport of lactose into the cell, and (22) __________________ whose function is unknown. All of these 3 genes are transcribed together to make a single (23) ________, which then is translated to make the 3 (24) ________________. Located close to the cluster of genes is a(n) (25) _____________ sequence for the binding of RNA polymerase. Another gene located away from the operon is called the (26) ______________ gene and makes a regulatory protein that is referred to as the lac (27) ______________ protein. This active protein binds to the (28) ______________ sequence that is located between the promoter and the cluster of genes. It blocks the path of RNA polymerase and thereby blocks the transcription of the genes that code for the 3 (29) ______________.

A. When lactose is present in the medium: Some of lactose enters the cell and is converted to allolactose by the enzyme (30) ________________ → allolactose acts as a(n) (31) ________________ → it binds to the lac repressor protein made by the (32) _______________ gene and inactivates it → the inactive repressor cannot bind to the (33) _____________ sequence → RNA polymerase binds to the (34) __________________ region and transcribes the cluster of genes (35) ______, ______, and ______ → a single (36) _______ is formed → 3 (37) ____________ are formed for the metabolism of lactose → bacteria is able to produce energy.

B. When lactose is absent in the medium: Absence of lactose means absence of (38) ________________ in the cell → the lac repressor protein is active and binds to the (39) ______________ sequence → RNA polymerase is blocked at the (40) __________________ region → no transcription of the cluster of genes (41) _______, _______, and _______ → no lactose metabolism (42) ____________ are formed.

This system where the presence of lactose, a(n) (43) ____________, increases the expression of the genes is referred to as the (44) ________________ operon. It is also called the (45) ______________ gene regulation because an active repressor turns off the cluster of genes. There is also a(n) (46) ______________ gene regulation system known that regulates lac (47) ______________. This system makes sure that if both, lactose and glucose are present, the cell uses the more preferred (48) ____________ and therefore does not make the 3 enzymes needed for (49) ____________ metabolism. However, if lactose is present and glucose is absent, this system ensures that the 3 (50) ____________ are made to metabolize lactose.

Positive gene regulation of lac operon: lac (51) ____________ is also regulated by the (52) ______________ gene regulation to ensure that when glucose is present in the medium, the bacterium preferentially uses it to produce (53) ____________. Lactose, being a(n) (54) ________________, requires additional processing before it can be used by the bacterium.

A. When lactose is present and glucose is absent: Lactose is converted to (55) ______________ → it binds to the (56) ________________ protein → (57) ______________ is inactivated. An inactive regulatory activator protein called (58) __________ is made by the bacterium → it is activated by (59) ________ → it then binds to the (60) _______ site → it promotes the binding of RNA polymerase to the (61) ______________ sequence → the cluster of genes is transcribed to make the 3 (62) ______________ for lactose metabolism.

B. When lactose and glucose are present: Glucose enters the bacterial cell and leads to the inactivation (63) ______________ cyclase → level of (64) _______ goes down → (65) _______ is not activated → RNA polymerase cannot bind to the (66) ________________ region → the cluster of genes is not transcribed and no (67) ________________ are made for lactose metabolism.

This modification of lac operon is called the (68) ______________ gene regulation because CAP ensures synthesis of the 3 enzymes needed for lactose metabolism.

trp operon: Tryptophan is a(n) (69) ______________ ____________ that is made by *E. coli* when it is absent in the medium.

The cluster of 5 genes (70) ________ - _______ code for the enzymes required for tryptophan biosynthesis. Preceding these genes are the (71) ________________ sequence for the RNA polymerase binding and (72) ________________ sequence for repressor binding. Located away from the operon is the (73) __________________ gene that makes the inactive repressor protein.

A. When tryptophan is absent in the medium: (74) ____________________ gene makes the inactive repressor → it cannot bind to the (75) __________________ sequence → RNA polymerase binds to the (76) ________________ sequence → the cluster of genes (77) __________-__________ are transcribed → 5 enzymes needed for the biosynthesis of (78) __________________ are made.

B. When tryptophan is present in the medium: Tryptophan enters the cell → it acts as a(n)

(79) ________________ and binds to the inactive repressor → repressor is activated and binds to the

(80) ______________ sequence → it blocks the transcription of the cluster (81) _________-__________ → 5

enzymes cannot be made → *E. coli* does not make (82) ________________.

This system where the metabolite tryptophan acts as a(n) (83) ________________ and combines with the

repressor to shut off the system is referred to as the (84) ________________ operon.

trp (85) ______________ is referred to as the (86) ______________ gene regulation systems where the

repressor turns off the gene expression.

Matching

Match each of the following terms with its correct definition.

87.	____	Lac operon model	A. A gene that makes the repressor protein
88.	____	Operon	B. An operon where the metabolite molecule represses or decreases the expression of the cluster of genes
89.	____	Promoter sequence	C. Proposed by Jacob and Monod to explain gene expression in *E. coli* for lactose metabolism
90.	____	Regulatory gene	D. Regulation mechanism where the active repressor turns off the gene expression
91.	____	Repressor protein	E. Regulation mechanism where a catabolite activator protein ensures the turning on or turning off of the genes
92.	____	Operator sequence	F. Refers to a cluster of genes and associated DNA sequences
93.	____	Inducer	G. DNA sequence where RNA polymerase binds
94.	____	Inducible operon	H. A chemical that turns on or induces the cluster of genes in an operon
95.	____	Repressible operon	I. A metabolite molecule combines with the repressor protein to shut off the operon
96.	____	Corepressor	J. Protein that is made by the regulatory gene in active or inactive form and binds to the operator to block the transcription of the cluster of genes
97.	____	Negative gene regulation	K. A sequence where the repressor protein binds and is present between the promoter and the cluster of genes
98.	____	Positive gene regulation	L. An operon where the metabolite molecule enhances or increases the expression of the cluster of genes
99.	____	CAP	M. Sequence next to promoter sequence where activated CAP binds
100.	____	CAP site	N. Catabolite activator protein that binds to the CAP site and helps in the binding of RNA polymerase to the promoter

Complete the Table

101. Prokaryotic cells have different types of operons and gene regulation mechanisms. Complete the following table with information.

Inducible Operon	*Repressible Operon*
Negative Gene Regulation	*Positive Gene Regulation*

16.2 Regulation of Transcription in Eukaryotes [pp. 335–342]

Section Review (Fill-in-the-Blanks)

Unlike prokaryotes, in eukaryotes genes are not organized into (102) ______________. They do contain (103) ____________-coding genes and closely placed (104) ______________ sequences. There are two major categories of regulation in eukaryotes: (105) ________-term regulation to respond to frequent changes in the environment; (106) ________-term regulation for development and differentiation in the organism. Unlike prokaryotes, eukaryotic cells not only regulate transcription, but they also regulate (107) ______-transcription, translation, and (108) _____-translation.

Chromatin control: In eukaryotes, DNA is wrapped around a cluster of proteins called the (109) ____________. Each unit structure called the (110) ______________ is made of two molecules of each of the four (111) ____________—H2A, H2B, H3 and H4. Another (112) ____________, H1 links these structural units together to further compact DNA. The complex of DNA and (113) ______________ is referred to as chromatin. If chromatin is tightly packed, the genes are (114) ________________ due to unavailability of the (115) ______________ sequence for RNA polymerase binding. On the other hand, when a gene is to be activated, chromatin changes, a process called chromatin (116) ______________, where a protein complex called the (117) ______________ complex displaces the (118) ________________ in a nucleosome, loosens DNA, and allows access to promoter sequence.

Transcription initiation control: Promoter region contains the (119) _______ box that helps in binding of the (120) ________________ factors which then recruit RNA polymerase II. The (121) __________________ ____________ complex is formed by (122) ________________ factors and RNA polymerase II binding at the promoter region. This complex then allows RNA polymerase II to unwind (123) ______ and transcribe it. Activators and (124) ________________ bind to the promoter (125) _____________ elements and promoter itself to promote the looping of (126) _______ and enhance transcription. On the other hand, (127) _________________ are proteins that negate the action of activators to block or reduce (128) ____________________.

Coordinated control: Even though eukaryotes do not have (129) ______________ as prokaryotes do, they do have a mechanism to coordinate the transcription of all the genes related to a specific function. As in hormone actions, all the genes that are under the regulation of a specific hormone have the same specific (130) ________________ sequences associated with them so that the (131) ______________-receptor complex can switch all of them at the same time.

DNA methylation control: Methylation is a process of adding (132) _________ group to (133) ___________ bases in DNA. This process takes place in the (134) ______________ sequences for hemoglobin genes in most body cells and turns off the genes. Therefore no transcription or synthesis of (135) ________________ can take place. This process of turning off the genes is called (136) ______________. On the other hand, in cells forming the red blood cells, enzymes remove the (137) ___________ group and hemoglobin genes are transcribed to make (138) ________________. The same applies to the Barr body, which is one of the two (139) _____ chromosomes in female somatic cell that is highly (140) ________________ to turn off all the genes.

Matching

Match each of the following terms with its correct definition.

141. ____ Histones
142. ____ Chromatin remodeling
143. ____ Nucleosome
144. ____ TATA box
145. ____ Transcription factors
146. ____ Transcription initiation complex
147. ____ Promoter proximal elements
148. ____ Activators
149. ____ Coactivator/mediator
150. ____ Hormone-receptor complex
151. ____ Methylation
152. ____ Silencing

A. It is part of eukaryotic promoter sequence that is recognized by the transcription factors necessary for RNA polymerase binding
B. Proteins that help in the binding of RNA polymerase to promoter region
C. Refers to proteins around which DNA is wrapped to form the nucleosome
D. It is a unit structure of chromatin formed when DNA wraps around 8 units of histones
E. Refers to activation of genes by loosening up chromatin
F. Proteins that bind to the promoter proximal elements and enhance transcription
G. Formed by binding of transcription factors and RNA polymerase II at the promoter sequence
H. A sequence located before the promoter sequence for the binding of activators that enhances transcription
I. Formed by the hormone combining with the receptor located in its target cell
J. A large multiprotein complex that loops DNA by joining activators
K. The process of methylating cytosines in DNA to turn off the genes
L. It is a process of adding methyl group to cytosine bases in DNA

Complete the Table

153. Eukaryotic cells have different types of gene regulation mechanisms at the transcriptional level. Complete the following table giving a short description of each type.

Mechanism	*Description*
Chromatin control	
Transcription initiation control	
Coordinated control	
DNA methylation control	

16.3 Posttranscriptional, Translational, and Posttranslational Regulation [pp. 342–345]

Section Review (Fill-in-the-Blanks)

Posttranscriptional control: This level of regulation is after (154) _________ is made.

In eukaryotic cells a longer transcript called the (155) ___________ is made. This pre-mRNA must undergo (156) _____________ to remove (157) _________ and splice different combinations of (158) __________ together to form different types of shorter transcripts that are now referred to as (159) _________. Different types of proteins can now be made from these different types of (160) __________.

In an unfertilized animal egg, (161) _________ is blocked from translation by the binding of (162) _____________ proteins. Once the egg is fertilized, these (163) ______________ proteins are removed in order to begin translation.

In mammary glands, the half-life of (164) __________, which is translated to make the milk protein casein, is extended by peptide hormone prolactin. This results in larger synthesis of (165) __________, the protein in milk.

Recently a new type of RNA was discovered during developmental stages of organisms, called the micro-RNA or (166) __-RNA that become a double stranded (167) ________ with the help of an enzyme dicer. The double stranded RNA is then degraded to single stranded RNA which can bind to mRNA to block their (168) ___________________.

Translational control: During early embryo development stages of animals, the spliced (169) __________ has 3'poly(A) tail that can be increased or decreased in length to increase or decrease translation.

Posttranslational control: Once a protein is made, its function can be regulated by chemically modifying, processing, or its degradation. Histones can be chemically modified by acetylation which then affects the structure of (170) ________________ and compactness of chromatin.

Many of the digestive enzymes are made as inactive, larger (171) ______________ that are broken down to form smaller, active (172) ___________ after their secretion into the digestive canal.

Some of the short-lived proteins in the cell have small proteins called the (173) _______________ attached that make them prone to proteasome digestion.

Matching

Match each of the following terms with its correct definition.

174. ____ Splicing
175. ____ Masking proteins
176. ____ mi-RNA
177. ____ Poly(A) tail
178. ____ Ubiquitin
179. ____ Proteasome

A. Micro-RNA that can bind to mRNA and block their translation
B. Small protein that attaches to the some of the short-lived proteins of the cell encouraging their quick breakdown
C. Process of removing exons and joining introns in a pre-mRNA to form an mRNA
D. Proteins that bind to mRNA to block translation in an unfertilized egg
E. A short chain of adenines attached to 3'end of mRNA whose length affects the translation process
F. Enzyme that breaks down proteins

Complete the Table

180. Eukaryotic cells have different types of gene regulation mechanisms in addition to the transcriptional level. Complete the following table giving a short description of each type.

Mechanism	*Description*
Posttranscriptional control	
Translational control	
Posttranslational control	

16.4 The Loss of Regulatory Control in Cancer [pp. 345–348]

Section Review (Fill-in-the-Blanks)

Cell division is regulated by specific (181) __________, which if mutated may allow the cells to continue dividing and form (182) __________. The cells revert from being specialized to embryonic form, a process that is called (183) __________. If the dividing cells remain as a single mass, they are referred to as (184) __________. If the cells separate and spread to other tissues and organs, the tumor is called (185) __________. This is also referred to as (186) __________. The spreading of malignant tumors is called (187) __________.

In normal cells, (188) ________-__________ code for proteins that stimulate cell division. In cancer cells, these genes are converted to (189) __________ by mutation, translocation, or viral infection. A gene in a normal cell that inhibits cell division is called the (190) ________-__________ gene. This gene, TP53, is an example of such a gene that codes for a protein (191) ______ and loss or mutation of this gene is involved in development of cancer.

Matching

Match each of the following terms with its correct definition.

192. ____ Mutation
193. ____ Benign tumor
194. ____ Malignant tumor
195. ____ Dedifferentiation
196. ____ Metastasis
197. ____ Proto-oncogenes
198. ____ Oncogenes
199. ____ TP53
200. ____ p53

A. Spreading of malignant tumors
B. The tumor whose cells separate and spread to other tissue and organs
C. A change in base sequence in DNA
D. The tumor that remains in its original site
E. The process by which differentiated cells become embryonic
F. Genes that regulate cell division in normal cells
G. Modified proto-oncogenes that stimulate the cell to become cancerous
H. A protein that is coded by the tumor-suppressor gene and stops cell division in normal cells
I. Tumor-suppressor gene that inhibit cell division in normal cells and is altered in cancer cells

Complete the Table

201. Complete the following table to compare terms connected to tumors.

Benign tumors	*Malignant tumors*
Proto-oncogenes	*Oncogenes*
Differentiation	*Dedifferentiation*

SELF-TEST

1. Which of the following terms refer to a collection of genes and sequences that regulate gene expression in prokaryotes? [p. 330]
   a. promoter
   b. operator
   c. operon
   d. repressor

2. To which region does the enzyme RNA polymerase bind? [p. 330]
   a. operator
   b. promoter
   c. regulatory
   d. inducer

3. In lac operon, lactose acts as a/an [p. 331]
   a. repressor
   b. corepressor
   c. activator
   d. inducer

4. In trp operon, tryptophan acts as a/an [p. 332]
   a. repressor
   b. corepressor
   c. activator
   d. inducer

5. In eukaryotes, RNA polymerase binds directly to the promoter sequence. [p. 336]
   a. True
   b. False

6. In both, prokaryotic and eukaryotic cells, gene regulation takes place mostly at the transcription level. [p. 335]
   a. True
   b. False

7. Which of the following are universally involved in packing DNA to form chromatin in eukaryotes? [p. 335]
   a. repressors
   b. activators
   c. histones
   d. inducers

8. Hormones are able to exert their affect on several genes due to the presence of the same ________________ sequence before all those genes. [p. 341]
   a. regulatory
   b. promoter
   c. operator
   d. repressor

9. Which of the following group is added to some of the bases of DNA in order to inactivate certain genes? [p. 341]
   a. methyl
   b. acetyl
   c. phosphate
   d. carbonyl

10. TATA box is part of the ______________ sequence that is recognized by the transcription factors before RNA polymerase can bind. [p. 336]
   a. operator
   b. promoter
   c. regulatory
   d. inducer

11. An unfertilized egg stores a lot of mRNA whose translation is blocked until after the sperm has fused with it. The proteins that block mRNA are called the ______________ proteins. [p. 342]
   a. repressor
   b. masking
   c. activator
   d. enzyme

12. Which of the following is involved in translational control of eukaryotes? [p. 344]
   a. ubiquitin
   b. 3'poly(A)
   c. acetylation of histones
   d. proteasome

13. Which of the following tumor spreads from one tissue to another tissue or organ? [p. 345]
   a. benign
   b. malignant

14. Which of the following gene stimulates a normal cell to become cancerous? [p. 346]
   a. proto-oncogene
   b. oncogenes

15. Which of the following refers to spreading of the cancer cells? [p. 345]
   a. dedifferentiation
   b. benign
   c. metastasis
   d. mutation
   e. oncogenes

INTEGRATING AND APPLYING KEY CONCEPTS

1. An unfertilized egg stores a lot of mRNA that is blocked from being translated. Explain how the mRNA is inhibited and how it is activated after sperm fuses.
2. All body cells have the same set of genes. Explain how red blood cells are able to make hemoglobin while the other body cells do not.
3. Explain mechanisms that transform proto-oncogenes to oncogenes.

17

BACTERIAL AND VIRAL GENETICS

CHAPTER HIGHLIGHTS

- Genetic recombination takes place in bacteria by three mechanisms:
    - Conjugation—direct transfer of DNA from one cell to another.
    - Transformation—absorption of DNA released by other disintegrating bacterial cells.
    - Transduction—transfer of DNA from one cell to another via a virus.
- Conjugation in *Escherichia coli* helped scientists map bacterial genes.
- Viruses have a very simple structure:
    - Nucleic acid core—DNA or RNA that carry limited genetic information.
    - Protein coat—forms a shell to enclose the nucleic acid.
- Viruses multiply inside a living host cell by using mostly the machinery of the host.
- Bacterial viruses called the bacteriophages, multiply by two mechanisms:
    - Lytic cycle—where the virus kills the host cell.
    - Lysogenic cycle—where the viral nucleic acid becomes incorporated into the host's genetic material and it multiplies with the host.
- Prokaryotic and eukaryotic genome have DNA sequences called the transposable elements that can move from one site in the genome to another.
- Transposable elements cause changes in the genome and affect the expression of the genes.

STUDY STRATEGIES

- This chapter is simpler if you have a good understanding of prokaryotic and viral structure.
- Remember, the goal of this chapter is to understand:
    - Multiplication of bacteria and viruses.
    - Genetic recombination.
    - Transposable elements in prokaryotic and eukaryotic cells.
- Concentrate on the key concepts and various pathways.

TOPIC MAP

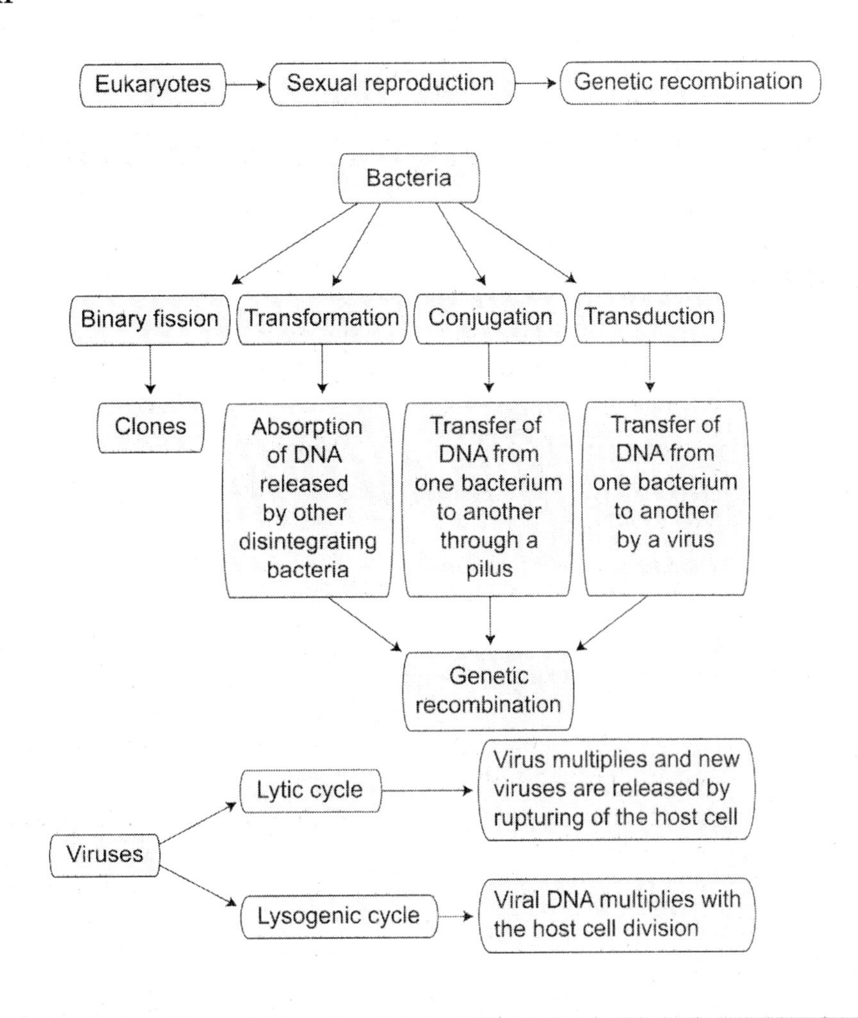

INTERACTIVE EXERCISES

Why It Matters [pp. 351–353]

Section Review (Fill-in-the-Blanks)

(1) ______________ ______________ named the bacterium that causes diarrhea as (2) __________________ ________. This bacterium was later found to be common in the large intestine of healthy humans and was renamed as (3) __________________ ______. This bacterium has been used extensively in research, including the study of viruses that attack bacteria, called (4) _________________ or simply (5) _________. Facts learned from research on bacteria, prokaryotic organisms, can usually be applied to (6) _____________ organisms.

17.1 Gene Transfer and Genetic Recombination in Bacteria [pp. 353–360]

Section Review (Fill-in-the-Blanks)

Most eukaryotic organisms reproduce (7) ______________ by (8) ____________ type of division that becomes a source of genetic recombination. Although bacteria reproduce (9) ______________, they have mechanisms to add new (10) ______ to their own (11) _______. Wild variety of bacteria can be grown on (12) ______________ medium that contains water, (13) ___________, and (14) ___________. When bacteria multiply, they produce genetically identical cells that are referred to as the (15) ___________.

Two scientists, Joshua (16) _______________ and Edward (17) ___________ experimented to demonstrate genetic recombination in bacteria. They exposed wild variety of *E. coli* bacteria to X rays or UV that act as (18) _____________ and cause genetic (19) ______________. New strains of bacteria emerged that are referred to as (20) ______________. The scientists determined that these new varieties required additional nutrients and are not able to grow on (21) ______________ medium. They mixed some of these varieties, plated them on (22) _____________ medium, and developed new genetic recombinants. Bacteria, which contain a single, (23) _____________ DNA molecule, connect to other bacterial cells by a tubelike structure called the (24) _________. A copy of DNA from the (25) __________ cell passes into a(n) (26) _______________ cells—a process called (27) __________________.

The donor cells, referred to as (28) _____ cells, have F genes in their (29) ___________. This gene allows them to make a pilus and conjugate with the recipient cells that lack this (30) __ gene. The recipient cells are called the (31) _____ cells. Once a copy of the (32) ___________ has been transferred from the donor cell to the recipient cell, the recipient cell that used to be (33) ____ cell, now gains the (34) ____ gene and becomes (35) ____ cell. This is the simplest type of (36) ___________ reproduction or genetic (37) ______________ demonstrated in bacteria.

A bacterial cell that has F genes integrated into its main chromosome is called (38) _____ cell. When it conjugates, its main (39) _______________ starts replication in the middle of its (40) ___ gene and the copy starts moving through the connecting bridge, the (41) ______. However, the connection breaks before the entire copy is transferred, and the recipient cells remains as (42) ____ cell.

Two scientists, Francois (43) ___________ and Elie (44) _______________ experimented with the conjugation duration to demonstrate genetic (45) _________________ in bacteria. Their technique was extensively used to map genes on bacterial chromosome, called the (46) ___________ _______. The circular arrangement of the genes in bacteria demonstrated that bacterial chromosome is (47) _______________.

Bacteria contain other plasmids: (48) ____ plasmid that has the antibiotic resistance gene.

In addition to conjugation, there are two other mechanisms for genetic (49) _________________ in bacteria.

Transformation: The scientist Fred (50) _______________ discovered that a nonpathogenic *Streptococcus pneumoniae* can become (51) ___________________ after being exposed to dead pathogenic bacteria. In 1944 (52) _____________, __________________ and _________________ found that the transforming factor was (53) _______. This transformation allows the nonpathogenic form to develop an extra protective (54) ______________, be able to survive phagocytosis, and cause the disease. Those species of bacteria that can naturally absorb (55) ______ from their surroundings have a(n) (56) _______ binding protein on their cell surface. The absorbed (57) ______ causes genetic (58) __________________ and become transformed. *E. coli*, which does not naturally absorb (59) ______, and does not naturally undergo transformation, can be experimentally made to absorb (60) _____. This experimental technique is referred to as (61) ______________ ___________________.

This can be achieved by giving cells a treatment of (62) _______________ ions and quick heat shock. More commonly, repeated pulses of electrical shocks, a procedure known as (63) _________________, are given to the bacterial cells. The application of this procedure is to introduce (64) _______ with desired genes.

Transduction: Scientists Joshua (65) ________________ and Norton (66) ______________ discovered transduction, where a bacteriophage, while multiplying inside a bacterium, picks up a piece of bacterial (67) _____ instead of or along with its own (68) _____. The new bacteriophage infects a new bacterium and transfers (69) _____ from the previous bacterial host—genetic (70) ________________ takes place.

To detect genetic recombination in bacteria, Joshua (71) _____________ and Esther (72) _______________ developed (73) ____________ plating technique. In this technique, bacteria are grown on (74) _________________ medium that is minimal medium plus a full range of nutrition added to it. This medium supports all the varieties—normal as well as the (75) _________________ that lack the ability to make one or more nutritional requirements. After the colonies develop, a sterile (76) ________________ block is pressed on the plate to transfer some of the bacteria from each colony. The (77) _______________ block is then pressed again on a new plate containing minimal medium or minimal medium with a specific nutrient added to it. This technique allows scientists to study (78) ________________ and specific chemical pathways they lack.

Matching

Match each of the following scientists with their major contribution.

79. ____	Theodore Escherich	A. A cycle of reactions that generates ATP, NADH, and $FADH_2$
80. ____	Joshua Lederberg and Edward Tatum	B. Demonstrated genetic recombination in bacteria through controlled conjugation
81. ____	Francois Jacob and Elie Wollman	C. Discovered conjugation in bacteria
82. ____	Avery, McCarty, McCleod	D. Demonstrated transformation in bacteria
83. ____	Fred Griffith	E. Developed replica plating technique
84. ____	Joshua Lederberg and Norton Zinder	F. Proved that DNA is the transforming factor
85. ____	Joshua Lederberg and Elie Lederberg	G. Demonstrated transduction

Match each of the following terms with its correct definition.

86. ____	Genetic recombination	A. Simple medium that contains water, sugar, and salt
87. ____	Minimal medium	B. Where bacterial cells make a physical connection to pass part of their genetic material from one to another
88. ____	Clone	C. Bacterial cells that have F gene incorporated into the main chromosome
89. ____	Auxotroph	D. Development of new combinations of genetic information to generate genetic variability
90. ____	Sex pilus	E. Genetically identical cells
91. ____	Conjugation	F. Bacterial cells that have F gene incorporated in the plasmid
92. ____	Hfr cells	G. A tubelike structure that bacteria develop to connect one cell to another
93. ____	F+ cells	H. Mutants that require additional nutrients added to the minimal medium
94. ____	F- cells	I. Refers to determining the location of the genes on a chromosome
95. ____	Genetic map	J. Plasmid that has the resistance gene
96. ____	R plasmid	K. Plasmid with fertility (F) gene
97. ____	F plasmid	L. Cells that lack F gene and act as recipient cells
98. ____	Transduction	M. Where a bacterium absorbs DNA released by other dead bacteria
99. ____	Transformation	N. Where a virus transfers DNA from one bacterium to another
100.	Replica plating technique	O. A medium that has a full complement of nutrients added to support auxotrophs
101.	Complete medium	P. A technique where bacterial colonies are transferred to a sterile velveteen and then grown in special media to identify auxotrophs

Complete the Table

102. Bacterial cells divide to produce clones. Complete the following table to describe three mechanisms by which bacteria achieve genetic recombination.

Mechanism of Recombination	Brief Description
A. Transformation	
B. Conjugation	
C. Transduction	

True/False

Mark if the statement is true or false. If the statement is false, justify your answer in the line below each statement.

103. ______________ When bacteria divide, they produce clones.

____

104. ______________ All varieties of bacteria are known to absorb DNA and transform.

____

105. ______________ During conjugation, DNA is transferred from the donor to the recipient. Eventually all bacteria will become F+.

____

17.2 Viruses and Viral Recombination [pp. 360–362]

Section Review (Fill-in-the-Blanks)

Viruses have a nucleic acid (106) _________ and a protective protein (107) __________. Some animal viruses also have a(n) (108) ______________, a plasma membrane layer that they derive from their host cell. They inject their (109) ___________ acids into the host cell and use the host cell for enzymes and chemicals necessary for their (110) ________________.

Viral nucleic acids can be (111) _______ or (112) _______. They code for viral (113) _________ proteins and (114) _____________ proteins that help in connecting to the host cell.

Bacteriophages are (115) ___________ that attack (116) _______________. Those phages that kill their host cell while multiplying are called (117) _____________ bacteriophages, while those that enter the host cell remain inactive and multiply with the host are called (118) ________________ phages.

A. Virulent bacteriophages: T-even phages are (119) _______________phages. They have a protein head that is packed with its (120) ______. It also has a protein tail that has (121) _________________ proteins to attach to the host bacterium. The tail contracts and viral (122) _____ is injected into the bacterium while the viral (123) ________ remains outside. Viral genetic material uses bacterial machinery to make enzymes to breakdown bacterial (124) ________________ and DNA (125) ___________ to make 100–200 copies of viral (126) _______. The host machinery is also used to make proteins to make new viral (127) __________ and (128) ____________. The replicated viral (129) ______ is packed into the assembled coat. Finally, an enzyme is made to rupture bacterial cell and new (130) ______________ are released. This cycle that virulent bacteriophage follow is called the (131) __________ cycle.

B. Temperate bacteriophage: Lambda phages, that attack *E. coli,* are (132) ______________ phages. This phage injects its (133) _______ into the bacterium. Viral (134) ______ becomes incorporated into bacterial (135) __________________. This stage of dormancy of a virus is now referred to as a(n) (136) ____________ and it replicates with the division of the host cell. This pathway followed by temperate bacteriophage is called the (137) _______________ cycle. Under certain environmental conditions, the incorporated (138) _________ can switch to (139) _______ cycle, that is, it becomes active, starts multiplying and ruptures the host cell to release hundreds of new viruses.

Matching

Match each of the following terms with its correct definition.

140. ____ Virulent phages
141. ____ Temperate phages
142. ____ Viral envelope
143. ____ Recognition proteins
144. ____ Lytic cycle
145. ____ Lysogenic cycle
146. ____ Prophage

A. An extra layer of membrane that some viruses carry from their host cell
B. Cycle followed by a temperate virus where its genetic material multiplies with the host cell division
C. Bacteriophage that follows lytic cycle
D. Bacteriophage that follows lysogenic cycle
E. The stage of a temperate phage where it becomes incorporated into the host chromosome
F. Proteins that help the phage attach to the host bacterium
G. Cycle followed by virulent phage where it multiplies and new viruses are released by rupturing of the host cell

Complete the Table

147. Complete the following table to describe the mechanisms of multiplication in bacteriophages.

Bacteriophage type	Type of Multiplication	Brief Description
A. Virulent		
B. Temperate		

Short Answer

148. Differentiate between lytic and lysogenic cycles. ____

____

____

149. Differentiate between a phage and a prophage. ____

____

____

17.3 Transposable Elements [pp. 362–368]

Section Review (Fill-in-the-Blanks)

Segments of DNA that can move from one place to another are called (150) ________________ elements (TEs) or (151) ________________ genes. The location where they move is called the (152) ________________ site. The movement is referred to as (153) ____________________. These segments of DNA cause (154) __________ changes such as elimination of a function or increase/decrease in gene expression by inserting in a(n) (155) ________.

In bacteria, DNA can be (156) __________________ from one location in the chromosome to another place in its chromosome, or between its chromosome and its plasmids, or between two (157) ________________.

There are two types of TEs in bacteria: (158) ________________ sequences (IS) that contains only the gene for the enzyme, (159) __________________, responsible for cutting or pasting TE from DNA. The second type is called (160) ________________ that has the transposase gene and additional genes such as (161) ________________ resistance gene.

The geneticist (162) _____________ __________________ made the first discovery of (163) __________________ elements in corn and received a Nobel prize for her work. Now TE are known in prokaryotes as well as in (164) ____________________.

In eukaryotes, there are two types of TEs: (165) ________________ that are similar to bacterial transposons for a cut and paste mechanism. The other type of eukaryotic TEs are the (166) ____________________ that not only cut and paste but also get transcribed to make (167) ___________________ RNA. An enzyme (168) ________________ ____________________, which is coded in retrotransposon, is used to reverse transcribe (169) ____________________ RNA and make DNA copies of retrotransposons. These DNA copies are then inserted in other locations of the chromosomes. If certain cellular genes are inserted in a(n) (170) ______________________, these genes can be frequently moved from one chromosome to another, making them abnormally active as in certain cancers.

Certain viruses, such as HIV, are called (171) __________________, which have been known to carry molecules of an enzyme, (172) __________________ ______________________. This enzyme is used for making complementary (173) ______ as part of their replication which can become incorporated into the host nucleus as (174) ______________.

Matching

Match each of the following terms with its correct definition.

175. ____ Transposable elements
176. ____ Transposition
177. ____ Target site
178. ____ Transposase
179. ____ Retrotransposons
180. ____ Reverse transcriptase
181. ____ Provirus
182. ____ Retrovirus

A. Refers to the location where a transposable element moves

B. TEs found in eukaryotes that not only cuts and pastes but also has the ability to make copies of TEs to be inserted in different locations

C. Segments of DNA that can move from one place to another

D. Process of moving segments of DNA from one place to another

E. Enzyme that copies RNA to make complementary DNA

F. State of virus where its DNA is inserted into eukaryotic genome

G. RNA viruses that carry reverse transcriptase

H. Enzyme that can do cut and paste of TEs

True/False

Mark if the statement is true or false. If the statement is false, justify your answer in the line below each statement.

183. ____________ Only bacteriophages are known to stay dormant inside the host cells for a number of days while viruses that attack eukaryotic cells are unable to do that.

____

184. ____________ Temperate viruses follow lysogenic cycle but they can switch to lytic cycle.

____

185. ____________ Transposable elements were first discovered in prokaryotes.

____

186. ____________ Transposable elements have the sequence to make transposase for cutting and pasting DNA.

____

SELF-TEST

1. All of the following cause genetic variation in bacterial populations except [p. 353]
   a. binary fission.
   b. transduction.
   c. conjugation.
   d. transformation.

2. Which of the following is the specific term used for bacterial varieties that emerge after exposure to X-rays or UV? [p. 353]
   a. wild
   b. temperate
   c. virulent
   d. auxotroph

3. The structure that connects two bacterial cells during transfer of DNA is called [p. 354]
   a. a plasmid.
   b. a nucleoid.
   c. a pilus.
   d. flagella.

4. Which of the following bacteria would have the F gene as part of their main chromosome? [p. 357]
   a. F-
   b. Hfr
   c. F+

5. During conjugation between an Hfr and an F- bacteria, [p. 357]

a. some of the genes from Hfr are combined with the DNA of F-.

b. F- becomes Hfr.

c. F- become F+.

d. genes from F- are transferred to Hfr.

6. Which of the following refers to a virus transferring DNA from one bacterium to another? [p. 358]

a. binary fission

b. transduction

c. conjugation

d. transformation

7. All varieties of bacteria are able to undergo transformation in nature. [p. 358]

a. True

b. False

8. All viruses have a membrane around them. [p. 360]

a. True

b. False

9. When a virus multiplies inside the host cell, it uses its own enzymes for replication. [p. 360]

a. True

b. False

10. During which cycle does a bacteriophage DNA becomes incorporated inside the host genome? [p. 361]

a. Lytic cycle

b. Lysogenic cycle

11. During which cycle does a virus rupture the host cell in order to be released? [p. 361]

a. Lytic cycle

b. Lysogenic cycle

12. A RNA virus usually carries an enzyme called [p. 365]

a. RNA polymerase.

b. DNA polymerase.

c. reverse transcriptase.

d. transposase.

13. A provirus is [p. 365]

a. a virus that contains DNA.

b. a virus that has RNA.

c. a virus whose DNA is incorporated into the host genome.

d. a virus that is assembled and ready to be released.

14. TE in bacteria can move DNA segments [p. 363]

a. from one location in its main chromosome to another location in the chromosome.

b. from a plasmid to the main chromosome.

c. from one plasmid to another plasmid.

d. Any of the above

15. TE are characteristic of prokaryotes and are rarely found in eukaryotes. [p. 364]

a. True

b. False

INTEGRATING AND APPLYING KEY CONCEPTS

1. In AIDS, the disease symptoms are expressed years after the initial infection of HIV. Explain what happens after the infection and during the long gap between infection and occurrence of the symptoms.

2. You may have heard that antibiotic abuse has led to the development of resistant varieties of bacteria. Explain what is involved in this process and how resistant varieties are different from normal varieties of bacteria.

3. When a patient gets a bacterial infection, the number of bacteria goes up gradually. However, when a patient gets a viral infection, after a gap period, the number of viruses increases very quickly. How can you explain this difference?

18

DNA TECHNOLOGIES AND GENOMICS

CHAPTER HIGHLIGHTS

- DNA is cloned/copied in a bacterium with the help of:
  - Restriction endonucleases that make cuts in DNA.
  - Bacterial plasmids that allow DNA segments to be inserted into bacteria for multiplication.
- DNA can also be cloned in vitro using the polymerase chain reaction (PCR).
- DNA technologies are used for:
  - Testing human genetic diseases.
  - Resolving forensic and paternity cases.
  - Distinguishing individuals of one species from another.
  - Altering genes in a cell or an organism to correct genetic disorders.
  - Inserting genes in plants to improve crops.
- Developments and use of DNA technologies is monitored by national and world agencies.
- DNA sequencing of certain species was completed by the Human Genome Project (HGP).
- DNA sequencing and analysis is important for understanding the:
  - Functions of genes.
  - Role of other parts of the genome in regulating the genes.
  - Phylogenetic relationships between species.

STUDY STRATEGIES

- In order to understand this chapter, you must review DNA structure, its multiplication and organization in prokaryotic and eukaryotic cells.
- This chapter has extensive terminologies. Trying to memorize, rather than understanding the terms is the most common mistake.
- Follow each step of the described procedures.

TOPIC MAP

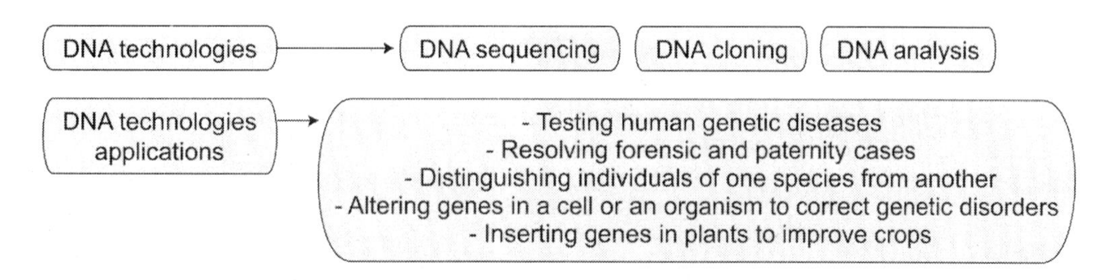

INTERACTIVE EXERCISES

Why It Matters [pp. 371–372]

Section Review (Fill-in-the-Blanks)

(1) ________________ refers to any technique that is used to make changes or improvements in living organisms or their products. DNA (2) ________________ is using techniques to isolate, analyze and manipulate DNA. Genetic (3) ________________ is to use technologies to manipulate genes that benefit us.

18.1 DNA Cloning [pp. 372–379]

Section Review (Fill-in-the-Blanks)

DNA (4) ____________ is to make multiple copies of a DNA segment. These multiple copies are used to understand how genes (5) ______________ and how they are regulated. They can be inserted into a bacterium to allow the bacterium to transcribe and translate to make large amounts of (6) ______________.

Methods used for DNA cloning: In the first method to clone DNA, eukaryotic genetic material containing the (7) __________ of interest is extracted from the human cells → treated with (8) ________________ enzyme, EcoRI, that cut this DNA into fragments with (9) ____________ ends → fragments with specific (10) ________ are isolated. At the same time, plasmids are isolated from (11) ____________ → plasmids are treated with the same type of (12) ________________ enzymes to open them → DNA fragments with specific (13) ________ of interest are inserted → enzyme DNA (14) __________ is used to seal the cut ends → (15) ______________ plasmids containing the specific gene are formed → plasmids are allowed to be absorbed by a culture of bacteria, a process called (16) __________________ → bacteria containing the (17) ______________ plasmid with the specific gene are identified using a nucleic acid probe, a technique called the DNA (18) ________________ → specific bacteria of interest are isolated → these bacteria are cultured in large numbers → specific gene or (19) ______________ made using these genes can now be isolated. A genomic (20) ____________ is a collection of plasmids with DNA fragments inserted into them as clones.

In the second method to clone DNA, (21) ________, the transcript that has already been processed to remove the non-coding sequences, (22) __________ and has the message for the protein of interest is isolated from the eukaryotic cell → enzyme (23) ______________ ____________________, isolated from a retrovirus, is used to make a single stranded DNA that is complementary to the (24) _______ the transcript → enzyme DNA (25) ________________ is used to make the second DNA molecule, complementary to the first → the resulting molecule of DNA is called the (26) __________________ DNA or cDNA. This cDNA is modified and inserted into a (27) __________ and added to the (28) ________ library.

<u>Methods used to make multiple copies of DNA fragments</u>: The technique developed in 1983 that makes copies of specific DNA sequences in a very short time period is called the (29) _______________ chain reaction or (30) _________. Isolated DNA fragments are heated → complementary strands (31) _______________ → the strands are cooled while DNA (32) _____________ are added to start DNA synthesis → enzyme DNA (33) __________________ and the four nucleotides are added → (34) __________________ DNA is made along the single DNA strands → two copies of (35) ________ molecules are formed. This procedure is repeated several time s to make multiple copies of (36) __________. A sample of the copies is analyzed on agarose (37) ______ _______________________.

Matching

Match each of the following terms with its correct definition.

38. ____	Biotechnology	A. To make multiple copies of a DNA segment
39. ____	DNA technology	B. Plasmid in which genes of interest are inserted
40. ____	Genetic engineering	C. Refers to any technique used to make improvements in living organisms or their products
41. ____	DNA cloning	D. Special enzymes isolated from bacteria to cut DNA at specific sites
42. ____	Bacterial plasmids	E. Short, single-stranded ends of a DNA fragment created by special restriction enzymes
43. ____	Restriction enzymes	F. Complementary DNA that is formed by copying mRNA by reverse transcriptase
44. ____	DNA ligase	G. Small, circular molecules of DNA found in certain bacteria in addition to the main chromosomes
45. ____	Recombinant plasmid	H. A technique used to separate DNA, RNA or proteins mixtures in a gel that is subjected to electrical field
46. ____	DNA hybridization	I. Polymerase Chain Reaction technique used to make multiple copies of small amounts of DNA
47. ____	Sticky ends	J. A collection of recombinant plasmids that contain cloned genes to be used for research and analysis
48. ____	Transformation	K. Using techniques to isolate, analyze and manipulate DNA
49. ____	Genomic library	L. To use technologies to manipulate genes that benefit us
50. ____	cDNA	M. Enzymes that seals open ends of DNA molecules
51. ____	PCR	N. A short single strand of DNA that is often used in DNA technologies to start DNA synthesis
52. ____	DNA primer	O. Absorption of DNA by bacteria to create genetic recombination
53. ____	Gel electrophoresis	P. A radioactively labeled DNA probe is used to bind to its complementary sequence on DNA in order to identify specific genes

Short Answer

54. Many different restriction enzymes have been identified in bacteria. How does enzyme EcoRI diferent from the others?

____

____

55. What is reverse transcriptase? What role does it play in DNA cloning? _______________

____

____

18.2 Applications of DNA Technologies [pp. 379–390]

18.3 Genome Analysis [pp. 390–398]

Section Review (Fill-in-the-Blanks)

(56) ________ cloning and amplification has lead to new research, new products and new treatments. Following are some of the applications of these techniques:

A. Testing of genetic diseases: DNA from individuals that have the normal or (57) ____________ genes are treated with (58) ______________ enzymes → DNA fragments, called the (59) ______________ ____________ __________ ________________ or RFLPs are formed → fragments are separated and the bands are compared using (60) ______________ _________ technique.

B. Identification of individuals for forensics, paternity cases and species comparisons: Genomic (61) ________ samples from individuals are collected → (62) __________ ____________ ____________ or STR loci from each sample are amplified using the (63) ________ technique → sample are separated into bands using (64) ________ __________________ → bands in different samples are compared/analyzed to pick similarities and dissimilarities in the genome.

C. Production of large amounts of proteins: Specific human genes (for insulin, growth hormone) are inserted into a bacterial (65) ____________ → which is then absorbed by the bacterium → a (66) ______________ bacterium is formed that has the DNA from human genome → bacteria starts making the specific (67) ____________ using the new gene.

D. Gene therapy: Normal genes are introduced into somatic cells that have the defective gene, (68) ___________ gene therapy, or into embryonic cells, (69) ________-__________ gene therapy.

E. Using other animals to produce human proteins of interest: The specific human gene is inserted into the (70) ___________ of the fertilized egg of the sheep → the (71) _____________ new sheep makes the protein.

F. Cloning of animals: Enucleated (72) _______ cell of a sheep and diploid somatic cell from the mammary gland are fused → the fused cell is allowed to grow and form a multicellular (73) _____________ → this is implanted into the (74) ____________ → Dolly was born.

G. Genetic engineering of plants: Genes are introduced into plants to increase (75) ____________________ to pests and diseases, greater (76) ________________ to heat and drought, and crops with higher (77) ________________ value. Bacterium *Rhizobium radiobacter* that inserts its (78) ____ plasmid into plant cells and causes tumors formation is often used to insert specific (79) __________ into plant cells.

Guidelines listed by (80) _____________ ________________ of ______________, NIH addresses the concern that (81) ________ technologies can be used towards negative research.

Genome analysis can be divided into two parts: (82) ________________ genomics that is the actual sequence of the genome; (83) __________________ genomics that is the location and function of the recognized genes. (84) ________________ combines biology with computers and mathematics to analyze DNA and study its evolutionary relationships. Sequencing most commonly is done by whole-genome shotgun method- the entire (85) ______________ of an organism is broken down randomly into thousands of smaller pieces → each piece is (86) ___________ and sequenced → the information is compiled by the computer to build the entire sequence.

Human Genome Project was to sequence the DNA of humans and other selected organisms. It showed that human genome is made of 3 billion (87) ________-___________. There are 20,000 – 25,000 (88) ______________-_____________ genes that occupy only (89) ____% of the entire genome. About (90) ____% of the genome has repeated sequences that have no known function. (91) _______________ refers to the complete set of proteins that can be expressed by an organism's genome and their functional interactions.

Matching

Match each of the following terms with its correct definition.

92. ____ RFLP

93. ____ Southern blot Analysis

94. ____ Short tandem repeats

95. ____ Somatic gene therapy

96. ____ Germ-line gene therapy

97. ____ Ti plasmid

A. Where normal genes are inserted into somatic cells to correct a genetic disorder

B. Tumor inducing plasmid that is inserted by certain bacteria to tumors in plants

C. Refers to different length fragments of DNA after restriction enzyme digestion called the restriction fragment length polymorphisms

D. A technique where DNA is digested with restriction enzyme, fragments are separated by gel electrophoresis and transferred to a filter paper, and analyzed using labeled probes

E. Where normal genes are inserted into fertilized egg to correct a genetic disorder

F. Short sequences in the non-coding regions of DNA repeated in series that are used to identify individuals through DNA fingerprinting

Match each of the following scientists with their contribution to cellular respiration.

98. ____ Paul Berg and Stanley Cohen

99. ____ Kary Mullis

A. Developed a technique where DNA is digested with restriction enzyme, fragments are separated by gel electrophoresis and transferred to a filter paper, and analyzed using labeled probes

B. Received Nobel Prize for the development of DNA cloning technique using bacterial plasmids and restriction enzymes

100.	____	Edward Southern	C. First one to demonstrate somatic gene therapy in humans to correct adenosine deaminase deficiency required for WBC maturation
101.	____	Sir Alec Jefferys	D. First ones to try gene therapy by inserting rat growth hormone gene in into the fertilized egg of a mouse
102.	____	Richard Palmiter and Ralph Brinster	E. Received Nobel Prize for the development of the PCR technique to amplify DNA
103.	____	French Anderson	F. Invented DNA fingerprinting technique to distinguish between individuals
104.	____	Ian Wilmut and Keith Campbell	G. Received Nobel Prize for the development of DNA sequencing technique
105.	____	Walter Gilbert and Allan Maxam	H. Cloned sheep using the genetic material of an adult somatic cell
106.	____	J. Craig Venter and Celera Genomics	I. First to sequence the entire genome of a bacterium using the shotgun method

Short Answer

107. Starting from collecting the cheek cells samples, outline the steps followed in a paternity case. ______________

____

____

108. Outline the procedure to make large amounts of insulin using bacterial culture. ______________________

____

____

Mark if the statement is true or false. If the statement is false, justify your answer by explaining in the line below each statement.

109. ______________ Normal genes introduced in the somatic cells with defective gene will eventually be passed on to the next generation.

____

110. ______________ A tobacco plant with a firefly gene is called a transgenic plant.

____

111. ______________ Human genome has about 100,000 protein coding genes.

____

112. ______________ All restriction enzymes leave sticky ends on DNA fragments.

____

113. ______________ Enzyme reverse transcriptase is present in prokaryotic and eukaryotic cells.

____

SELF-TEST

1. Which of the following would be considered as transgenic? [p. 376]
   a. Bacterial plasmid with antibiotic resistance gene
   b. Bacterial chromosome with lac operon
   c. Human cell with growth hormone gene
   d. Bacterial plasmid with insulin gene

2. cDNA is made by copying [p. 376]
   a. RNA
   b. DNA
   c. proteins
   d. plasmids

3. In PCR, which of the following is used as the primer? [p. 379]
   a. RNA
   b. DNA
   c. proteins
   d. plasmids

4. Which of the following enzyme is used in PCR? [p. 379]
   a. RNA polymerase
   b. DNA polymerase
   c. reverse transcriptase
   d. DNA ligase

5. Gel electrophoresis is used to separate [p. 379]
   a. DNA
   b. RNA
   c. protein
   d. Any of the above

6. Which of the following most accurately describes sticky ends? [p. 374]
   a. Bacterial plasmids with insulin gene
   b. mRNA isolated from human cells
   c. DNA fragments with single stranded ends
   d. Proteins that stick to DNA

7. Over 50% of the human genome consists of repeated sequences that have no apparent function. [p. 395]
   a. True
   b. False

8. Only about 2% of the human genome has the protein-coding sequences. [p. 394]
   a. True
   b. False

9. The one and only purpose of Human Genome Project was to sequence human genome. [p. 391]
   a. True
   b. False

10. In cloning of the sheep, the scientists used the nucleus from [p. 387]
    a. somatic cell
    b. an egg
    c. a zygote
    d. a sperm

11. Which of the following does the enzyme reverse transcriptase use as a template? [p. 376]

a. RNA

b. DNA

c. proteins

d. plasmids

12. Enzyme reverse transcriptase is obtained from. [p. 376]

a. bacteria

b. virus

c. humans

d. plants

13. Which of the following scientist/s developed PCR? [p. 378]

a. Kary Mullis

b. Edward Southern

c. Ian Wilmut and Keith Campbell

d. Sir Alec Jefferys

14. Which of the following scientist/s was the first to clone sheep? [p. 387]

a. Kary Mullis

b. Edward Southern

c. Ian Wilmut and Keith Campbell

15. Gene therapy has been successful in animals other than humans. [p. 386]

a. True

b. False

INTEGRATING AND APPLYING KEY CONCEPTS

1. Somatic gene therapy is being experimented in humans but germ-line gene therapy is not permitted in humans. Explain.

2. Genetically engineered food products are slowly coming into our market. What is the rationale for the skepticism in our community to buy these products?

3. Human Genome Project came up with much smaller numbers of protein coding genes in human genome than the estimate of the scientists. How do you explain production of many more types of proteins than the number of genes?

19

DEVELOPMENT OF EVOLUTIONARY THOUGHT

CHAPTER HIGHLIGHTS

- Evidence supporting the fact of evolution was accumulating in Darwin's time, and it was gradually gaining acceptance among scientists.
- Darwin's thinking was influenced by observations he made during his travels on the *Beagle* and by his intellectual predecessors and contemporaries.
- Evidence from many areas of biology supports the fact of evolution.
- The key components of evolution by natural selection are:
    - Organisms "overproduce" offspring, but populations remain relatively stable in size.
    - Individuals in populations vary, and the variations are heritable.
    - The individuals with the most favorable traits do most of the reproducing, and those traits are inherited by their offspring, becoming more common in the population.
- Relatively recent advances in evolutionary biology have been incorporated into the "modern synthesis."

STUDY STRATEGIES

- Have a solid understanding of the population concept. Interactions among members of a population are intraspecific.

INTERACTIVE EXERCISES

Why It Matters [pp. 401-402]

Section Review (Fill-in-the-Blanks)

The scientist who proposed the idea of evolution by natural selection independently of Charles Darwin was (1) _____.

19.1 The Recognition of Evolutionary change [pp. 402 – 405]

Section Review (Fill-in-the-Blanks)

The area of biology that studies the form and variety of organisms in their natural environment is called (2) _____. Prior to Darwin's era, biologists were more concerned with (3) _____, which sought to reconcile observations about nature with biblical teachings. Much effort was also spent on classifying organisms, the branch of biology known as (4) _____.

As Europeans began exploring other parts of the world and discovering new kinds of organisms, a new branch of biology, (5) _____, arose, concerned chiefly with the worldwide distribution of plants and animals. At the same time (6) _____, scientists who compared the anatomical structure of organisms, began making note of similarities and differences between them, e.g., (7) _____ structures that are apparently useless in organisms today but had important functions in their ancestors.

The discovery of (8) _____, mineralized parts of ancient organisms, also contributed to the development of evolutionary thought. Their relative ages could be determined by their position in the horizontal layering of sedimentary rocks in the earth's crust. This layering is referred to as (9) _____, and the study of the ancient organisms found in those layers is called (10) _____. One of the leading zoologists of the time, Georges Cuvier, theorized that the different fossils found in different layers were the result of sudden, local shifts in the environment, e.g., severe flooding. His theory became known as (11) _____.

By the beginning of the 19th century, scientists were becoming less concerned with reconciling their observations with scripture and sought more objective explanations for natural phenomena. Thus, the Scottish geologist, James Hutton, proposed that the earth's geologic features were the result of physical processes that acted over periods of time much longer than could be accounted for in the Bible. His theory is referred to as (12) _____, and it allowed for the long time spans required of evolution. Later in the 19th century, Charles Lyell, an English geologist, argued that the physical processes that shaped the Earth's surface are the same processes and occur at the same rate as those that occurred in antiquity. This theory, called (13) _____, implied that the age of the Earth was much greater than previously believed.

Choice

For each of the following statements, choose the most appropriate concept from the list below.

a. catastrophism b. gradualism c. natural history d. uniformitarianism e. taxonomy

14. _____ Belief that the same physical processes that shape the Earth's geological features today also operated in prehistoric times
15. _____ Branch of biology concerned with classification of organisms
16. _____ Branch of biology that examines the form and variety of organisms in their natural environment
17. _____ Idea that geologic and topographic features of the Earth change very slowly
18. _____ Periodic disasters leading to localized extinction of populations

19.2 Darwin's Journeys [pp. 405 – 411]

Section Review (Fill-in-the-Blanks)

Among the important observations that Darwin made during the voyage of the *Beagle* were: (19) _____ he collected in an area usually resembled living organisms that Darwin observed in that same region. Another observation was that animals from very different South American environments were similar in form but were very different from European animals that lived in similar environments. He deduced that the similarities between the South American species were the result of their descent from a(n) (20) _____. He also took note of and realized the significance of the great variability in the shape of the (21) _____ of the finches he collected on different islands of the Galapagos.

Darwin's thoughts on evolution were influenced by his intellectual predecessors, including (22) _____, an English economist concerned with the ability of agricultural output to keep pace with human population growth, as well as his own experiences, including the controlled breeding of domestic animals to produce desirable traits. He called this process (23) _____. A similar process that occurred in nature, without human intervention, and resulted in certain traits becoming more common in natural populations, he called (24) _____. Those traits that increased the number of offspring produced by individuals who possessed them are said to be (25) _____. Darwin also realized that related organisms that face different environmental challenges could become more different over time, a process we call (26) _____. Darwin referred to this evolutionary diversification of species as (27) _____.

Matching

Match each of the following persons with their contribution to Darwin's thinking about evolution by natural selection.

28.	____	Aristotle	A. Changes in the earth's geological features occur slowly over long periods of time
29.	____	Georges Cuvier	B. Geologic processes and rates of change have remained constant throughout the earth's history
30.	____	James Hutton	C. Inheritance of acquired characteristics
31.	____	J. B. de Lamarck	D. Periodic "catastrophes" lead to localized extinction of species
32.	____	Charles Lyell	E. Populations grow at a rate that outpaces increases in resources
33.	____	Thomas Malthus	F. *Scala Naturae*, the ladder of life

19.3 Evolutionary Biology since Darwin [pp. 411 – 415]

Section Review (Fill-in-the-Blanks)

Since Darwin published *On the Origin of Species*, scientists have learned much more about evolution and the world in general. Remarkably, almost all of these discoveries are consistent with Darwin's theory. One of the first was actually a rediscovery of the work of Gregor Mendel on inheritance of characteristics in 1900. In the early part of the 20th century, geneticists and mathematicians applied Mendel's findings to Darwin's work and developed the branch of biology known as (34) _____. Several decades later, additional data from other fields of biology were used to create a unified theory of evolution called (35) _____ (also known as "neo-Darwinism"). Among other things, this unified theory described the relationship between two levels of evolutionary change: The first involves small-scale genetic change in a population in response to a subtly changing environment. This is referred to as (36) _____. The second deals with large-scale change that may result in the formation or extinction of species. This is referred to as (37) _____.

Despite its power and importance, the theory of evolution by natural selection is misinterpreted by many people today. One example of these misunderstandings is the idea of (38) _____, which implies that natural selection is goal oriented.

Choice

For each of the following statements, choose the branch of biology most likely to use the tool or method from the list below.

a. comparative embryology b. comparative molecular biology c. comparative morphology
d. historical biogeography e. paleobiology

39. _____ The study of homologous structures

40. _____ Examination of similarities and differences in macromolecules such as proteins and DNA in different species.

41. _____ The study of similarities and differences in development between species.

42. _____ The study of fossilized organisms

43. _____ The study of the distribution of organisms around the world

Building Vocabulary

Prefixes	Meaning
bio-	life, living things
macro-	large, long, great, excessive
ortho-	straight, regular
paleo-	ancient, prehistoric

Prefix	Suffix	Definition
_____	-evolution	44. Large scale changes in the history of a species
_____	-geography	45. World distribution of plants and animals
_____	-biology	46. The study of extremely old, often extinct organims
_____	-genesis	47. Progressive, goal-oriented evolution

True/False

If the statement is true, write a "T" in the blank. If the statement is false, make it correct by changing the underlined word(s) and writing the correct word(s) in the answer blank.

48. _______________ Natural selection is based on differences in <u>number of surviving offspring</u> among individuals.

49. _______________ A whale's flipper and a bat's wing are <u>analogous</u> structures.

50. _______________ Embryological similarity between different species often reflects a <u>common ancestry</u>.

51. _______________ A unified theory of evolution which incorporates evidence unknown in Darwin's time is called <u>orthogenesis</u>.

SELF-TEST

1. The Greek philosopher ________ is considered the first student of natural history. His ideas on classification of organisms still influenced thinking in Darwin's time. [p. 402]

a. Archimedes

b. Aristotle

c. Euclid

d. Plato

2. Which of the following structures is NOT homologous with the others? [p. 403]

a. a bat's wing

b. a fly's wing

c. a human arm

d. a whale's flipper

3. Fossils collected from the same geologic stratum are probably similar in ________. [p. 403]

a. age

b. appearance

c. size

d. structure

4. Jean Baptiste de Lamark proposed ________ as a mechanism of evolution. [p. 404]

a. DNA analysis

b. the inheritance of acquired characteristics

c. the principle of use and disuse

d. b and c

5. Which of the following is a key component of Darwin's theory? [p. 410]

a. descent with modification

b. inheritance of acquired characteristics

c. special creation

d. spontaneous generation

6. Populations of the same species occupying different environments may exhibit ________ over time. [p. 410]

a. evolutionary convergence

b. evolutionary divergence

c. orthogenesis

d. stratification

7. Microevolution may be synonymous with ________. [p. 412]

a. adaptation

b. gradualism

c. historical biogeography

d. mutationism

8. Comparing the ________ of organisms can often yield information about evolutionary relationships. [pp. 413 - 414]

a. embryology

b. macromolecules

c. morphology

d. all of the above

9. Which of the following statements is FALSE? [p. 414]

a. Humans represent the pinnacle of evolution.

b. The DNA of humans and chimpanzees will be more similar than the DNA of humans and chickens.

c. Dinosaurs and birds have a common ancestor.

d. Evolution is neither progressive nor goal-directed.

10. Which of the following statements is TRUE? [p. 414]

a. Evolution bears on every aspect of biology.

b. Humans represent the apex of evolution and, thus, have stopped evolving.

c. The study of natural selection allows one to predict the outcome of evolution.

d. Our understanding of evolution is complete and evolutionary biology is no longer an area of active research.

INTEGRATING AND APPLYING KEY CONCEPTS

1. *Unanswered Questions:* Many genetic traits (e.g., Huntington's disease) are clearly maladaptive. Based on your knowledge of genes and their expression, think of reasons why natural selection has not totally eliminated these deleterious alleles from human populations.

2. *Insights from the Molecular Revolution:* Describe the artificial selection experiments of John J. Toole and draw parallels between them and natural selection.

3. *Focus on Research:* Some people are puzzled by the fact that Darwin interrupted his work on evolution to spend 8 years studying barnacles. Others have speculated that he spent so much time on barnacles because he was reluctant to publish his work on evolution. In fact, Darwin did not publish his work until he was in danger of being "scooped" by Wallace. Speculate on why Darwin might have procrastinated in publishing.

20

MICROEVOLUTION: GENETIC CHANGES WITHIN POPULATIONS

CHAPTER HIGHLIGHTS

- Individuals in populations vary, and the variations may be caused by differences in their genetic makeup, environmental factors, or the interaction of the two.
- All populations have a genetic structure that can be described, and mathematical models can be used to test hypotheses relating to the evolution of those populations.
- The Hardy-Weinberg principle describes how populations will achieve equilibrium if certain assumptions are met.
- Three modes of natural selection (directional selection, stabilizing selection, and disruptive selection) have differing effects on phenotypic variation within populations.
- Sexual selection may lead to the evolution of showy structures or display behavior and sexual dimorphism.
- A variety of factors maintain genetic and phenotypic variation in populations despite the effects of stabilizing selection and genetic drift.
- Natural selection leads to the evolution of adaptive traits, although there are constraints on this evolution.

STUDY STRATEGIES

- Understand the sources of variation in populations and be able to devise experiments that can identify the basis of those variations.
- Be familiar with the Hardy-Weinberg principle and know the conditions that must exist for a population to be in genetic equilibrium.
- Know the three modes of natural selection and their effect on the phenotypic variation of a population.
- Be able to discuss the various factors that maintain genetic and phenotypic variation in populations.
- Understand the concept of adaptation and remember that several factors constrain the evolution of adaptations.

INTERACTIVE EXERCISES

Why It Matters [pp. 419-420]

Section Review (Fill-in-the-Blanks)

(1) _____ is a drug used to fight bacteria that enter through the skin and which saved many lives after the Cocoanut Grove nightclub fire in 1942. It's overuse led to the evolution of (2) _____ strains of *Staphylococcus.*

20.1 Variation in Natural Populations [pp. 420-423]

Section Review (Fill-in-the-Blanks)

Any heritable change in the genetic structure of a population is called (3) ______. A population consists of a group of individuals of the same (4) ______ that live together and interact with each other. The individuals in most populations differ with regard to external and internal structure and function, that is, they show (5) ______ variation.

If those variations are generally small and incremental, they are said to be (6) ______; if variations occur in two or more discrete states, e.g., blood type in humans, they are said to be (7) ______. The existence of these discrete variations is an example of a(n) (8) ______.

True/False

If the statement is true, write a "T" in the blank. If the statement is false, make it correct by changing the underlined word(s) and writing the correct word(s) in the answer blank.

9. ______________ Phenotypic variation in individuals may have an <u>environmental</u> as well as a genetic basis.

10. ______________ The <u>behavior</u> of animals may be influenced by genetic differences between individuals.

11. ______________ Only <u>environmentally</u> based variation is subject to evolutionary change.

12. ______________ Differences in the DNA of organisms <u>always</u> produces phenotypic differences.

13. ______________ Sexual reproduction "shuffles the genetic deck" and results in offspring that differ from both parents. Scientists estimate that in humans there are more than <u>10^{10}</u> possible combinations of alleles in every gamete.

14. ______________ Height in humans is a trait that varies <u>quantitatively</u>.

Complete the Table

Complete the following table by filling in the appropriate description of a listed transport process or by naming the transport process based on the listed description.

Process	*Description/Function*
Polymorphism	15.
Microevolution	16.
17.	A technique which can identify different forms of a protein and which is useful in analyzing genetic differences between individuals.
Population	18.

20.2 Population Genetics [pp. 423-425]

Section Review (Fill-in-the-Blanks)

The sum of all alleles at all gene loci in all individuals in a population is referred to as its (19) _____. The relative abundance of each allele is its (20) _____, while the proportion of individuals in a population possessing a specific genotype is that population's (21) _____. The (22) _____ principle describes the conditions under which a population achieves (23) _____, a state where the genetic structure of a population is not evolving. Since natural populations are often difficult to manipulate experimentally, much of what we know about them comes from observational data. Study of these populations is often based on (24) _____ models which predict what would happen if a specific factor had no effect.

Short Answer

List the five conditions which must be met for a population to be in Hardy-Weinberg equilibrium.

25. ____________________

26. ____________________

27. ____________________

28. ____________________

29. ____________________

20.3 The Agents of Microevolution [pp. 425-434]

Section Review (Fill-in-the-Blanks)

If any of the conditions necessary for a population to be in Hardy-Weinberg equilibrium are not met, microevolution (a change in allele frequencies over a number of generations) will occur. (30) _____ is a heritable change in a gene or chromosome and may cause changes in allele frequencies. Likewise, (31) _____ caused by immigrants entering a population or emigrants leaving it will change the genetic structure of the population. Changes in allele frequencies may also occur due to chance events, an effect called (32) _____, which is minimized in very large populations. Occasionally populations are reduced to very small sizes due to disease, drought, or other stresses. Alleles may be eliminated from such small populations resulting in a (33) _____, which reduces genetic variation. A similar reduction in variation can occur when a small number of individuals start a new population, a phenomenon called (34) _____. The process of (35) _____ results in certain individuals with specific, heritable traits producing more offspring than individuals lacking these traits. This difference in reproductive success is reflected in an individual's (36) _____, a measure that reflects the number of surviving offspring an individual produces compared to other members of the population.

Three modes of natural selection can cause changes in the phenotypic variation of a population. They are (37) _____, in which individuals possessing traits at the extremes of the phenotypic continuum are selected against and phenotypic variation is, thus, reduced. Under other circumstances, only one end of the distribution of a trait is selected against, and the mean value of that trait changes. This type of selection is called (38) _____. A third mode, (39) _____, involves selection against the most common phenotype and both extremes of the distribution are favored. The result is increased genetic variability and may result in the evolution of a polymorphism.

Another selective process may result in the evolution of showy structures, complex courtship behaviors, large body size, or structures used in competition with other members of the same sex. This is termed (40) _____ selection and may result in (41) _____, a condition wherein males and females differ in size or appearance.

Choice

For each of the following statements, choose the most appropriate concept from the list below.

a. directional selection b. disruptive selection c. intersexual selection d. intrasexual selection
e. stabilizing selection

42. _____ A horse breeder mates the fastest males with the fastest females.

43. _____ Babies with very low or very high birth weights suffer higher than average mortality.

44. _____ During droughts cactus finches with very long <u>or</u> very deep bills enjoy higher survival than those with average sized bills.

45. _____ Female bowerbirds prefer males with extremely long tails.

46. _____ Very large elephant seals win battles with smaller males. The winner of one of these encounters attains dominant status and gets to mate with females.

20.4 Maintaining Genetic and Phenotypic Variation [pp. 435-437]

20.5 Adaptation and Evolutionary Constraints [pp. 437-440]

Section Review (Fill-in-the-Blanks)

The evolution of (47) _____ helped to maintain gene tic variability in populations by hiding recessive alleles in heterozygotes from the effects of natural selection. (48) _____ occur when circumstances allow for two or more phenotypes to be maintained in a population. For example, (49) _____, the condition where heterozygotes enjoy a higher relative fitness than either homozygote, keeps individuals carrying two different alleles numerous. In some instances, rare phenotypes in a population enjoy an advantage due to (50) _____. Many biologists believe that much variation in populations has no measurable effect on fitness. These variations are said to be (51) _____. Adaptive traits are products of natural selection that increase the (52) _____ of individuals possessing them. The accumulation of these traits over time is (53) _____.

Choice

For each of the following statements, choose term that best describes the situation from the list below.

a. adaptation b. frequency dependent selection c. heterozygote advantage
d. neutral variation hypothesis e. selection in varying environments

54. _____ A change in plumage resulting from natural selection that makes a mouse more difficult to be seen by predators.

55. _____ A mutation changes base sequence in DNA, but the protein it codes for is unchanged.

56. _____ Different shell patterns in snails are favored in different environments.

57. _____ Homozygous individuals have lower fitness than heterozygotes.

58. _____ Rare or unusual phenotypes in a population enjoy higher than average reproductive success.

SELF-TEST

1. How does diploidy maintain genetic variability in a population? [p. 435]
   a. It doubles the number of gametes produced by individuals.
   b. It hides recessive alleles from natural selection.
   c. It increases mutation rate.
   d. It promotes genetic drift.

2. An individual's phenotype is the result of ________. [p. 422]
   a. environmental factors
   b. genetic factors
   c. quantitative factors
   d. a and b

3. Only ________ variation in phenotypes is subject to evolutionary change. [p. 422]
   a. acquired
   b. anatomic
   c. genetic
   d. environmental

4. ________ creates new alleles while ________ shuffles existing alleles into new combinations, producing novel genotypes and phenotypes. [p. 422]
   a. artificial selection; stabilizing selection
   b. founder effect; nonrandom mating
   c. gene flow; genetic drift
   d. mutation; sexual reproduction

5. Population bottlenecks are a cause of ________ [p. 428]
   a. gene flow
   b. genetic drift
   c. mutation
   d. polymorphism

6. The number of surviving offspring an individual produces compared to other members of a population is that individual's ________. [p. 430]

a. allele frequency

b. gene flow

c. heterozygote advantage

d. relative fitness

7. Which of the following will not alter allele frequencies in future generations? [p. 425]

a. genetic drift

b. mutation

c. natural selection

d. nonrandom mating

8. Which of the following will reduce genetic variability in a population? [p. 431]

a. directional selection

b. disruptive selection

c. sexual selection

d. stabilizing selection

9. The evolution of showy structures and behaviors in males of many species is the result of ________. [p. 433]

a. directional selection.

b. disruptive selection.

c. sexual selection.

d. stabilizing selection

10. Which of the following statements is TRUE? [p. 437-440]

a. Biologists are in agreement that all genetic change has fitness implications.

b. In the absence of human intervention, natural selection produces organisms perfectly adapted to all conditions.

c. No organism is perfectly adapted; evolutionary compromises must always be made.

d. Traits evolved for a specific function cannot be co-opted for other purposes.

INTEGRATING AND APPLYING KEY CONCEPTS

1. *Unanswered Questions:* The relationship between recombination rate and genetic variation is complex. Summarize the current state of knowledge regarding this relationship, including what we know about selective sweep and background selection.

2. *Insights from the Molecular Revolution:* Investigators studying genetic variation in humpback whales were surprised to learn that two of the three populations that survived until whale hunting was outlawed retained relatively high levels of variability. Would you expect to obtain similar results if you were studying a bird species with a comparatively short generation time? Explain why or why not.

3. *Focus on Research:* If a population geneticist finds that a population is not in Hardy-Weinberg equilibrium, he or she will probably want to determine which of the required assumptions is violated. What sorts of things might he or she do to determine whether each condition is met?

21

SPECIATION

CHAPTER HIGHLIGHTS

- There are three ways that biologists commonly think of the concept of "species." They are the morphological, biological, and phylogenetic species models.
- Pre- and postzygotic reproductive isolating mechanisms maintain species integrity.
- Most biologists recognize three ways in which new species arise: sympatric, parapatric, and allopatric speciation.
- The accumulation of genetic variation between allopatric populations can lead to speciation.
- Polyploidy is a mechanism by which sympatric speciation may occur in plants.
- Speciation may be a result of chromosomal alterations.

STUDY STRATEGIES

- Learn the various ways that biologists define "species" but be aware of the limitations of each and understand that there is considerable overlap between the different definitions.
- Understand the importance of reproductive isolation between species in maintaining species integrity and familiarize yourself with the various pre- and postzygotic isolating mechanisms.
- Know the three generally recognized models of speciation and that all of the models include some mechanism by which reproductive isolation can evolve.
- Be familiar with the genetic mechanisms of speciation and note that, again, they involve the evolution of reproductive isolating mechanisms.

TOPIC MAPS

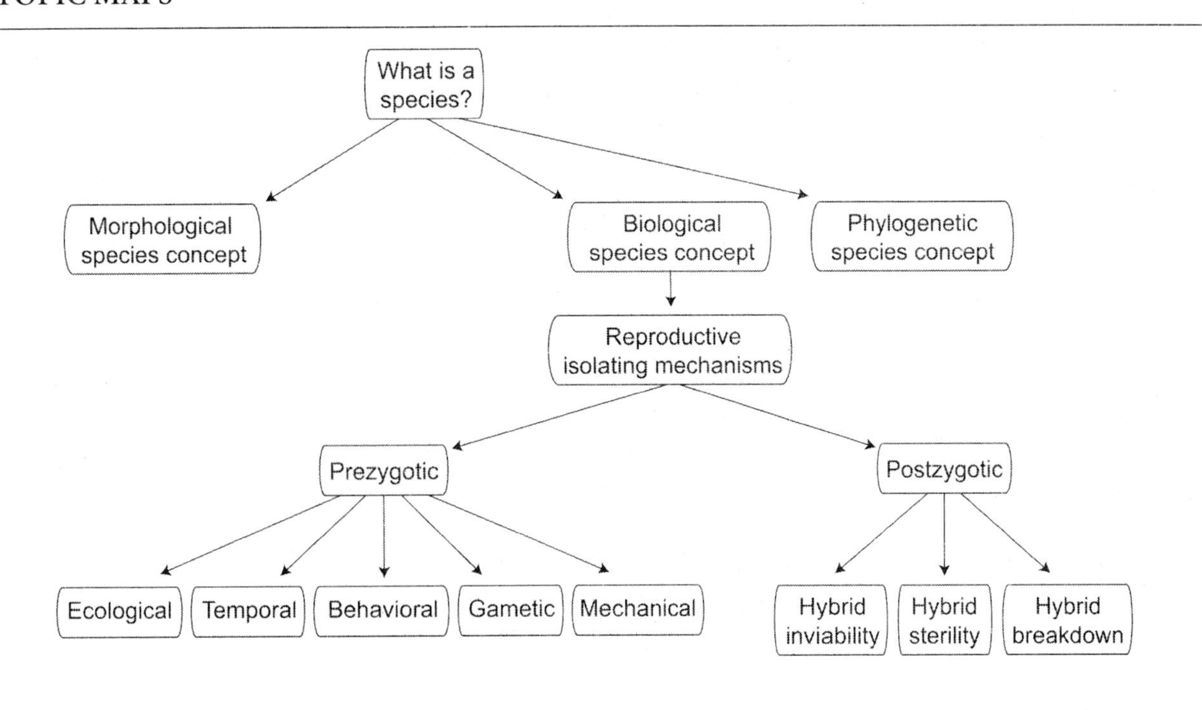

INTERACTIVE EXERCISES

Why It Matters [pp. 443-444]

Section Review (Fill-in-the-Blanks)

In his study of birds of paradise, Ernst Mayr became interested in (1) _____, the mechanism by which new species arise.

21.1 What Is a Species? [pp. 444-447]

Section Review (Fill-in-the-Blanks)

There are several ways that biologists define "species." One of those is based on the visible anatomical features of a group. This is referred to as the (2) _____ concept. A second is based on the ability of different groups of organisms to breed in nature and produce fully fertile offspring. This is the (3) _____ concept. Yet another model uses morphological and genetic sequence data to construct an evolutionary tree. The populations that represent the smallest branches of this tree are grouped together as a single species. This represents the (4) _____ concept.

Species usually show geographic variation, i.e., populations in different regions exhibit some degree of genetic and phenotypic variation. These populations are referred to as (5) _____ or, sometimes, races. Depending on circumstances, these populations may have a geographical distribution that allows for gene flow between adjacent populations but not directly between distant populations. These populations are called (6) _____ species. Similarly, species which occupy a wide geographical range often exhibit some genetic and phenotypic variation as they adapt to local conditions. This variation is often gradual and continuous, and the traits represent a (7) _____.

Choice

For each of the following statements, choose the most appropriate concept from the list below.

a. biological species concept b. morphological species concept c. phylogenetic species concept

8. ____ Based on anatomical characters
9. ____ Based on closely clustered populations on an evolutionary tree constructed from morphological and genetic evidence
10. ____ Based on individuals reproducing fully fertile offspring in nature

21.2 Maintaining Reproductive Isolation [pp. 447-449]

Section Review (Fill-in-the-Blanks)

A(n) (11) _____ is a biological trait that prevents the gene pools of two species from mixing. They may be (12) _____ if they prevent fertilization of an egg or (13) _____ if they operate after fertilization has occurred. Among the mechanisms that can prevent fertilization are (14) _____, where individuals occupy different habitats. (15) _____ occurs when individuals reproduce at different times. One of the most common mechanisms is (16) _____, wherein various displays, vocalizations, and other signals indicate species identity to potential mates.

In some instances, the reproductive structures of males and females are only compatible with other members of the same species, a situation called (17) _____. Finally, (18) _____ occurs when the reproductive cells of different species are incompatible.

Mechanisms that prevent mixing of genes from different species after fertilization include (19) _____ in which embryonic development fails due to different genetic instructions from the two parents. Sometimes development proceeds and produces otherwise healthy offspring that cannot make functional gametes. This is an example of (20) _____. (21) _____ occurs when hybrids are fertile and can breed with other hybrids and either parental species, but F_2 offspring exhibit reduced survival or fertility.

Complete the Table

Fill in the blanks in the following table with the proper term or description.

Prezygotic Mechanisms	(22) _____	Individuals occupy different habitats.
	Temporal isolation	(23) _____
	Behavioral isolation	(24) _____
	Mechanical isolation	(25) _____
	(26) _____	Incompatiblity between gametes of different species
Postzygotic Mechanisms	(27) _____	Abnormal embryonic development
	(28) _____	Hybrid offspring do not make functional gametes
	Hybrid breakdown	(29)

21.3 The Geography of Speciation [pp. 449-454]

21.4 Genetic Mechanisms of Speciation [pp. 454-460]

Section Review (Fill-in-the-Blanks)

Species can arise by several mechanisms. If a population is subdivided by a physical barrier to gene flow, the mechanism of speciation is said to be (30) _____. An example of this type of speciation might be a small number of colonizers founding populations on a group of nearby islands. During the period of separation, reproductive isolation mechanisms may evolve and the populations can become discrete, but closely related species. These species make up a (31) _____. Sometimes the barrier between newly evolved species is removed, but the evolution of prezygotic isolating mechanisms is not complete. Individuals from the two populations are capable of meeting each other, interbreeding, and producing fertile offspring in areas called (32) _____. These areas are usually small, and other mechanisms maintain reproductive isolation between most members of the two species. Additionally, the hybrids' offspring may have lowered fitness and selection will favor individuals that mate only with members of their own species, a phenomenon known as (33) _____.

Under some circumstances, such as a rapid and sharp discontinuity in environmental conditions, speciation may occur between adjacent populations that would otherwise be capable of unrestricted gene flow. This is (34) _____ speciation. There is not a universal agreement about the whether this form of speciation is a reality.

It is also possible for speciation to occur within a continuously distributed population. This is (35) _____ speciation. It might occur in insects in which a mutation changes the preferred plant species upon which most of their life cycle takes place. This isolation on a different plant species could result in the development of a subspecies called a(n) (36) _____. (37) _____, a genetic mechanism resulting in an individual receiving extra sets of chromosomes, is another mechanism by which a new species may arise within a preexisting population. In (38) _____, gametes containing the 2n number of chromosomes are produced. This poses problems for most animal species, but in plants these (39) _____ gametes may be capable of self-fertilizing its own ova. The resultant offspring may be fertile but reproductively isolated from the parent species because of the difference in chromosome number. Finally, new species may arise when two closely related species hybridize and chromosome number doubles in the hybrid offspring. This type of speciation is called (40) _____ and allows the tetraploid hybrids to either self-fertilize or mate with other tetraploid individuals.

Building Vocabulary

Prefixes	Meaning
allo-	different
auto-	self
para-	beside
sym-	together

Prefix	Suffix	Definition
_____	-patric	41. A type of speciation that occurs as a result of a sudden discontinuity in environmental condition
_____	-polyploidy	42. The production of unreduced gametes; it may lead to speciation
_____	-polyploidy	43. Hybrid offspring become polyploid; it may lead to speciation
_____	-patric	44. A type of speciation that occurs within one continuously distributed population
_____	-patric	45. Speciation that requires geographical isolation of subpopulations

---

SELF-TEST

1. Two populations that can and do interbreed in nature and produce fully fertile offspring would exemplify the ________. [p. 445]

a. biological species concept

b. morphological species concept

c. phylogenetic species concept

d. all of the above

2. Which of the following species concepts relies mainly on visible anatomical characters to define species? [p. 444]

a. biological species concept

b. morphological species concept

c. phylogenetic species concept

d. all of the above

3. Local variants of species that interbreed where their distributions overlap are called _________. [p. 445]

a. races

b. clines

c. subspecies

d. and c

4. Traits that vary along an environmental gradient represent a ________. [p. 446]

a. cline

b. hybrid zone

c. reproductive isolating mechanism

d. subspecies

5. The elaborate courtship displays of many animal species have evolved as ________. [p. 448]

a. behavioral isolating mechanisms

b. ecological isolating mechanisms

c. mechanical isolating mechanisms

d. temporal isolating mechanisms

6. In some cases, the genitalia of closely related insect species do not fit and, thus, prevent successful mating. This is an example of ________. [p. 448]

a. behavioral isolation

b. ecological isolation

c. hybrid inviability

d. mechanical isolation

7. Mules, the hybrid offspring of horses and donkeys, are sterile. This is the result of ________ [p. 449]

a. a postzygotic isolating mechanism

b. a prezygotic isolating mechanism

c. allopolyploidy

d. clinal variation

8. An earthquake causes the course of a river to change and separates a subpopulation of mammals from the main body of the population. Over time this isolated population accumulates enough genetic variation to be considered a new species. This is an example of ________. [p. 450]

a. allopatric speciation

b. autopolyploidy

c. parapatric speciation

d. sympatric speciation

9. Polyploidy is a mechanism that can allow ________ speciation to occur. [p. 454]

a. allopatric

b. nongenetic

c. parapatric

d. sympatric

10. Which of the following statements is FALSE? [p. 456]

a. Autopolyploidy is common in animals.

b. Chromosomal rearrangements may foster speciation.

c. Speciation may occur between adjacent populations.

d. Sympatric speciation may occur within a continuously distributed population.

INTEGRATING AND APPLYING KEY CONCEPTS

1. *Unanswered Questions:* Most of the small number of "speciation genes" that have been identified so far affect postzygotic reproductive isolating mechanisms. Speculate on why this might be so and why it may be more difficult to identify genes that contribute to the evolution of prezygotic reproductive isolating mechanisms.

2. *Insights from the Molecular Revolution:* A comparison of base pairs of human and chimpanzee DNA reveals differences at only 1.2% in their respective genomes. Humans and chimpanzees belong to different genera. Given these facts, are you surprised by the results of H. D. Bradshaw's work that indicates that reproductive isolation in different monkey-flower species may be determined by as few as 8 gene loci? Explain why or why not.

3. *Focus on Research:* Hampton Carson's work on fruit flies indicates that flies from older islands in the Hawaiian archipelago colonized the young island of Hawaii at least 19 different times. Review his research and speculate on how sexual selection could have hastened the evolution of prezygotic isolating mechanism and, hence, speciation.

22

PALEOBIOLOGY AND MACROEVOLUTION

CHAPTER HIGHLIGHTS

- Macroevolution is simply microevolution occurring over vast periods of time resulting in large-scale change in morphology and species diversity.
- Much of the evidenced for macroevolution comes from the analysis of fossils by paleobiologists. Paleobiology requires extensive knowledge of the comparative anatomy of animals and plants.
- The earth's history includes many changes in its physical environment and, as a result, changes in the distribution of organisms.
- Convergent evolution results in distantly related organisms adapting in similar ways to similar selection pressures.
- There is disagreement among evolutionary biologists regarding the rate of macroevolution.
- Increases in biodiversity (speciation) and decreases (extinction) have occurred frequently in evolutionary history.
- In recent years, the study of macroevolution has emphasized evolutionary developmental biology (evo-devo).

STUDY STRATEGIES

- Macroevolution is essentially the same as microevolution. The large-scale changes of macroevolution are simply the result of many small-scale changes that have accumulated over long periods of time.
- Know that the earth is a dynamic planet and changes in its physical properties (e.g., climatic change, movement of continents) result in changes in species composition and diversity. New species evolve and existing species go extinct.
- While much of our knowledge of paleobiology is based on analysis of fossils, understand that, for many reasons, the fossil record paints an incomplete picture of evolutionary history.
- Understand why convergent evolution results in superficial similarities among distantly related species.
- Be familiar with both sides of the debate over evolutionary rates and the evidence that supports them.
- Evolutionary developmental biology (evo-devo) is an important new tool to help us understand macroevolution. Know how changes in the timing of various developmental events can lead to evolutionary change.

INTERACTIVE EXERCISES

Why It Matters [pp. 463 – 464]

Section Review (Fill-in-the-Blanks)

In 1796, Georges Cuvier made the then-startling suggestion that (1) _____ represented species that had gone extinct, something not thought possible at that time. Cuvier is generally considered to be the founder of two branches of biology, (2) _____ and (3) _____. Although Cuvier never expressed support for Darwin's theory of evolution by natural selection, today his work is considered important to the concept of (4) _____, large-scale changes in the anatomy and diversity of organism throughout the earth's history.

22.1 The Fossil Record [pp. 464-469]

22.2 Earth History, Biogeography and Convergent Evolution [pp. 469-471]

Section Review (Fill-in-the-Blanks)

(5) _____ can be formed in a number of ways, but most commonly they are the result of mineralization of parts of organisms. There are a number of methods used to estimate the relative age of fossils, but one technique, (6) _____, allows paleobiologists to determine the age of fossils rather precisely by measuring the amount of a radioactive isotope in a fossil and knowing the (7) _____ of that isotope.

Biogeography, the distribution of living and ancient organisms over the face of the earth is also important to the understanding of macroevolution. In the last 100 years scientists have determined that the position of the earth's great land masses has changed over its history. This is due to the fact that the earth's crust is divided into irregularly shaped segments that float on its semisolid mantle, a concept known as (8) _____. As these segments move, the continents that sit atop them slowly change their positions relative to each other, a phenomenon called (9) _____.

Historical biogeographers attempt to determine how the biogeography of organisms has developed over evolutionary time spans. Many species live in all suitable habitats throughout a geographical area. They are said to have a (10) _____ that requires no special historical explanation. But other organisms occupy widely separated areas, a phenomenon known as (11) _____. Two factors can create disjunct distributions: (12) _____, which is the movement of organisms away from their place of origin; the second is (13) _____, the fragmentation of a continuous distribution caused by external factors. An example of these external factors is the breakup of the supercontinent, Pangaea. It resulted in the isolation of the continents as we know them today. Each of them has a different mix of organisms and led to the evolution of unique (14) _____ on each. Alfred Russel Wallace used them to define the six (15) _____ that we still recognize today. Some of these biogeographical regions, e.g., Australia, are occupied by unique (16) _____ species that occur nowhere else on Earth. Sometimes distantly related species that are biogeographically isolated evolve similar features. Thus, many fish and aquatic mammals are torpedo shaped because that form facilitates movement in water. These similarities are the result of (17) _____.

True/False

If the statement is true, write a "T" in the blank. If the statement is false, make it correct by changing the underlined word(s) and writing the correct word(s) in the answer blank.

18. ______________ Fossils form when organisms are buried under sediment or are preserved in oxygen-rich conditions.

19. ______________ The fossil record represents a virtually complete record of life throughout evolutionary history.

20. ______________ Historical biogeography attempts to explain the geographical distribution of organisms.

21. ______________ The similarities between American cacti and African spurges are the result of vicariant evolution.

22. ______________ The half-life of ^{14}C is approximately 5600 years. A piece of wood with 1/4 of the ^{14}C found in living specimens is roughly 5600 years old.

23. ______________ The theory of plate tectonics helps explain continental drift and historical biogeography.

24. ______________ The redfin darter (a small fish) is found only in the rivers of the Ozarks, thus it is endemic to the region.

25. ______________ The ring-billed gull is found in all coastal regions of the continental U.S. It's distribution is disjunct.

22.3 Interpreting Evolutionary Lineages [pp. 473-477]

22.4 Macroevolutionary Trends in Morphology [pp. 477-480]

Section Review (Fill-in-the-Blanks)

The evolutionary transformation of an existing species into a new one is called (26) _____. It does not increase the number of existing species. In contrast, (27) _____ is the evolution of two or more descendant species from a single ancestral species. The result is an increase in the number of existing species. The tempo of macroevolutionary change is disputed by biologists. The (28) _____ hypothesis contends that speciation is the result of the slow, continuous accumulation of small changes at a steady rate. A competing explanation, the (29) _____ hypothesis, suggests that most species endure long periods during which they change only little, then undergo brief periods of cladogenesis and rapid change. The two hypotheses are not necessarily mutually exclusive.

Some evolutionary lines exhibit a trend toward greater size and complexity, while others tend to exhibit increased morphological novelty. Steven Stanley has proposed that the trend toward great size in some lineages may be a result of (30) _____, a mechanism akin to natural selection. His hypothesis is based on the idea that the evolutionary success of a species is measured by the number of species which have descended from it. Thus, if large species leave more descendant species than small ones, the trend will be toward more large species. Along with an evolutionary increase in size, we often observe an increase in morphological complexity.

Morphological novelties may appear suddenly in the fossil record. Several mechanisms have be postulated to explain this observation. One is (31) _____, a trait that is adaptive in one context also having benefits in a different context. For example, feathers and winglike forelimbs may have evolved among some dinosaurs as a means of capturing prey and conserving body heat. Later they became further modified and used for flight. (32) _____ is another mechanism which may contribute to speciation based on differences in growth rate of specific anatomical structures. For example, the skulls of newborn chimps resemble those of newborn humans while the skulls of adults are more different. The reason for this is that growth rates of certain parts of the skulls differ. By the time skull development is complete, significant changes are apparent. Yet another mechanism leading to differences in species is a change in the timing of developmental events, or (33) _____. Thus, some species of amphibians retain gills, a juvenile trait, into adulthood. In this case, a gene responsible for metamorphosis may have mutated.

Building Vocabulary

Prefixes	Meaning
allo-	different
clado-	a branch
hetero-	different
paedo-	child

Prefix	Suffix	Definition
_____	-morphosis	34. Retention of juvenile characteristics into adulthood
_____	-chrony	35. Changes in timing of developmental events
_____	-metric	36. Different relative sizes of structures in related species
_____	-genesis	37. Evolution of two or more descendant species from a common ancestor

Matching

Match each of the following persons with the correct concept.

38.	____	Edward Drinker Cope	A.	An evolutionary trend toward greater size in a lineage.
39.	____	Niles Eldredge and Stephen Jay Gould	B.	Punctuated equilibrium
40.	____	Othniel C. Marsh	C.	Phylogeny of horse species
41.	____	John Ostrom	D.	Preadaptation
42.	____	Steven Stanley	E.	Species selection

22.5 Macroevolutionary Trends in Biodiversity [pp. 480-483]

Section Review (Fill-in-the-Blanks)

The number of extant (currently living) species on the Earth represents its (43) _____. It may fluctuate over time for a variety of reasons. One is (44) _____, where a lineage rapidly evolves into a number of species with different requirements and preferences. An example is the 13 species of Galapagos finches. They have adapted different structures and behaviors to exploit different food sources, yet they all evolved from a single ancestral species. This does not increase the number of existing species. This speciation occurred when the ancestral species was carried from the mainland of South America to the Galapagos and found itself in an unfilled (45) _____. Thus without competing warbler species present on the islands, a line of finches evolved characteristics that allowed them to feed on insects in a manner similar to that of the true warblers.

Just as speciation can increase the number of extant species, (46) ______ results in the elimination of species when its last member dies. For any species, there is a nonzero probability that it will one day disappear in this manner as environments change. If these environmental changes are slow, species will usually disappear at a low rate called the (47) ______. However, at some points in the Earth's evolutionary history, large numbers of species have died out over relatively short periods of time an event know as (48) ______.

True/False

If the statement is true, write a "T" in the blank. If the statement is false, make it correct by changing the underlined word(s) and writing the correct word(s) in the answer blank.

49. ______________ The <u>background rate of extinction</u> is the result of small scale, relatively slow changes in the environment.

50. ______________ Mass extinctions are caused by <u>long periods of environmental stability</u>.

51. ______________ It has been estimated that the <u>vast majority</u> of species that have ever existed have also gone extinct.

52. ______________ The trend in evolutionary history is for biodiversity to <u>increase at a constant rate</u>.

22.6 Evolutionary Developmental Biology [pp. 483-488]

Section Review (Fill-in-the-Blanks)

The study of changes in regulatory genes that affect development and that can create morphological changes is called (53)______. These studies show that alterations in the timing or sequence of various developmental processes can lead to new forms. If the changes are due to changes in genes that code for transcription factors that bind to regulatory sites on DNA, they are referred to as (54) ______ genes. Several hundred of these genes are found in most animals and are called the "genetic tool-kit." The (55) ______ family of genes is an example. They are a set of genes that control the overall body plan of animals that possess them. All the genes in this family contain a nucleotide sequence, the (56) ______, which codes for a protein component called a(n) (57) ______ that functions as a transcription factor that can either turn a regulatory gene on or off.

Choice

For each of the following descriptions, choose the correct gene or gene family from the list below.

a. *Hox* b. *Pax-6* c. *Pitx1*

58. ____ Regulates genes that control the development of light sensing organs in a wide variety of animals

59. ____ Regulates genes that control the development of spines on the pelvic fins of sticklebacks

60. ____ Regulates genes that control where appendages will form on an animal's body

---

SELF-TEST

1. Fossils are most likely to form when [p. 464]

a. aquatic invertebrates are exposed to the air and are preserved by dehydration.

b. hard parts of organisms are buried under sediments in anaerobic conditions.

c. soft parts of organisms are buried under sediments in aerobic conditions

d. volcanic activity buries organisms in lava and ash.

2. Radiometric dating is based on the ________ of various radioactive ________. [p. 468]

a. decay rate; isotopes

b. fusion rate; ores

c. lifespan; isomers

d. wavelength; emissions

3. Continental drift is responsible for the ________ of some populations. [p. 469]

a. continuous distribution

b. disjunct distribution

c. dispersal

d. preadaptation

4. Which of the following in NOT one of Wallace's six biogeograhpical realms? [p. 471]

a. Australian

b. Ethiopian

c. Nearctic

d. Occidental

5. The evolution of the horse lineage is an example of ________. [p. 473]

a. abiogenesis

b. anagenesis

c. cladogenesis

d. pangenesis

6. Periods of stability followed by periods of rapid evolutionary change are cornerstones of ________. [p. 475]

a. convergence

b. gradualism

c. heterochrony

d. punctuated equilibrium

7. Cope's Rule states that there is a trend toward increasing ________ within evolutionary lineages. [p. 478]

a. allometric growth
b. paedomorphosis
c. size
d. vicariance

8. The evolution of species to fill unoccupied roles in the environment is called ________. [p. 480]

a. adaptive radiation
b. convergence
c. evo-devo
d. species selection

9. The change in relative size of various anatomical features of humans through development is an example of ________. [p. 479]

a. adaptive radiation
b. allometric growth.
c. paedomorphosis.
d. preadaptation.

10. Homeotic genes serve as a "tool kit" that ________. [p. 484]

a. links regulatory genes on a single chromosome
b. recycles worn out cellular components
c. regulates the timing and sequence of developmental processes
d. repairs mutations that occur in structural genes

INTEGRATING AND APPLYING KEY CONCEPTS

1. *Unanswered Questions:* Based on what you have learned about homeotic genes, why is it not surprising that organisms with serially repeating segments are most likely to show extreme organ displacement, e.g., substitution of legs for antennae in *Drosophila*? Arthropods, which exhibit this pattern of segmentation, are the most diverse group in the animal kingdom. How might these two facts relate to each other?

2. *Insights from the Molecular Revolution:* The results of research done by Paulo Sordino and his colleagues indicates that tetrapod digits are a morphological novelty. Can you think of any further experiments or observations that could determine whether or not fishes have the potential to develop digits?

3. *Focus on Research:* The evolution of placental mammals resulted in the extinction of most marsupials in most part of the world. Why are marsupials so common in Australia? Knowing what you know about changes in the fauna of North and South America after establishment of a land bridge between the two continents, what do you think would be the fate of Australian marsupials if placentals from other parts of the world were introduced to without regulation?

23

SYSTEMATIC BIOLOGY: PHYLOGENY AND CLASSIFICATION

CHAPTER HIGHLIGHTS

- The study of systematics aims to classify species and reconstruct their evolutionary history.
- Organisms are organized into a hierarchical classification system.
- Morphological characters provide clues about evolutionary relationships.
- Cladistics uses shared derived characters to trace evolutionary history.
- Molecular characteristics have clarified many evolutionary relationships.

STUDY STRATEGIES

- The study of phylogeny and classification is important for understanding the evolutionary relationships among organisms and the origins of their structure and function. Therefore, take time now to learn these basic concepts.
- This chapter contains some familiar material and some unfamiliar material. Go slow. Take one section at a time, then work through the companion section(s) in the study guide.
- Practice drawing phylogenies on paper.

TOPIC MAP

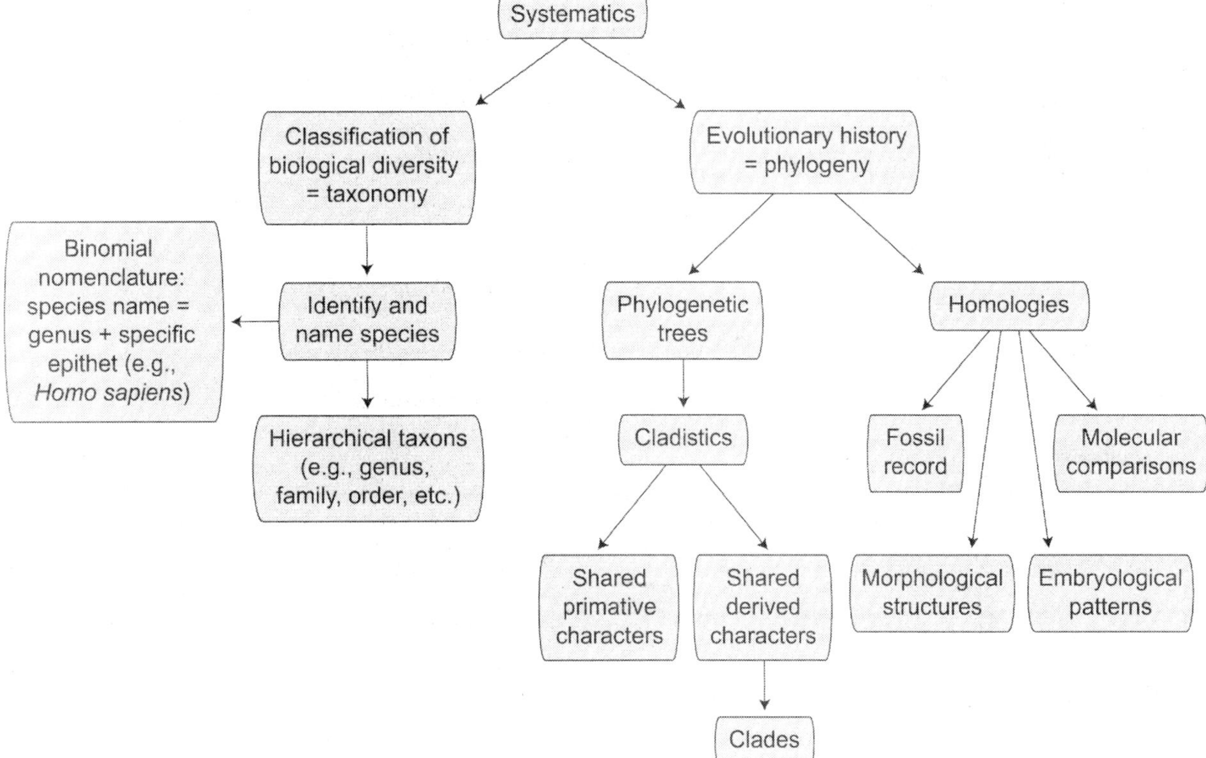

INTERACTIVE EXERCISES

Why It Matters [pp. 491-492]

Section Review

The branch of biology that studies the diversity of life and its evolutionary relationships is called (1) ____.

23.1 Systematic Biology: An Overview [pp. 492-493]

23.2 The Linnaean System of Taxonomy [pp. 493-494]

23.3 Organismal Traits [pp. 494-495]

Section Review

One goal of systematics is to reconstruct the (2) ____, or evolutionary history of organisms; such relationships are illustrated using (3) ____. The second goal is the identification and naming of species, which is called (4) ____. A(n) (5) ____ is an arrangement that reflects the relatedness of organisms. (6) ____ is the system of nomenclature in which species are given a two-part name or (7) ____. (8) ____ is the first part of the species name, and the (9) ____ is the second part. A(n) (10) ____ is an arrangement of organisms into increasingly inclusive categories. A(n) (11) ____ is a group of related genera. Similar families are grouped into (12) ____, orders into (13) ____, classes into (14) ____, and phyla into (15) ____. All life on earth is classified into three (16) ____. A particular category of the taxonomic hierarchy is a(n) (17) ____.

Sequence

Arrange each of the following taxa into the correct sequence, from the least inclusive group to the most inclusive.

18. ____ a. family
19. ____ b. phylum
20. ____ c. genus
21. ____ d. order
22. ____ e. domain
23 ____ f. class
24. ____ g. kingdom

Identify

List the species name, genus name, and specific epithet for each of the following:

25. *Oncorhynchus mykiss*: ____________________

26. *Xenopus leavis*: ____________________

27. *Homo sapiens*: ____________________

Matching

Match each of the following with its correct definition.

28. ____	classification	A.	the evolutionary history of a group of organisms
29. ____	taxon	B.	an illustration that identifies likely evolutionary relationships among species
30. ____	binomial nomenclature	C.	the science devoted to naming and classifying organisms
31. ____	phylogeny	D.	an arrangement of organisms into hierarchical groups
32. ____	taxonomy	E.	a Latinized two-part name
33. ____	taxonomic hierarchy	F.	the system in which species are assigned a two-part name
34. ____	phylogenetic tree	G.	an arrangement of organisms into ever more inclusive groups
35. ____	binomial	H.	the group of organisms at a given taxonomic level

23.4 Evaluating Systematic Characters [pp. 495-497]

23.5 Phylogenetic Inference and Classification [pp. 497-501]

23.6 Molecular Phylogenetics [pp. 501-508]

Section Review

(36) ____ are similarities that result from shared ancestry. Phenotypic similarities that evolved independently in different lineages are called (37) ____. Some characters evolve slowly while others evolve rapidly in (38) ____. As a result, every species displays old forms of traits, or (39) ____, and new forms of traits, or (40)____. In (41) ____, characters identified in the group under study are compared to distantly related species not included in the analysis.

(42) ____ include a single ancestral species and all of its descendents. (43) ____ include species from separate evolutionary lineages. (44) ____ include an ancestor and some but not all of its descendents. The (45) ____ is the notion that species can be traced to a single common ancestor. The (46) ____ holds that because a particular evolutionary change is rare, it is unlikely that the change evolve more than once in the same lineage. (47) ____ groups together species that share both ancestral and derived character.

(48) ____ is a classification approach based solely on evolutionary relationships. In this approach, only species that share derived characters and form a monophyletic lineage are grouped together in a (49) ____. Trees that illustrate hypothesized evolutionary branching are called (50) _____. (51) ____ is a cladistic system of identifying and naming clades.

Differences in DNA sequences of two species can serve as a(n) (52) ____. A phylogenetic tree based on rRNA divides living organisms into three primary lineages or (53)____: Bacteria, Archaea, and Eukarya.

Building Vocabulary

Prefixes	Meaning
mono-	single, alone
poly-	many, much
para-	next to, associated with

Prefix	Suffix	Definition
_____	-phyletic	54. Taxon in which all of the of subgroups have a common ancestor
_____	-phyletic	55. Taxon consisting of a common ancestor and some but not all of its descendents
_____	-phyletic	56. Taxon consisting of several evolutionary lines and not including a common ancestor

Choice

For each of the following characteristics, choose the most appropriate term from the list below.

a. homoplasies b. homologies

57. ____ similarities that result from shared ancestry

58. ____ similarities that evolved independently in different lineages

For each of the following characteristics, choose the most appropriate term from the list below.

a. ancestral characteristics b. derived characteristics

59. ____ traits that have remained unchanged in a species

60. ____ traits not present in ancestors

Match the taxonomic approach with its basis for classification.

a. cladistics b. traditional evolutionary systematics

61. ____ shared ancestral characters and shared derived characters

62. ____ shared derived characters only

For each of the following, choose the most appropriate taxonomic principle from the list below.

a. principle of monophyly b. principle of parsimony

63. ____ The idea that species can be traced to a single common ancestor

64. ____ The idea that the same change evolved more that once in a lineage is extremely unlikely

Matching

Match each of the following terms with its correct definition.

65.	____	cladogram	A. a phenomenon that describes how some characters evolve slowly and others evolve rapidly
66.	____	outgroup comparison	B. a technique that compares a group under study to a distantly related species
67.	____	molecular clock	C. a group that possesses a unique set of derived characters
68.	____	mosaic evolution	D. a tree that illustrates a hypothesized sequence of evolutionary branchings
69.	____	clade	E. a system that identifies and names clades rather than familiar taxonomic levels
70.	____	PhyloCode	F. differences in DNA sequences that marks their time of divergence

SELF-TEST

1. A taxon that comprises related genera. [p. 493]
   a. class
   b. family
   c. species
   d. order

2. Systematics encompasses the study of ____. [p. 494]
   a. the development of organisms
   b. the evolutionary history of organisms
   c. the function of organisms
   d. the identification and naming of organisms

3. In the binomial name for rainbow trout, *Oncorhynchus mykiss*, the species name is _______. [p. 493]
   a. *Oncorhynchus*
   b. *mykiss*
   c. *Oncorhynchus mykiss*

4. Phenotypic similarities that evolved independently in different lineages are called ________. [p. 495]
   a. homologies
   b. evolutionary mosaics
   c. feathers
   d. homoplasies

5. The fossil record and embryological evidence indicates that these are derived characters. [p. 497]
   a. vertebral column
   b. four-chambered heart
   c. two-chambered heart
   d. insects with four walking legs
   e. insects with six walking legs

6. Cladistics would consider which of the following when classifying organisms? [p. 499]
   a. shared ancestral characters
   b. shared derived characters
   c. a combination of ancestral and derived characters
   d. monophyletic lineage

7. Molecular systematists conduct phylogentic analyses using which kinds of molecules? [p. 501]
   a. Protein
   b. DNA
   c. RNA
   d. carbohydrate
   e. lipid

8. Molecular phylogentics has identified which three major lineages? [pp. 506-508]

a. Prokarya

b. Eukarya

c. Bacteria

d. Protista

e. Archaea

9. The hyomandibula, a bone that braced the lower jaw in early jawed fishes, is homologous to ____. [p. 496]

a. incus

b. malleus

c. stapes

d. mandibal

e. tibia

For questions 10 and 11, refer to the following figure.

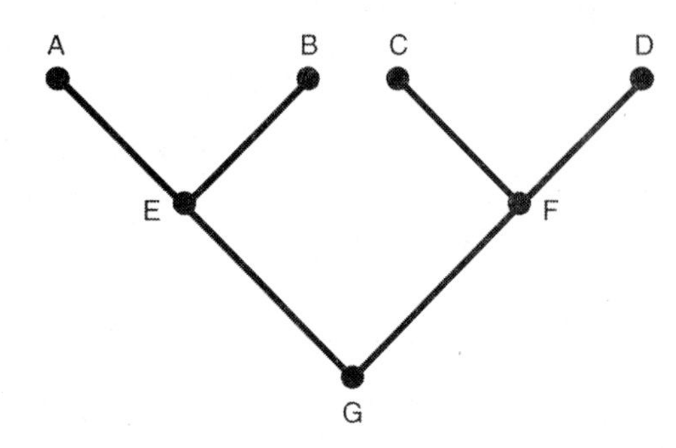

10. Species C and D are part of monophyletic taxon. Which is the ancestor for this taxon? [p. 498]

a. species E

b. species F

c. species A

d. species B

11. Species B and C are part of a taxon that is best described as ____. [p. 498]

a. monophyletic

b. polyphyletic

c. paraphyletic

INTEGRATING AND APPLYING KEY CONCEPTS

1. Two different but related 9-amino acid hormones--oxytocin and arginine vasopressin—occur in the brain of mammals. How can comparisons of nanopeptides among different groups of animals lead to misconceptions about the evolution of the genes that encode these peptides?

24

THE ORIGIN OF LIFE

CHAPTER HIGHLIGHTS

- Organic and inorganic molecules that make up and support living things arose spontaneously under primitive conditions of earth and its atmosphere.
- Random aggregation of these molecules resulted in chemical reactions that eventually supported life.
- Primitive cells were consequently formed by the aggregation of organic molecules.
- These primitive protocells had to:

  —Develop the pathways to produce energy for supporting cellular functions.

  —Develop the pathways to be able to store information for protein synthesis, multiply, and pass information from one generation to another.

  —Organize the molecules to form a prokaryotic cell.
- The endosymbiotic hypothesis to explain the evolution of eukaryotic cells suggests that:

  —Eukaryotic cells evolved from ancestral prokaryotic cells.

  —Some of the organelles, such as mitochondria and chloroplasts, developed from the ingested prokaryotic cells.
- Multicellular organisms probably evolved by:

  —Cells of the same species aggregating to form colonies.

  —Differentiation of the cells in a colony to perform various specialized functions.

STUDY STRATEGIES

- This chapter is fairly simple but requires the imagination to understand the atmospheric and earth conditions that existed at the time life may have evolved.
- Remember the goal of this chapter is to understand:

  —How organic molecules spontaneously evolved.
  —How prokaryotic cells evolved.
  —How eukaryotic cells evolved.
  —How living things became multicellular.
- Concentrate on the key concepts and hypotheses that are proposed.

TOPIC MAP

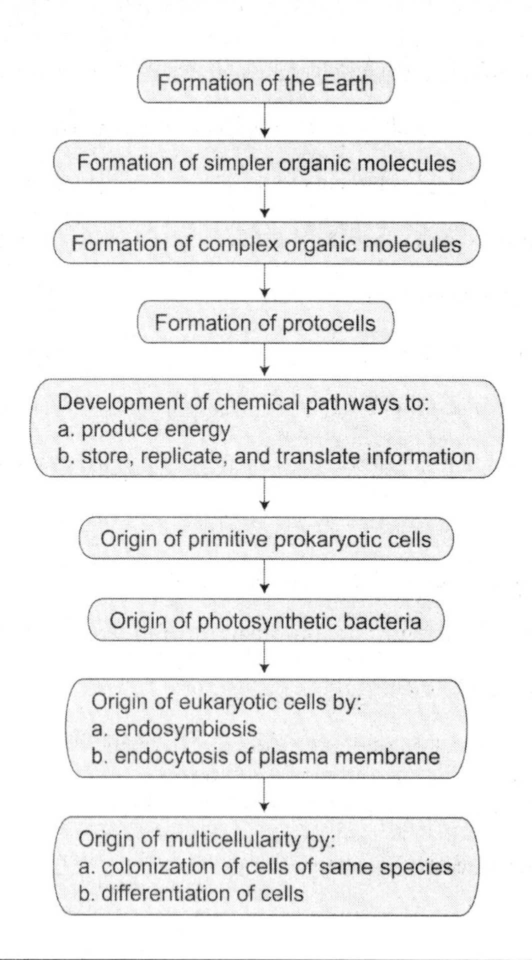

INTERACTIVE EXERCISES

Why It Matters [pp. 511–512]

Section Review (Fill-in-the-Blanks)

The Big Bang Theory was proposed in 1947 by (1) ________________ ________________ to explain the origin of universe. The (2) ________ and (3) ________ released during the explosion condensed to form the sun and the planets that revolve around it, including the (4) __________. Earth evolved about (5) ________ billion years ago, and the oldest fossil record of cell-like structure is about (6) _______ billion years old.

24.1 The Formation of Molecules Necessary for Life [pp. 512–515]

Section Review (Fill-in-the-Blanks)]

Present day cells are characterized by a(n) (7) ________________ that forms a boundary, (8) ___________ acids that code the genetic information, a mechanism that allows the cell to produce regulatory (9) __________, and metabolic reactions to produce (10) ____________. Laboratory experimentation has provided substantial support for the idea that life originated (11) ______________________ from the chemical and physical processing of nonliving matter present on the earth.

Hypothesis for the formation of Earth and its atmosphere: Astronomers have suggested that our solar system started forming about (12) ______ billion years ago. Dust and gas started to compress with the help of force to pull the heavier (13) ________________ elements to the center and lighter silicates, carbides, and sulfides of the (14) ________________ elements to the surface. As the (15) ________ was lost through radiation, earth started to cool down and the atmosphere was formed by some of the (16) ________ and (17) ________. The sun warmed the cooling earth enough to maintain liquid (18) ____________ on the surface as rivers, lakes, and seas. Evaporating (19) _________ added moisture and erupting (20) ________________ added various types of gases to the atmosphere. Eventually sunlight and lightning promoted chemical reactions that led to the formation of (21) _______________ chemicals.

Hypotheses for the origin of organic chemicals: A. Oparin-Haldane hypothesis assumes that the earth's (22) ________________ was very different at the time when life was originating on earth. It also proposes that the (23) _____________ from lightning and storm, and the absence of (24) _____________ in the atmosphere played key roles in the formation and concentration of organic chemicals. Absence of oxygen and microorganisms also prevented the breakdown of (25) ____________ chemicals. This resulted in formation of the (26) "______________" soup that encouraged random aggregation of chemicals and chemical reactions characteristic of life.

Direct experimental support came in the 1950s from the lab of (27) ____________ ____________ and (28) _______________ __________. They created a(n) (29) ________________ atmosphere by placing gases and (30) __________ ____________ in a closed container. They then sparked high voltage through the mixture and within a weeks time were able to form (31) _______________ chemicals such as urea, amino acids, and lactic, formic, and acetic acids. Later they were able to also form more complex (32) ________________ molecules such as purines, pyrimidines, sugars, and phospholipids. B. Scientists also believe that life may have evolved in a(n) (33) ______________ atmosphere found in the hydrothermal vents of the sea floor. Japanese scientists, (34) ________________ ___________________ and his colleagues, simulated thermal vents in the lab and were able to generate amino acids that polymerized to form short (35) ___________________. C. Some scientists have proposed that organic chemicals are extraterrestrial and arrived on earth through a(n) (36) ______________ or a(n) (37) _______________.

Matching

Match each of the following scientists with their contribution to origin of organic chemicals.

38. ____	George Lemaitre	A. Provided experimental evidence to support origin of organic molecules.
39. ____	Aleksandr Oprarin and J. B. S. Haldane	B. Proposed the Big Bang Theory to explain the origin of universe.
40. ____	Stanley Miller and Harold Urey	C. Proposed the hypothesis to explain how organic chemicals were formed on earth.

41. ____ Koichiro Matsuno and his colleagues — D. Provided experimental evidence to support origin of life near the hydrothermal vents.

Match each of the following terms with its correct definition.

42. ____ Big Bang Theory — A. Submarine volcanoes in the sea floor that emit superheated mineral-rich water.

43. ____ Oparin-Haldane Hypothesis — B. Refers to oceans and other water bodies with very high concentrations of organic chemicals.

44. ____ Prebiotic soup — C. Explains the origin of the universe.

45. ____ Hydrothermal vents — D. Explains how organic chemicals formed on earth.

Complete the Table

46. There are three major theories to explain the origin of organic chemicals. Give a short description of each.

Hypotheses	*Short description*
A. Oparin-Haldane	
B. Hydrothermal Vents	
C. Extraterrestrial Origin	

24.2 The Origin of Cells [pp. 515–519]

Section Review (Fill-in-the-Blanks)

Monomers such as amino acids and nucleotides polymerized by (47) ________________ of water or by (48) __________________ synthesis. Once the larger molecules were made, they randomly assembled, surrounded by a(n) (49) ______________ to form primitive (50) _____________.

Hypotheses for the origin of protocells: A. Absorption into clays: This hypothesis proposes that protocells may have originated in between the layers of (51) _______ where water, ions, and (52) ______________ chemicals become trapped. Scientists (53) ________ __________ and (54) ______________ __________ provided the experimental support for this hypothesis. B. Lipid bilayer assembly: R. J. (55) ___________ hypothesized that protocells may have formed by spontaneous assembly of (56) ________ bilayer. Scientist (57) _________ ____________ provided the experimental support for this hypothesis.

Transition from (58) ____________ to the formation living cells must involve development of two major characteristics: A. Ability to generate energy: This required the development of (59) __________, where electrons are removed from a chemical and energy is released to support other reactions; and (60) ____________, where electrons are added to a chemical and energy is added to form larger chemicals. To support stepwise release of energy, a(n) (61) _____________ transport mechanism with intermediate (62) ____________ carriers developed. Also, (63) _______ may have developed to store energy for immediate use. B. Ability to store, reproduce, and translate information: In present-day cells, information is passed from DNA to (64) _______ and then to proteins. This process is catalyzed by several (65) ____________. Scientists were faced with the dilemma—which evolved first: nucleic acids or the (66) ________________? According to RNA-first hypothesis, (67) ________ acted as the first genes and as enzymes called the (68) __________________. From these precursor RNA molecules, different types of (69) _________ developed to provide specific functions. (70) _______ may have developed later and was subsequently chosen as a chemical to store information. According to the Protein-first hypothesis, (71) ____________ acted as enzymes and catalyzed the formation of the nucleic acids, (72) _______ and (73) _________.

Once large (74) _______________ molecules developed, they were enclosed in membranes formed spontaneously by in a pool of (75) _____________. At this point (76) ______________ may have formed.

Hypothesis for the origin of prokaryotic cells: (77) __________________ slowly organized themselves into self-replicating prokaryotic cells. Scientist (78) _____________ __________________ proposed that the first photosynthetic prokaryotes may have evolved about 3.5 billion years ago and may have used (79) _________ as the electron donor. As a result, no (80) _______________ is released. At some point, photosynthetic prokaryotes shifted to using the more common chemical, (81) ___________, as an electron donor. By splitting this donor chemical, (82) ___________ was released into the atmosphere. This allowed some prokaryotes to develop electron transfer system and to start using (83) _____________ as an electron acceptor. Possible evidence of the first photosynthetic cyanobacteria was found in the rocks, (84) ______________________.

Matching

Match each of the following scientists with their contribution to origin of organic chemicals.

85. ____ Noam Lahav and Sherwood Chang
86. ____ R. J. Goldacre
87. ____ David Deamer
88. ____ Richard Dickerson

A. Provided experimental support for the origin of protocells by lipid bilayer assembly.
B. Provided experimental support for the origin of protocells by absorption into clays.
C. Proposed the origin of protocells by lipid bilayer spontaneous assembly.
D. Proposed the origin of photosynthetic bacteria.

Match each of the following terms with its correct definition.

89. ____ Dehydration synthesis
90. ____ Absorption into clays
91. ____ Lipid bilayer assembly
92. ____ RNA-first hypothesis
93. ____ Protein-first hypothesis
94. ____ Ribozymes
95. ____ Stromatolites

A. Proposes that protocells developed between the layers of clay.
B. Proposes that the first genes and enzymes were RNA molecules.
C. Where smaller subunits are joined to form a larger molecule by removal of water.
D. Proposes that lipids spontaneously form bilayer.
E. RNA that acts as an enzyme.
F. Proposes that the key organic molecules were protein molecules.
G. Rocks that contained fossils of ancient photosynthetic cyanobacteria.

Short Answer

96. Why is the hypothesis lipid bilayer assembly more attractive than the hypothesis of absorption into clays for the formation of protocells? ______

______

______

97. What evidence supports the RNA-first hypothesis over the Protein-first hypothesis? ______

______

______

98. Why do scientists believe that the ancient atmosphere lacked oxygen and was a reducing environment? ______

______

______

Sequence

99. Arrange the following steps of that may have been followed for the origin of life.

A. Aggregation of complex organic molecules inside a membrane bound protocell

B. Abiotic synthesis of organic molecules

C. Assembly of complex organic macromolecules

D. Reorganization of chemicals to form a prokaryotic cells

______ ______ ______ ______

True/False

Mark if the statement is true or false. If the statement is false, justify your answer in the line below each statement.

100. ______ Photosynthetic bacteria that used water as an electron donor were the first living organisms to evolve.

______

101. ______ Oxygen was present in the earth's atmosphere from the time of its origination.

______

24.3 The Origins of Eukaryotic Cells [pp. 519–523]

Section Review (Fill-in-the-Blanks)

Present day eukaryotic cells have a nucleus that is separated from the (102) ______ and numerous membrane-bound (103) ______.

Scientist (104) ______ ______ suggested the endosymbiotic hypothesis for the evolution of eukaryotic cells. According to this hypothesis, an anaerobic (105) ______ cell engulfed a(n) (106) ______ prokaryotic cell. This resulted in permanent partnership for mutual gain and is referred to as (107) ______. The engulfed bacteria became the (108) ______ of the cell, an organelle typically found in eukaryotic cell. The ancestral anaerobic cell was using organic electron (109) ______ and after endosymbiosis, it was able to use (110) ______ as the electron acceptor.

Similarly, a nonphotosynthetic (111) _______________ cell engulfed a(n) (112) ___________________ prokaryotic cell. The engulfed bacteria continued to photosynthesize and became a(n) (113) _________________, typically found in plant cells.

If the endosymbiotic hypothesis is true for the origination of mitochondria and chloroplast, these organelles must have characteristics in common with the (114) ___________________ cells. Scientists found several similarities, such as: presence of single, circular (115) __________; code for (116) _________________ RNA; and (117) _________ of the ribosomes. The earliest fossil of a eukaryotic cell is about (118) _______ billion years old. For some of the other membrane structures, such as the nuclear envelope, endoplasmic reticulum, and Golgi complex, researchers believe that they were formed by the (119) _____________ membrane forming invaginations as in endocytosis.

Some scientists believe that the domain (120) ______________ may have been the ancestor of eukaryotic cells because they have the same basic structure as bacteria but also have some of the characteristics of (121) __________________ cells.

The first eukaryotic organism must have been (122) ______________________. The first multicellular eukaryote must have evolved about (123) ________ - __________ million years ago. The unicellular eukaryotic cells must have formed (124) _________________, followed by the (125) ___________________ of the cells.

Short Answer

126. What kind of evidence is used to support endosymbiotic hypothesis? ___________________________

____

____

127. Which group of organisms is hypothesized as the ancestor of eukaryotic cells? ____________________

____

____

True/False

Mark if the statement is true or false. If the statement is false, justify your answer in the line below each statement.

128. ______________ ER, Golgi, and nuclear membranes developed by endosymbiosis.

____

129. ______________ All multicellular eukaryotes developed from the same unicellular organism.

____

130. ______________ Mitochondria and chloroplasts have a single, circular DNA as prokaryotes.

____

SELF-TEST

1. Which of the following must be a characteristic of all living things? [p. 512]
   a. They must be multicellular.
   b. They must photosynthesize.
   c. They must replicate and translate their genetic information.
   d. They must be able to produce all their organic and inorganic chemicals.

2. Which of the following must be the possible composition of the atmosphere to support synthesis of organic chemicals (according to the Oparin-Haldane hypothesis)? [p. 513]
   a. H_2O, H_2, CH_4, and NH_3
   b. O_2, H_2 and H_2O
   c. O_2, N_2, H_2, and H_2O
   d. CH_4, NH_3 and O_2

3. Absence of which of the following in the primitive atmosphere is critical to the Oparin-Haldane hypothesis? [p. 513]
   a. O_2
   b. N_2
   c. H_2O
   d. H_2

4. Which of the following supports the origin of organic chemicals in the hydrothermal vents of the ocean floor? [p. 515]
   a. Water is needed for the synthesis of organic chemicals.
   b. High temperature is needed for synthesis of organic chemicals.
   c. Reducing condition surrounds a pool of chemicals needed for the synthesis of organic chemicals.
   d. High temperature supports the survival of living things.

5. Which of the following must have been the most critical molecule for the storage of information and support of chemical reactions? [p. 517]
   a. DNA
   b. Ribozyme
   c. Proteins
   d. Ribosome

6. Which of the following is a major difference between prokaryotic and eukaryotic cells? [p. 519]

a. Prokaryotic cells have a plasma membrane unlike eukaryotic cells.

b. Eukaryotic cells have membrane-bound organelles unlike prokaryotic cells.

c. Eukaryotic cells have DNA unlike prokaryotic cells.

d. Prokaryotic cells have ribosomes unlike eukaryotic cells

7. Which of the following structures may have originated by endosymbiotic mechanism? [p. 519]

a. Plasma membrane and ER

b. ER and nuclear membrane

c. Mitochondria and chloroplast

d. Ribosomes and cell wall

8. Which of the following structures may have originated by endocytosis mechanism?[p. 520]

a. Plasma membrane and ER

b. ER and nuclear membrane

c. Mitochondria and chloroplast

d. Ribosomes and cell wall

9. Earth is estimated to have formed ______ billion years ago. [p. 511]

a. 4.6

b. 3.5

c. 2.2

d. 1.0

10. The oldest fossil of a eukaryote is estimated to be ________ billions years old. [p. 521]

a. 4.6

b. 3.5

c. 2.2

d. 1.0

11. The oldest fossil of a prokaryote found in the stromatolites is estimated to be ________ billions years old. [p. 521]

a. 4.6

b. 3.5

c. 2.2

d. 1.0

12. Why is reducing atmosphere proposed to be critical for the synthesis of organic chemicals? [p.513]

a. Presence of oxygen would break down the pool of organic chemicals through oxidation.

b. Presence of oxygen would encourage the synthesis of organic chemicals.

c. Presence of oxygen would encourage making of water molecules.

d. Presence of oxygen would create a reducing environment.

13. Eukaryotes may have evolved from a common ancestral line shared with Archaeans. [p. 521]

a. True

b. False

14. All of the following are key characteristics that support endosymbiotic origination of mitochondria from a prokaryote except [pp. 520-521]

a. presence of single circular DNA molecule.

b. presence of smaller ribosomes.

c. similar coding of rRNA.

d. DNA sequence is similar to nuclear DNA.

15. Which of the following is true for the experiments done by Miller-Urey? [p. 514]

a. They were able to make organic chemicals required by living things in a reducing environment.

b. They were able to use cold temperature to create organic chemicals.

c. They were able to make organic chemicals in the presence of oxygen.

d. They were able to make cells in their apparatus.

e. Their experiments did not require water.

INTEGRATING AND APPLYING KEY CONCEPTS

1. Knowing the structure of phospholipids, explain how they spontaneously form membrane structure.

2. Even if water was found in the atmosphere or the crust, what would scientists look for if they wanted to know whether life existed on that planet?

3. Why would scientists not expect creation of new life on the planet earth as they hypothesize to have happened billions of years ago?

25

PROKARYOTES AND VIRUSES

CHAPTER HIGHLIGHTS

- Prokaryotes:
  - Have a very simple structure.
  - Are metabolically most diverse.
  - Mostly reproduce asexually.
- The domain Bacteria:
  - Contains commonly found bacteria.
  - Includes many that cause diseases.
- The domain Archaea:
  - Is mostly found in extreme environments.
  - Shares characteristics with common bacteria as well as eukaryotes.
- Viruses:
  - Lack cellular structure.
  - Infect all groups of organisms.
  - Depend upon the host for all metabolic and reproductive activities.
- Viroids and prions:
  - Have structure simpler than viruses.
  - Are rare, but viroids infect animals and prions infect plants.

STUDY STRATEGIES

- This chapter is fairly simple.
- Since the organisms discussed here are extremely small, understanding requires some imagination.
- Focus on the detail structure and their evolution.

TOPIC MAP

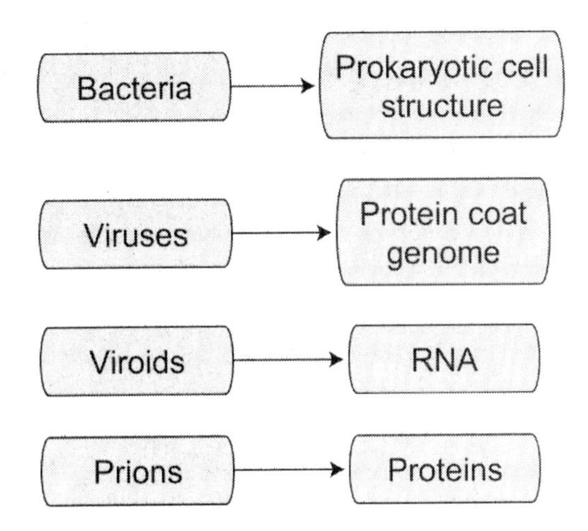

INTERACTIVE EXERCISES

Why It Matters [pp. 525–526]

Section Review (Fill-in-the-Blanks)

Escherichia coli is a bacterium commonly found in human large intestine and belongs to the group of smallest organisms called (1) ________________. This group is divided into two domains: (2) ______________ and (3) ________________. They are extremely small and some of them cause (4) ____________. They are found in diverse environmental conditions and have the most diverse (5) _______________ capabilities.

Viruses are (6) ___________ than prokaryotes that infect all groups of organisms but are considered as (7) _____-living.

25.1 Prokaryotic Structure and Function [pp. 526–534]

Section Review (Fill-in-the-Blanks)

Prokaryotes are small and lack membrane-bound (8) ________________. Even though they reproduce asexually, they get genetic variability through (9) _____________, (10) _______________, (11) _________________, and (12) _________________.

Prokaryotes have a cell (13) _______ and a plasma (14) ________________ that enclose cytoplasm. They have a single, (15) ______________ DNA that is referred to as the prokaryotic (16) _________________ and is located in the (17) _____________ region of the cytoplasm. In addition to this DNA, some prokaryotes also have additional pieces of DNA called (18) ____________.

Prokaryotes have different shapes: spherical are called (19) ________; rod shaped are called (20) __________; comma-shaped are called (21) _________; and corkscrew shaped are called (22) _____________.

The cell wall of bacteria is made of (23) __________________. The physician (24) _______ ______________ _______ developed a special staining procedure called the (25) ________ staining technique that allowed him to divide bacteria into two types: Gram (26) ____ and Gram (27) ____. The staining difference is due to the difference in the (28) _______ ________ of the bacteria. Gram (29) ____ have a thick (30) _________________wall whereas Gram (31) ____ have a thin layer of (32) _________________ and an outer (33) _________________. This outer membrane contains LPS, the (34) ___________________, and protects the cell from bursting, from killing by the immune system, and from antibiotic affects. Most pathogenic bacteria are Gram (35) ____.

Many bacteria have a protective (36) ___________________ slime coat. If attached to the cell, it is called the (37) __________, or if loosely associated with the cells, it is called the (38) __________ ____________.

Many bacteria have one or more hairlike structures called the (39) ________________ attached to their cell wall. This structure is quite different from those present in eukaryotic cells, but functionally, it allows them to (40) __________.

Some bacteria also have rigid tubelike structures called the (41) ________ that allows them to attach to their host or help in transfer of some of their (42) ________ to another bacterial cell for genetic recombination.

Prokaryotes are metabolically very diverse and they can be divided into four types based on the source of (43) ______________ and carbon: (44) __________________ are the bacteria that obtain energy by oxidizing inorganic substances and use CO_2 as their carbon source; (45) ___________________ are the bacteria that obtain energy by oxidizing organic molecules and use already prepared organic chemicals as their carbon source; (46) ______________________ are the bacteria that use light as their energy source and CO_2 as their carbon source; and (47) _____________________ are the bacteria that use light as their energy source and organic chemicals as their carbon source.

Bacteria that use oxygen in their metabolic process for producing energy and are called (48) ____________. Many of these bacteria cannot grow without oxygen and are called (49) _____________ aerobes.

Some bacteria produce (50) _____________ without using oxygen and are called (51) _________________. Some of these bacteria are poisoned by the presence of oxygen and are called (52) _____________ anaerobes.

Finally, there are bacteria that can produce (53) _____________ in the presence or absence of (54) __________; they are referred to as the (55) _______________ anaerobes.

Some prokaryotes are able to reduce atmospheric nitrogen to form nitrate or ammonia, a process that is only done by (56) ________________ and is called nitrogen (57) _________________. This form of nitrogen is then used by all other living organisms to make organic chemicals such as (58) ____________ acids and nucleotides.

Prokaryotes normally reproduce asexually by (59) _____________ fission that forms new cells that are genetically (60) _________________. Some prokaryotes may exchange DNA by a process called (61) ________________ that requires two cells to join through a tubelike structure called (62) ________. Some bacteria also form (63) ___________________ that are formed inside the parent cell and can survive through unfavorable heat, water, and chemical conditions.

Matching

Match each of the following terms related to prokaryotic cell structure and function with its correct definition.

64. ____	Organelles	A. Additional pieces of DNA that have supplemental genes in some prokaryotes
65. ____	Plasmid	B. A polysaccharide slime coat attached to the cells
66. ____	Prokaryotic chromosome	C. Unit structures in the cells that have a specific structure and function
67. ____	Pilus	D. The single circular DNA molecule that has most the required genes
68. ____	Flagella	E. A rigid tubelike structure that allows certain bacteria to attach to their host or transfer some of their DNA to another bacterial cell
69. ____	Capsule	F. A polysaccharide slime coat that is loosely associated with the cells
70. ____	Slime layer	G. A hairlike structure that is attached to the bacterial cell wall and helps the cell in movement
71. ____	Nitrogen fixation	H. A heat, water, and chemical resistant structure formed by some bacteria to survive through unfavorable environmental conditions
72. ____	Endospore	I. A hairlike structure that is attached to the bacterial cell wall and helps the cell in movement

Match each of the following terms related to prokaryotic cell shape with its correct definition.

73. ____	Coccus	A. Helical or corkscrew-shaped prokaryote
74. ____	Bacillus	B. Comma-shaped prokaryote
75. ____	Vibrio	C. Spherical prokaryote
76. ____	Spirilla	D. Rod-shaped prokaryote

Match each of the following terms related to prokaryotic mechanism to achieve genetic variability with its correct definition.

77. ____	Mutation	A. When DNA is passed from one bacteria to another through a physical contact to cause genetic recombination
78. ____	Transformation	B. A change in DNA sequence due to exposure to mutagens
79. ____	Transduction	C. When DNA is absorbed by bacterial cells from their environment to cause genetic recombination
80. ____	Conjugation	D. When a virus transfers DNA from one bacterium to another to cause genetic recombination

Match each of the following terms related to the differences in prokaryotic cell wall with its correct definition.

81. ____	Gram +	A. A layer present outside the thin layer of peptidoglycan in the wall of Gram- bacteria
82. ____	Gram -	B. Bacterial types that stain purple with the Gram staining technique and have a thick wall of peptidoglycan
83. ____	Peptidoglycan	C. Bacterial types that stain red with the Gram staining technique and have a thin wall of peptidoglycan and an outer membrane
84. ____	Outer membrane	D. A combination of lipids and polysaccharides that makes up the outer membrane of Gram- bacteria
85. ____	Lipopolysaccharides	E. Chemical that makes up the cell wall of bacteria

Match each of the following terms related to the mechanism by which the prokaryotes secure organic chemicals with its correct definition.

86. ____ Chemoautotrophs

87. ____ Chemoheterotrophs

88. ____ Photoautotrophs

89. ____ Photoheterotrophs

A. Use light as their energy source and organic chemicals as their carbon source

B. Use light as their energy source and CO_2 as their carbon source

C. Obtain energy by oxidizing inorganic substances and use CO_2 as their carbon source

D. Obtain energy by oxidizing organic molecules and use already prepared organic chemicals as their carbon source

Match each of the following terms related to the oxygen requirements of the prokaryotic cell with its correct definition.

90. ____ Aerobes

91. ____ Obligate aerobes

92. ____ Anaerobes

93. ____ Obligate anaerobes

94. ____ Facultative anaerobes

A. Do not require oxygen; in fact are poisoned by the presence of oxygen

B. Do not require oxygen

C. Use oxygen as the electron acceptor for cellular respiration

D. Cannot grow without oxygen

E. Use oxygen when present, otherwise produce energy anaerobically

Choice

For each of the following DNA forms, choose the most appropriate description.

a. Prokaryotic chromosome b. Plasmid

95. ____ ____ The main DNA molecule in all prokaryotes that has the genes required for most metabolic activities of the cell.

96. ____ ____ The additional pieces of DNA that are found in some bacteria that have special genes such as the antibiotic resistance gene and fertility gene.

For each of the following forms of bacteria, choose the most appropriate description.

a. Gram+ b. Gram-

97. ____ ____ Bacterial types that stain red with the Gram staining technique and have a thin wall of peptidoglycan and an outer membrane

98. ____ ____ Bacterial types that stain purple with the Gram staining technique and have a thick wall of peptidoglycan.

25.2 The Domain Bacteria [pp. 534–537]

25.3 The Domain Archaea [pp. 537–540]

Section Review (Fill-in-the-Blanks)

Based on (99) ________, (100) ________, and (101) ____________ sequences, prokaryotes have been divided into two domains: (102) ____________ and (103) ______________.

The domain Bacteria includes: (104) ______________, the Gram- purple bacteria; (105) ______________________, the Gram- photosynthetic green bacteria; (106) ______________, the Gram- blue-green bacteria; (107) ________-____________, the Gram+ chemoheterotrophic bacteria; (108) ______________, the spiral bacteria; (109) ________________, the Gram- bacteria that lack peptidoglycan.

Scientist Carl (110) ________ gave DNA and rRNA sequence evidence to separate Archaea from domain Bacteria. Archaea includes bacteria that mostly live in extreme environments and they are referred to as the

(111) _______________, although some live in normal environments and are called (112) ___________________.

Their unique features include differences in (113) ____________ membrane and cell (114) ________.

The Domain Archaea includes: (115) ____________________ that release methane,

(116) __________________ that live in high-salt areas, (117) ___________________ that live in high-temperatures,

and (118) __________________ that live in cold temperature.

Matching

Match each of the following terms related to Archaea with their correct definitions.

119. ____	Extremophile	A. Archaea that live in high-temperature environments
120. ____	Mesophile	B. Archaea that live in normal environments
121. ____	Methanogens	C. Archaea that live in extreme environments
122. ____	Halophiles	D. Archaea that live in high-salt environments
123. ____	Thermophiles	E. Archaea that live in cold environments
124. ____	Psychrophiles	F. Archaea that release methane

Short Answer

125. Give two features that are unique to Archaea. ____

____

____

126. Why was the discovery of thermophiles important for biochemical research? ___________________

____

____

25.4 Viruses, Viroids, and Prions [pp. 540–546]

Section Review (Fill-in-the-Blanks)

Viruses have nucleic acid, (127) ________ or (128) ________, (129) ___________-stranded or

(130) _____________-stranded. Nucleic acid is enclosed inside a protein coat called the (131) ____________. Only

some viruses come with a few molecules of (132) _______________ that will help their replication inside the host

cell. Even though viruses are known to infect most groups of organisms, they are considered (133) _______-living.

Viruses could be: (134) _____________ viruses that have proteins spirally arranged to form a rod shape,

(135) ______________ viruses that have triangular protein units joined to form a geometric sphere,

(136) ________________ viruses that have a membrane surrounding their basic structure, and

(137) _______________ viruses that have a head with nucleic acid and a tail that helps inject the viral genome into

the host.

Viruses that infect bacteria: They are called (138) _________________. They either go through (139)

___________ cycle where the virus multiplies inside the bacterium and kills the host cell to release new viruses, or they go through (140) ________________ cycle where the virus nucleic acid becomes incorporated into the bacterial chromosome as a(n) (141) __________ and multiplies with the bacterium.

Viruses that infect animal cells: Enveloped viruses inject both (142) _________ and (143) ___________ into the host cell. The (144) ___________ is then removed to release viral (145) ____________ into the host cell. Viruses that are not enveloped enter the host by (146) _______________. The viral (147) _____________ triggers the multiplication process and the new viruses are released by death of the host or the viruses take a part of the host plasma (148) _______________ as an envelope.

Viruses that infect plant cells: Due to the presence of cell wall, plant viruses enter the host cell through (149) __________ and inject (150) ________. They move from one cell to another through (151) _____________ that interconnect plant cells.

Viruses cannot be treat with antibiotics that were developed to treat (152) ____________. Most drugs used to treat viral disease target viral (153) _______ into the host cell or inhibit viral (154) _______________.

Viroids are simpler than viruses and are made of just naked (155) _______ strands. They have been known to infect (156) __________ and cause diseases.

Prions are (157) ________ particles that infect animals and cause diseases such as the mad cow disease, scrapie, and kuru.

Matching

Match each of the following terms with its correct definition.

158. ____ Capsid
159. ____ Helical viruses
160. ____ Polyhedral viruses
161. ____ Enveloped viruses
162. ____ Complex viruses
163. ____ Virus head
164. ____ Virus tail
165. ____ Bacteriophage
166. ____ Prophage
167. ____ Lytic cycle
168. ____ Lysogenic cycle
169. ____ Plasmodesmata
170. ____ Viroids
171. ____ Prions

A. Viruses that have a membrane around them
B. Part of a complex virus that helps to inject its genome into the host cell
C. A cycle followed by bacteriophage where the viral genome is incorporated into the bacterial chromosome and multiplies with it
D. RNA molecules that infect and causes diseases in plants
E. Protein coat of a virus
F. A virus made of triangular units joined together to form a geometric shape
G. Protein particles that infect animals and cause diseases
H. Cytoplasmic connections between plant cell
I. A cycle followed by a bacteriophage where the new viruses are released by rupturing of the bacterial cell
J. Part of a complex virus that contains viral genome
K. Viruses that are made of a head containing the genome and a tail to inject
L. A virus that attacks bacteria
M. A rod shaped virus
N. State of a bacteriophage when its genome is incorporated into the bacterial chromosome

Complete the Table

172. Complete the following table with the chemical composition of virus, viroids, and prions.

Structural Entity	Genome	Protein	Envelope
Virus			
Viroid			
Prion			

True/False

Mark if the statement is true or false. If the statement is false, justify your answer in the line below each statement.

173. ______________ Viruses are living organisms.

____

174. ______________ Viruses can multiply inside or outside a living host.

____

175. ______________ Antibiotics such as penicillin are prescribed to treat bacterial and viral diseases.

____

SELF-TEST

1. All of the following are considered non-living except [p. 526]
   a. bacteria.
   b. viruses.
   c. viroids.
   d. prions.

2. All of the following are found in a bacterial cell except [p. 527]
   a. a plasma membrane.
   b. a cell wall.
   c. ribosomes.
   d. a nucleus.

3. All of the following are true for prokaryotic DNA except: [p. 527]
   a. They have a single molecule of DNA.
   b. Their DNA is circular.
   c. Both, prokaryotic chromosome and plasmids are made of DNA.
   d. Their DNA is located inside a nucleus.

4. All of the following are true for Gram negative bacteria except: [p. 528]
   a. They have a thin peptidoglycan layer.
   b. They have an outer membrane.
   c. They often cause diseases.
   d. They stain purple with Gram staining technique.

5. Which of the following structure gives additional protection to some bacteria? [p. 528]
   a. plasma membrane
   b. nucleoid
   c. cell wall
   d. capsule

6. Eukaryotic and prokaryotic flagella are made of microtubules. [p. 530]
   a. True
   b. False

7. Which of the following type of bacteria are able to produce energy with or without oxygen? [p. 531]
   a. obligate aerobes
   b. obligate anaerobes
   c. facultative anaerobes

8. All of the following produce genetic variability in bacteria except [p. 534]
   a. binary fission.
   b. mutation.
   c. transformation.
   d. transduction.
   e. conjugation.

9. Which of the following part of bacteria contains lipopolysaccharides? [p. 528]
   a. inner plasma membrane
   b. cell wall
   c. outer membrane
   d. capsule

10. Which of the following types of bacteria use light as their energy source and CO_2 as their carbon source? [p. 531]
    a. Chemoautotrophs
    b. Chemoheterotrophs
    c. Photoautotrophs
    d. Photoheterotrophs

11. Which of the following type of bacteria obtain energy by oxidizing organic molecules and use already prepared organic chemicals as their carbon source? [p. 531]
    a. Chemoautotrophs
    b. Chemoheterotrophs
    c. Photoautotrophs
    d. Photoheterotrophs

12. Which of the following type of Archaea are typically found in freezing environments? [p. 540]
    a. Psychrophiles
    b. Halophiles
    c. Thermophiles
    d. Mesophiles

13. Which of the following type of Archaea are typically found in high-salt environments? [p. 540]
    a. Psychrophiles
    b. Halophiles
    c. Thermophiles
    d. Mesophiles

14. Which of the following are made of only RNA?

[p. 540]

a. viruses

b. viroids

c. prions

d. bacteria

15. Which of the following are made of cells?

[p. 526]

a. viruses

b. viroids

c. prions

d. bacteria

INTEGRATING AND APPLYING KEY CONCEPTS

1. Give a scientist's opinion on the origin of a virus.
2. Why do common antibiotics not work in killing viruses?
3. Explain how bacteria obtain genetic variation in spite of the fact that they mostly divide by binary fission.

26

PROTISTS

CHAPTER HIGHLIGHTS

- Protists are eukaryotes that mostly live in aquatic or moist environments.
- Protists are mostly unicellular, or some are simple multicellular.
- Protists can be autotrophic or heterotrophic.
- Protists reproduce asexually by mitosis and/or sexually by meiosis.
- Because of their diverse structure, metabolism, and reproduction, they are difficult to classify.
- Major groups:
    - Excavates—Single celled, flagellated, lack mitochondria, produce ATP anaerobically, heterotrophic, parasitic.
    - Discicristates—Single celled, flagellated, autotrophic or heterotrophic, disc-shaped mitochondrial cristae.
    - Alveolates—Single celled, flagellated or nonmotile, autotrophic or heterotrophic, specialized structures such as membrane-bound vesicles called the alveoli.
    - Heterokonts—Funguslike or bacteria-like or algaelike, autotrophic or heterotrophic, flagellated gametes.
    - Cerozoa—Single celled, autotrophic or heterotrophic, move or ingest food by producing stiff projections called the axopods, hard outer shell called tests.
    - Amoebozoa—Irregular cell shape, use lobose pseudopodia to move and engulf food particles, heterotrophic.
    - Archaeplastida—Multicellular, higher level of differentiation, autotrophic, closest to land plants.
    - Opisthokonts—Unicellular or colonial, heterotrophic, flagellated, closest to fungi and animals.
- Endosymbiotic theory used to explain the origin of chloroplast in protists.

STUDY STRATEGIES

- Remember the goal of this chapter is to understand what protists are and how they differ in their structure, metabolism, and reproduction.
- Due to their diversity, it is confusing when it comes to the classification of protists.
- Concentrate on the major groups first.
- Once the key characteristics are understood, you can then learn further divisions of each group.

TOPIC MAP

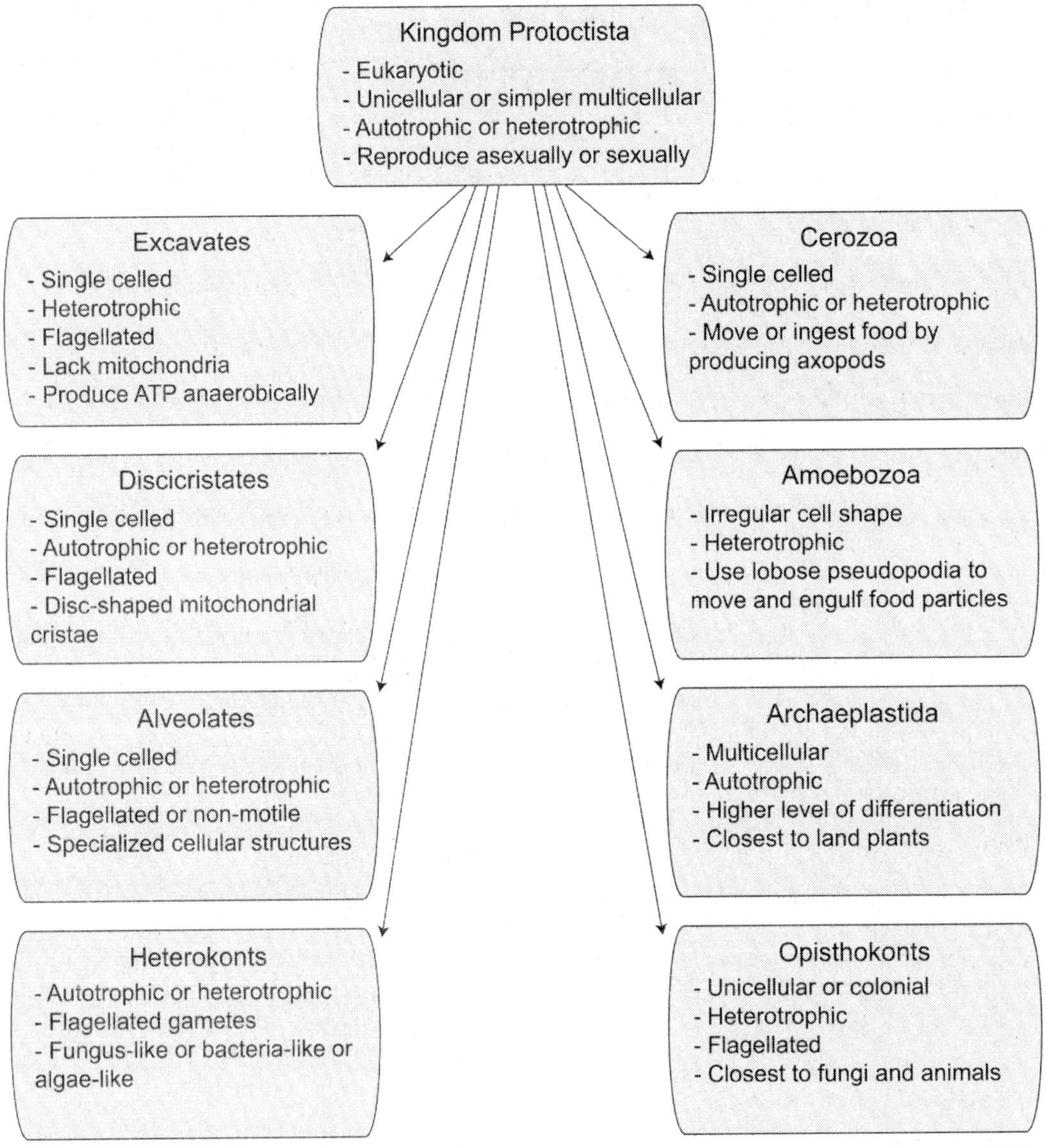

Kingdom Protoctista
- Eukaryotic
- Unicellular or simpler multicellular
- Autotrophic or heterotrophic
- Reproduce asexually or sexually

Excavates
- Single celled
- Heterotrophic
- Flagellated
- Lack mitochondria
- Produce ATP anaerobically

Discicristates
- Single celled
- Autotrophic or heterotrophic
- Flagellated
- Disc-shaped mitochondrial cristae

Alveolates
- Single celled
- Autotrophic or heterotrophic
- Flagellated or non-motile
- Specialized cellular structures

Heterokonts
- Autotrophic or heterotrophic
- Flagellated gametes
- Fungus-like or bacteria-like or algae-like

Cerozoa
- Single celled
- Autotrophic or heterotrophic
- Move or ingest food by producing axopods

Amoebozoa
- Irregular cell shape
- Heterotrophic
- Use lobose pseudopodia to move and engulf food particles

Archaeplastida
- Multicellular
- Autotrophic
- Higher level of differentiation
- Closest to land plants

Opisthokonts
- Unicellular or colonial
- Heterotrophic
- Flagellated
- Closest to fungi and animals

INTERACTIVE EXERCISES

Why It Matters [p. 549–550]

Section Review (Fill-in-the-Blanks)

Almost all of the protists live in a(n) (1) ________________ environment. They belong to the kingdom (2) ________________ and branched off during very early evolution of (3) ________________ organisms.

26.1 What Is a Protist? [pp. 550–553]

Section Review (Fill-in-the-Blanks)

Protists are the most (4) ______________ group on this planet. Almost all of the protists live in (5) ______________ or in moist soil. The majority of them are single celled, (6) ________________, but some are larger and multicellular. They have a(n) (7) ________________ cell structure, that is, they have membrane-bound

(8) ______________ and cytoskeleton. They can reproduce (9) ______________, or (10) ______________ by production of gametes. They share a number of characteristics with other eukaryotic kingdoms such as (11) ____________, ______________, and ______________. However, they differ from them in a number of ways. Their cell (12) _________ is chemically different from the Fungi. They lack the differentiation of the body into stem, leaves, roots, and seeds as in (13) ____________. They also lack the differentiation into nerve cells and a digestive tract, and collagen, the extracellular binding material common in (14) ______________.

Metabolically, most protists are aerobic and have the (15) ______________, the powerhouse of the cell. A few protists that lack this organelle produce ATP (16) ______________. Some protists are (17) ______________, that is, they obtain their organic nutrients by ingesting or absorbing. Ingested food particles are enclosed in an organelle, the (18) __________ vacuole, where it is digested with the help of (19) ______________. Other protists are (20) ______________, that is, they photosynthesize. The pigments and the process are very similar to the (21) _________.

Being aquatic, some protists have special structures such as (22) ______________ vacuoles that help in getting rid of excess water that diffuses into the cell. (23) ____________ and ____________ are hairlike structures that help some of the protists move. Others extend their cytoplasm or cell, forming (24) ______________ that help them in their motility.

Small photosynthetic protists that live together in ponds, lakes, and oceans are referred to as the (25) __________________. They produce organic nutrients and oxygen for the (26) ______________, the microorganisms, and aquatic animals.

Matching

Match each of the following terms with its correct definition.

27. ____ Food vacuole
28. ____ Contractile vacuole
29. ____ Phytoplankton
30. ____ Zooplankton
31. ____ Pseudopodia

A. Small photosynthetic protists that collectively live in ponds, lakes, and oceans
B. Heterotrophic microorganisms and animals that live in ponds, lakes, and oceans that depend on protists for nutrients and oxygen
C. Cytoplasmic and cellular extensions of protists that help them move
D. A membrane-bound organelle found in protists that contains engulfed food particles for digestion
E. A membrane-bound organelle that helps a protist get rid of excess water

Complete the Table

32. Complete the following table for the characteristics of protists.

Major Characteristics	*Description*
Habitat	
Cell Structure	
Metabolism	
Nutrition	
Reproduction	

26.2 The Protist Groups [pp. 553–572]

Section Review (Fill-in-the-Blanks)]

Protoctista is the most diverse (33) _______________ and hardest to classify into groups:

A. Excavates—Single celled, flagellated, lack (34) _______________, produce ATP by (35) _______________ respiration, heterotrophic, parasitic. *Giardia* causes intestinal infection; *Trichomonas* causes urinary and reproductive tract infection.

B. Discicristates—Single celled, flagellated, autotrophic or heterotrophic, (36) _________ -shaped mitochondrial cristae. *Euglena* that is photosynthetic; *Trypanosoma* causes sleeping sickness.

C. Alveolates—Single celled, flagellated or nonmotile, autotrophic or heterotrophic, specialized structures such as membrane-bound vesicles called the (37) _______________. *Paramecium* that has two nuclei, rows of cilia, a food vacuole, a contractile vacuole, and very elaborate sexual reproduction; Dinoflagellates, a photosynthetic species associated with coral reefs, have a cellulose shell and may glow— (38) _______________.

D. Heterokonts—Funguslike or bacteria-like or algaelike, (39) _______________ or heterotrophic, flagellated (40) _______________. Water molds that have fungal-like (41) _______________ but a cellulose wall; diatoms with a glassy (42) _____________ shell; golden algae that photosynthesize like (43) _______________ but have additional pigment called (44) _______________; brown algae that not only photosynthesizes, but also differentiates into leaflike (45) _____________, stemlike (46) _____________, and rootlike (47) _______________.

E. Cerozoa—Single celled, autotrophic or heterotrophic, move or ingest food by producing stiff projections called the (48) _______________, and have a hard outer shell called (49) __________. Radiolarians have microtubule-supported extensions called (50) _______________; Forams have an organic shell with (51) _____________ _______________; chloroarachniophytes are photosynthetic amoeba that also (52) _______________ food.

F. Amoebozoa—Irregular cell shape, use lobose (53) ______________ to move and engulf food particles, heterotrophic. *Amoeba* is single celled and reproduces only asexually by (54) ______________; *Dictyostelium* is a slime mold that is extensively used in research because it does form a(n) (55) ______________ body under unfavorable conditions.

G. Archaeplastida—Multicellular, higher level of differentiation, autotrophic, closest to land (56) ______________. Red algae have additional pigment called (57) ______________; green algae have the same pigments as land plants, little cellular (58) ______________ of body parts, and may have been ancestors for the land (59) ______________.

H. Opisthokonts—Unicellular or colonial, heterotrophic, flagellated, and closest to (60) ______________ (and) ______________. Choanoflagellates have a collarlike extension that surrounds the single (61) ______________.

Matching

Match each of the following terms with its correct definition.

62. ____	Bioluminescent	A. Refers to leaflike structures of algae
63. ____	Fucoxanthin	B. Refers to rootlike structures of algae
64. ____	Blades	C. Additional pigment present in red algae
65. ____	Stipes	D. Light released by Dinoflagellates when they are disturbed
66. ____	Holdfast	E. Additional pigment present in golden and brown algae
67. ____	Axopods	F. Refers to stemlike structures of algae
68. ____	Phycobilins	G. Slender strands of cytoplasmic extensions that are typically present in Cerozoa

Short Answer

69. In what respect are protists most diverse as compared to other kingdoms? ______________________________

____

____

70. What constitutes phytoplanktons and zooplanktons? ______________________________

____

____

71. Explain primary and secondary endosymbiosis for the evolution of plastids. ______________________________

____

____

SELF-TEST

1. Protists are [p. 549]
   a. prokaryotic.
   b. eukaryotic.

2. Most protists are [p. 549]
   a. aquatic.
   b. land forms.

3. All protists produce ATP aerobically. [p. 553]
   a. True
   b. False

4. Which of the following structures is involved in getting rid of excess water? [p. 552]
   a. Food vacuole
   b. Contractile vacuole
   c. Pseudopodia
   d. Axopods

5. Which of the following pigments gives brown and golden algae their color? [p. 561]
   a. Phycobilins
   b. Fucoxanthin
   c. Chlorophylls
   d. Carotenoids

6. Which of the following pigments gives red algae their color? [p. 567]
   a. Phycobilins
   b. Fucoxanthin
   c. Chlorophylls
   d. Carotenoids

7. Which of the following may have been the ancestors of animals? [p. 569]
   a. Opisthokonts
   b. Archaeplastida
   c. Excavates
   d. Amoebozoa
   e. Alveolates

8. Which of the following lack mitochondria? [p. 553]
   a. Opisthokonts
   b. Archaeplastida
   c. Excavates
   d. Amoebozoa
   e. Alveolates

9. To which of the following do Amoeba and slime molds belong? [p. 563]
   a. Opisthokonts
   b. Archaeplastida
   c. Excavates
   d. Amoebozoa
   e. Alveolates

10. The protists that have been studied very closely to understand the process of differentiation belong to the group [p. 564]

a. Opisthokonts.
b. Archaeplastida.
c. Excavates.
d. Amoebozoa.
e. Alveolates.

11. Primary endosymbiosis is where a non-photosynthetic eukaryotic cell engulfs a _________________ that becomes a permanent resident and transforms into a plastid, as in red-green algae and land plants. [p. 570]

a. nonphotosynthetic prokaryotic cell
b. photosynthetic prokaryotic cell
c. nonphotosynthetic eukaryotic cell
d. photosynthetic eukaryotic cell

12. Secondary endosymbiosis is where a non-photosynthetic eukaryotic cell engulfs a _________________ that becomes a permanent resident and transforms into a plastid in *Euglena.* [p. 571]

a. nonphotosynthetic prokaryotic cell
b. photosynthetic prokaryotic cell
c. nonphotosynthetic eukaryotic cell
d. photosynthetic eukaryotic cell

13. The protists that have complex cytoplasmic structures and move with the help of cilia belong to the group [p. 555]

a. Opisthokonts.
b. Archaeplastida.
c. Excavates.
d. Amoebozoa.
e. Alveolates.

14. Protoctista is a kingdom where organisms that are not plants, animals, fungi, or prokaryotes are just combined. [p. 551]

a. True
b. False

15. Protoctista includes organisms that reproduce asexually by mitosis or sexually by meiosis. [p. 552]

a. True
b. False

INTEGRATING AND APPLYING KEY CONCEPTS

1. What are the distinguishing characteristics of protists that separate them from organisms in other kingdoms?

2. Discuss why it is difficult to classify protists.

3. What is primary and secondary endosymbiosis? Give two examples to support this concept.

27

PLANTS

CHAPTER HIGHLIGHTS

- Plants are thought to have evolved from green algae some 450 million years ago.
- Adaptations displayed by early land plants include a waxy cuticle, lignified tissues, and internal chambers to protect the developing gamete.
- Later adaptations to land included development of root and shoot systems, vascular tissue, a shift from long-lived haploid (gametophyte) generations to a larger, long-lived diploid (sporophyte) generation, and a shift from homospory to heterospory with separate male and female gametophytes.
- Living nonvascular plants (bryophytes) include liverworts, hornworts, and mosses.
- Living seedless vascular plants include club mosses, ferns, and horsetails.
- Gymnosperms were the first seed plants. A seed forms when an ovule matures following fertilization and functions to protect and help disperse the embryonic sporophyte.
- Angiosperms are flowering seed plants. Ovules are protected by ovaries, which mature into fruit that nourishes the embryo and aids in seed dispersal.

STUDY STRATEGIES

- This chapter makes use of the concepts of evolution, cell structure, mitosis, meiosis, and sexual reproduction developed earlier. You should briefly review these concepts to make sure that you understand them.
- While many of the terms in this chapters may be familiar, others probably won't be, so don't be hasty in your study.
- This chapter is divided into sections. DO NOT try to go through them all in one sitting. Take one section at a time, then work through the companion section(s) in the study guide.
- Draw pictures of the various life cycles, being sure to note the ploidy of each generation and the characteristics of each group.

INTERACTIVE EXERCISES

Why It Matters [pp. 575–576]

Section Review

The (1) ____ has over 300,000 living species and ranges from mosses, horsetails, and ferns to conifers and flowering plants.

27.1 The Transition to Life on Land [pp. 576–581]

Section Review

Life on land presented numerous challenges to resist desiccation. Early plants made (2) ____, a thick polymer to prevent zygotes from drying out. Some land plants also evolved a waxy (3) ____ layer on their surface to reduce water loss. The presence of small openings or (4) ____ on this layer allowed for control of gas exchange and water evaporation. Land plants split into two groups: nonvascular plants or bryophytes, which lack internal transport vessels, and vascular plants or tracheophytes, which possess internal transport vessels. Vascular plants also have (5) ____, a polymer that strengths cell walls and allows plants to grow taller and stay erect on land, and a(n) (6) ____, a region of unspecialized dividing cells near the tips of shoots and roots. Because all land plants produce embryos, they are referred to as (7) ____. One type of vascular tissue is (8) ____, which distributes water and ions, and the other is (9) ____, which distribute sugars made during photosynthesis. (10) ____ anchor the plant to the substrate and absorb water and nutrients. (11) ____ are modified stems that also can anchor pants and absorb water and nutrients. The are vast underground (12) ____ and extensive above ground (13) ____ in vascular plants. As plants moved into drier habitats, their life cycles became more variable. Some haploid cells may be gametes (sperm or eggs) or (14) ____, which give rise to new haploid individuals asexually. Plants cycle between haploid and diploid phases in phenomenon called (15) ____. The diploid generation is a(n) (16) ____ and the haploid generation is a(n) (17) ____. Sporophytes eventually develop capsules called (18) ____, which produce spores. Some plants make only one type of spore and are (19) ____, others are (20) ____ and produce two types of spores: a smaller type that develops into a male gametophyte—a pollen grain, and a larger type that develops into a female gametophyte.

Building Vocabulary

Prefixes	Meaning
gamet(o)-	sex cells (egg, sperm)
spor(o)-	spore
homo-	same
hetero-	different

Suffix	Meaning
-angi(o)(um)	vessel, container
-phyte	plant
-spor(e)(ous)	of or relating to a spore

Prefix	Suffix	Definition
____	____	21. structure in which gametes are formed
____	____	22. structure in which spores are produced
____	____	23. the gamete-producing phase in the life cycle of a plant
____	____	24. the spore-producing phase in the life cycle of a plant
____	____	25. production of one type of spore
____	____	26. production of two different types of spores

Choice

For each of the following statements, choose the most appropriate term from list below.

a. vascular plant (bryophyte) b. nonvascular plant (tracheophyte)

27. ____ have internal transport vessels
28. ____ do not have internal transport vessels

For each of the following statements, choose the most appropriate term from list below.

a. xylem b. phloem

29. ____ vascular tissue that distributes sugars made during photosynthesis
30. ____ vascular tissue that distributes water and ions

For each of the following statements, choose the most appropriate term from list below.

a. sporophyte b. gametophyte

31. ____ the haploid generation of a plant
32. ____ the diploid generation of a plant

For each of the following statements, choose the most appropriate term from list below.

a. homosporous b. heterosporous

33. ____ a plant that makes two different kinds of spores
34. ____ a plant that makes only one type of spore

For each of the following statements, choose the most appropriate term from list below.

a. sporangium b. gametangium

35. ____ structure in which spores develop
36. ____ structure in which gametes develop

For each of the following statements, choose the most appropriate term from list below.

a. root system b. shoot system

37. ____ tissue specialized for the uptake of water and nutrients
38. ____ tissue specialized for the absorption of light and the uptake of CO_2 from air

Matching

Match each of the following structure with its correct definition.

39.	____	sporopollenin	A. thick polymer that protects reproductive spores
40.	____	rhizome	B. waxy layer on surface of plant
41.	____	spore	C. openings in surface of plant that control gas exchange and H_2O evaporation
42.	____	lignin	D. polymer that strengthens secondary cell walls
43.	____	stomata	E. region of unspecialized dividing cells near shoot and root tips
44.	____	alteration of generation	F. another name for land plants
45.	____	embryophyte	G. horizontal modified stem that can penetrate substrate
46.	____	apical meristem	H. gives rise to a new haploid individual asexually
47.	____	cuticle	I. life cycle that consists of haploid and diploid generations

27.2 Bryophytes: Nonvascular Land Plants [pp. 581–584]

27.3 Seedless Vascular Plants [pp. 584–590]

Section Review

Because they lack a system for conducting water, (48) ____ commonly grow in moist places or as (49) ____, growing independently (not as a parasite) on another organism. The gametes of bryophytes are sheltered within a layer of protective cells called a(n) (50) ____; that in which eggs form is a(n) (51) ____, and that in which sperm form is a(n) (52) ____. Liverworts, perhaps the first land plants, make up the phylum (53) ____ and have a simple gametophyte body called a(n) (54) ____. Some species can reproduce asexually, in which small cell masses or (55) ____ form on a thallus. Hornworts make up the phylum (56) ____ and mosses make up the phylum (57) ____. The haploid spores of mosses germinate on a wet surface and grow into a web of tissue called (58) ____.

Members of the phylum (59) ____ dominated North America and Europe during the Carboniferous period. Also abundant during this period were representatives of the phylum (60) ____, a group that includes ferns and horsetails. Modern lycophytes are small; their sporangia occur at the base of specialized leaves called (61) ____, which cluster to form a(n) (62) ____ or (63) ____. In most ferns, the fronds arise from (64) ____ positioned along a rhizome. Sporangia on the lower surface of some leaves cluster into a(n) (65) ____. Thick-walled cells in a layer called the (66) ____ encircle the sporangium.

Choice

For each of the following descriptions , choose the most appropriate term from the list below.

a. archegonia b. antheridia

67. ____ female gametangium

68. ____ male gametangium

For each of the following descriptions, choose the most appropriate term from the list below.

a. node b. sorus

69. ____ point on stem where fern leaves attach

70. ____ cluster of sporangia on under side of some leaves

For each of the following descriptions, choose the most appropriate term from the list below.

a. sporophylls b. strobilus

71. ____ cluster of leaves that form a "cone"

72. ____ specialized leaves on which sporangia are clustered

Complete the Table

Phylum	Representative(s)	Feature
73.	Liverworts	Flat, ribbonlike tissue
Anthocerophyta	Hornworts	74.
75.	Mosses	Threadlike gametophytes
Lycophyta	76.	77.
Pterophyta	78.	Familiar sporophyte body with fronds

Matching

Match each of the following type of junction with its correct definition.

79. ____ epiphyte
80. ____ annulus
81. ____ gametangium
82. ____ gemmae
83. ____ thallus
84. ____ protonema

A. life history involves growing independently on another organism
B. structure in which gametes form
C. gametophyte body of liverworts
D. small cell masses that give rise to new thalli asexually
E. filamentous web of tissue
F. row of thick-walled cells that encircles sporangium

27.4 Gymnosperms: The First Seed Plants [pp. 590–594]

27.5 Angiosperms: Flowering Plants [pp. 594–601]

Section Review

(85) ____ are conifers and their relatives. Sperm arise inside of a(n) (86) ____. (87) ____ is the transfer of pollen to female reproductive parts. A(n) (88)____ is a structure in which a female gametophyte develops. A(n) (89)____ forms when an ovule matures after a pollen grain reaches it and a sperm fertilizes the egg. The (90) ____ are restricted to warmer climates. There is only one living species in the phylum (91) ____, which grows in temperate forests in China. Conifers or phylum (92) ____ are the most abundant gymnosperms. Small male cones possess sporangia that undergo meiosis to form haploid (93) ____. The larger female cones have sporangia that undergo meiosis to form haploid (94) ____.

The most successful group of plants is the (95) ____. Defining and key reproductive structures of this group are (96) ____ and (97) ____, the latter of which nourishes the embryo and helps disperse seeds. Angiosperms are assigned to the phylum (98) ____, most of which are classified as (99) ____ and (100) ____. Four additional angiosperm groups have been recognized: the (101) ____, which are most closely related to monocots, and three groups considered to be (102) ____. A two-step (103) ____ process in the seeds of flowering plants is unique and improves reproductive success. The ovule containing a female gametophyte is enclosed within a protective (104) ____. Angiosperms (105) ____ with pollinators.

Choice

For each of the following descriptions, choose the most appropriate term from the list below.

a. gymnosperms b. angiosperms

106. ____ have "naked" seeds

107. ____ seeds develop in a vessels (carpel) formed by a modified leaf

For each of the following descriptions, choose the most appropriate term from the list below.

a. microspores b. megaspores

108. ____ produced in female cones and will form a gametophyte

109. ____ produced in male cones and will form a pollen grain

For each of the following descriptions, choose the most appropriate term from the list below.

a. monocots b. eudicots

110. ____ embryos possess a single leaf

111. ____ pollen grain has three grooves

For each of the following descriptions, choose the most appropriate term from the list below.

a. ovule b. ovary

112. ____ develops from a carpel and matures into a fruit

113. ____ sporophyte structure in which female gametophyte develops

Complete the Table

Group	Feature(s)
Cycadophyta	114.
115.	Only one living species, the ginkgo tree
Coniferophyta	116.

Group	Representative(s)
Monocots	117.
Eudicots	118.
119.	Magnolias, laurels, avocados
Basal angiosperms	120.

Matching

Match each of the following type of junction with its correct definition.

121. ____ pollen grain
122. ____ Anthophyta
123. ____ fruit
124. ____ coevolution
125. ____ flowers
126. ____ seed
127. ____ double fertilization
128. ____ pollination

A. male gametophyte that gives rise to sperm
B. the process of transferring pollen to female reproductive structures
C. structure formed after fertilization, includes embryo sporophyte, nutritive tissue, and protective cover
D. defining feature of angiosperms
E. nourishes angiosperm embryo and aids in seed dispersal
F. the interdependent evolution of two or more species
G. process by which seed of flowering plants give rise to an embryo and to nutritive tissue
H. the phylum name for angiosperms

SELF-TEST

1. Many factors contributed to the adaptive success of angiosperms, including ____. [pp. 597–598]
   a. more efficient transport of water and nutrients
   b. enhanced nutrition and physical protection of embryos
   c. enhanced dispersal of seeds
   d. requirement of water for sperm to fertilize egg

2. A pine tree has [pp. 594]
   a. sporophylls.
   b. megasporangia on female cones.
   c. separate male and female parts on the same tree.
   d. two sizes of spores in separate cones.
   e. female cones that are larger than male cones.

3. In gymnosperms, the pollen grain develops from _____. [p. 590]

a. the gametophyte generation

b. microspore cells

c. the male gametophyte

d. meiosis of cells in microsporangium

4. A strobilus is _____. [p. 586]

a. on a diploid plant

b. on a vascular plant

c. found on horsetails

d. on a haploid plant

5. The sporophyte generation of a plant [p. 579]

a. is haploid.

b. is diploid.

c. produces haploid spores by meiosis.

d. produces haploid spores by mitosis.

6. The spore case on ferns is _____. [p. 589]

a. formed by the haploid generation

b. usually on fronds

c. called sporangia

d. often arranged in a sorus

e. a precursor to the fiddlehead

7. Land plants are thought to have evolved from _____. [p. 576]

a. green algae

b. fungi

c. bryophytes

d. mosses

e. Euglena-like autotrophs

8. The gametophyte generation of a plant [p. 579]

a. is haploid.

b. is diploid.

c. produces haploid spores.

d. produces haploid gametes by mitosis.

e. produces haploid gametes by meiosis.

9. Plant sperm form in [p. 579]

a. diploid gametophyte plants.

b. haploid gametophyte plants.

c. haploid sporophyte plants.

d. antheridia.

e. archegonia.

10. Spores grow [p. 579]

a. into gametophyte plants.

b. into sporophyte plants.

c. into a haploid plant.

d. to form a plant body by mitosis.

e. to form a plant body by meiosis.

INTEGRATING AND APPLYING KEY CONCEPTS

1. *Evolution link*: What is the adaptive significance of flowers? What has made angiosperms so successful? What factors could limit or threaten their success?

2. *Insights from the molecular revolution*: What are the likely origins of the three distinct sets of genes in plants? How can markers such as the *rbcL* gene be used to confirm your hypothesis? How does a plant cell coordinate the activity of three sets of genes?

28

FUNGI

CHAPTER HIGHLIGHTS

- Fungi are heterotrophic organisms that get their nutrients by extracellular digestion and absorption.
- Fungi reproduce by making asexual and sexual spores.
- Kingdom Fungi is divided into major groups based mostly on the structures produced for sexual reproduction:
    - Chytridiomycota are mostly aquatic fungi that produce flagellated spores.
    - Zygomycota have aseptic, coenocytic hyphae and reproduce sexually by producing zygospore.
    - Glomeromycota are the mycorrhizae that are associated with plant roots and produce asexual spores at the tip of the hyphae.
    - Ascomycota produce ascocarp that contain saclike asci with 4-8 haploid ascospores.
    - Basidiomycota produce basidiocarp that contain saclike basidia with 4 haploid basidiospores.
    - Conidial or Imperfect fungi are those species where sexual reproduction has not been observed.
- Lichens are symbiotic combination of a mycobiont (fungal partner) and a photobiont (algal or cyanobacterial partner).

STUDY STRATEGIES

- Understand the major characteristics of the kingdom Fungi.
- Remember that the kingdom Fungi is divided on the basis of the type of sexual structure produced.
- Some memorization is required for the characteristics of species included in each phylum.

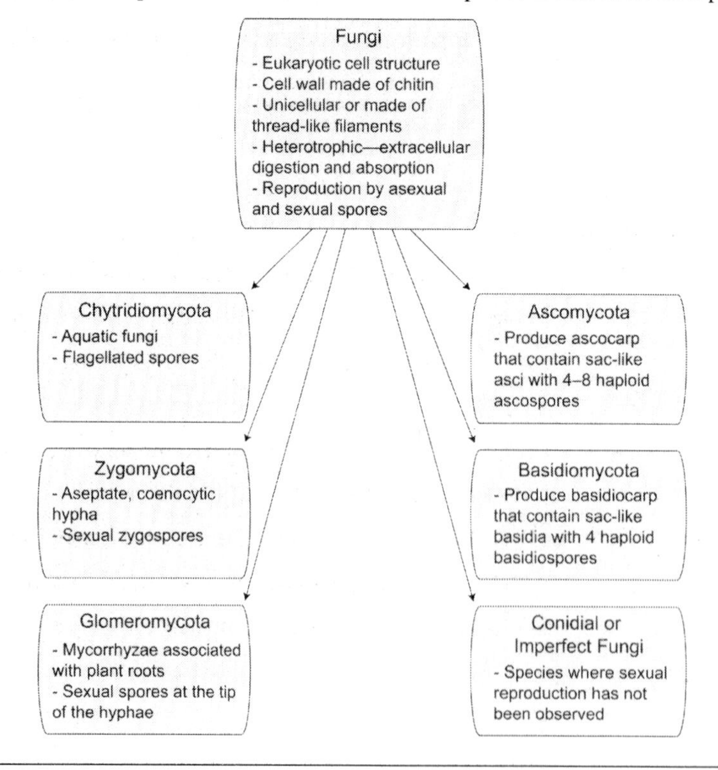

INTERACTIVE EXERCISES

Why It Matters [pp. 605–606]

Section Review (Fill-in-the-Blanks)

Fungi are the biggest players in (1) ________________ of dead organic matter, returning it to the atmosphere in the form of (2) ____________ ________________. They are therefore referred to as the (3) ______________.

28.1 General Characteristics of Fungi [pp. 606–610]

Section Review (Fill-in-the-Blanks)

Fungi may be single celled or (4) ______________. They have the (5) ____________ cell structure, and the cell wall is made of a polysaccharide called (6) ___________. Most fungi are made of threadlike structures called (7) ____________ and a mass of these filaments is referred to as (8) ________________.

Those fungi that live on dead organisms and break down organic chemicals to get their nutrients are called (9) __________________. Fungi that get nutrients from living plants and animals causing diseases are referred to as the (10) ________________, or form a partnership called (11) ______________. Either way, fungi release (12) ________________ for extracellular digestion and then absorb the nutrients.

During asexual reproduction, fungi produce large numbers of microscopic, haploid, asexual (13) ____________ at the tip of a hypha.

Sexual reproduction involves fusion of the cytoplasm of two hyphe or two gametes, a process called (14) _________________. This results in a new cell called a(n) (15) ________________ that contains two haploid nuclei. The nuclei fuse, (16) _________________, resulting in a diploid cell. This diploid cell undergoes meiosis to form 4 or 8 genetically different, haploid, sexual (17) _______________.

Matching

Match each of the following terms with its correct definition.

18. ____ Hyphe
19. ____ Mycelium
20. ____ Saprobes
21. ____ Parasites
22. ____ Symbiosis
23. ____ Spores
24. ____ Plasmogamy
25. ____ Dikaryon
26. ____ Karyogamy

A. Organisms that live on dead organisms to get nutrients
B. Threadlike structure that make up most fungi
C. Organisms that live on other living things and harm them
D. A mass of hyphe
E. Single cells that are produced by fungi as part of reproduction
F. Fusion of the two nuclei in a dikaryon
G. Where two organisms live together benefiting each other
H. Fusion of the cytoplasm of two hyphe or two gametes
I. A cell with two haploid nuclei

Complete the Table

27. Complete the following table by giving definitions of the terms used to describe different ways fungi get their nutrients.

Terms	Definitions
Saprobes	
Parasites	
Symbiosis	

28. Complete the following table by giving the major characteristics of fungi.

Terms	Characteristic of Fungi
Cell Type	
Cell Wall	
Number of Cells	
Mode of Nutrition	
Reproduction	

28.2 Major Groups of Fungi [pp. 610–620]

Section Review (Fill-in-the-Blanks)

Kingdom Fungi has been divided into (29) _____ phyla. A sixth group has been added as a temporary group that includes fungi whose (30) _____________ reproduction has not been demonstrated.

Chytridiomycota: The only fungi that produce (31) _________________ spores that swim. The spores are formed in a sac called (32) _________________. Most Chytrids are (33) ______________, although some are found in moist soil.

Zygomycota: These fungi typically have hyphe that lack cell partitions and are called (34) ________________ hyphe. The hyphae therefore has numerous nuclei in the continuous cytoplasm, which is referred to as (35) __________________. During sexual reproduction, the two mating types, (36) _____ and (37) _____, produce terminal (38) _________________ that contain numerous (39) _____________ nuclei. This is followed by fusion of the cytoplasm, (40) ________________, followed by fusion of the nuclei, (41) _________________. The cell wall of the new cell thickens to form a dormant spore called the (42) _________________. Under favorable conditions, the diploid nuclei undergo (43) _____________, and the spore germinates to form new mycelia.

Glomeromycota: These are fungi are symbiotically associated with the (44) __________ of 80–90% plants and are often referred to as the (45) ________________. They produce enzymes to enter the plant cells and form clusters of branching hyphe called (46) ________________. These structure allow the fungus to get (47) ____________ from the plant and in return provide dissolved (48) ________________ from the surrounding soil.

Ascomycota: These fungi reproduce asexually by producing exposed chains of (49) ____________ at the tip of specialized hyphe known as (50) ________________. Sexual reproduction involves plasmogamy of the female, enlarged tip of the hypha called the (51) __________________ and the male, smaller tip called the (52) ____________________ to produce a(n) (53) ________________ cell containing the two nuclei. The two nuclei fuse to produce a(n) (54) ______________ zygote, which then undergoes meiosis to form 4–8 (55) __________________ inside a sac called the (56) _________. Several of these sacs may be formed in a reproductive body called the (57) __________________.

Basidiomycota: The hyphae of these fungi form a tight cluster to form a reproductive body called (58) ________________. This structure consists of a stalk with a cap and develop numerous sacs called the (59) ______________. Each sac has a diploid nucleus that undergoes (60) ____________ to eventually form 4 haploid, exposed (61) ________________.

Imperfect fungi: Any fungal species whose (62) _____________ reproduction has not been demonstrated is temporarily housed in this group.

Matching

Match each of the following terms with their correct definition.

63. ____	Ascocarp	A. Asexual spore
64. ____	Mycorrhizae	B. The tip of – mating hypha of an Ascomycota that has the - mating nucleus
65. ____	Arbuscule	C. A cluster of mycorrhizae hyphae that allow the fungus to get sugars from the plant
66. ____	Sporangia	D. The reproductive body of the Ascomycota that bears sexual spores
67. ____	Conidium	E. A sac that contains 4–8 sexual spores of Ascomycota
68. ____	Conidiophore	F. The sexual spores of Ascomycota
69. ____	Ascogonium	G. The Glomeromycota associated with plant roots
70. ____	Anthredium	H. A terminal sac that is filled with asexual spores in Chytridiomycota and Zygomycota
71. ____	Ascospores	I. The specialized hypha of the Ascomycota that bear chains of asexual spores
72. ____	Ascus	J. The tip of + mating hypha of an Ascomycota that has the + mating nucleus

73. Complete the following table by giving the major characteristics of each phylum of the kingdom Fungi.

Phyla/Group	Major Characteristics
Ascomycota	
Zygomycota	
Basidiomycota	
Chytridiomycota	
Glomeromycota	
Imperfect Fungi	

74. Complete the following table by giving examples that will help you remember each phylum of the kingdom Fungi.

Phyla/Group	Major Characteristics
Ascomycota	
Zygomycota	
Basidiomycota	
Chytridiomycota	
Glomeromycota	

Short Answer

75. On what basis are the phyla formed in the kingdom Fungi? ______________________________

____

____

76. Why is Imperfect Fungi not given a phyla name? ______________________________

____

____

77. Considering the mechanism for getting nutrients, how do fungi differ from other living things? ____________

____

____

28.3 Fungal Associations [pp. 620–623]

Section Review (Fill-in-the-Blanks)

Some fungi form (78) ________________ partnerships with cyanobacteria, algae, and plants. Lichens are a combination of the fungal partner, the (79) ________________ and an algae or cyanobacteria partner, the (80) ________________. The fungal partner could be from phyla (81) ________________ or (82) ________________. The main body of a lichen is referred to as a(n) (83) ______________ that can multiply by breaking off, a form of (84) ______________ reproduction. Special asexual clusters called the

(85) ______________ are produced that have the fungal and algal cells. As sexual reproduction, depending upon the type of fungal partner a(n) (86) ______________ or (87) ______________ may be produced.

Matching

Match each of the following terms with its correct definition.

88. ____	Thallus	A. A cluster of fungal and algal cells used for asexual reproduction
89. ____	Soredia	B. The main body of the lichen
90. ____	Mycobiont	C. The photosynthetic partner of a lichen
91. ____	Photobiont	D. The fungal partner of a lichen

SELF-TEST

1. Fungi are [p. 605]
   a. prokaryotic.
   b. eukaryotic.

2. Fungi are [p. 605]
   a. autotrophic.
   b. heterotrophic.

3. All of the fungi produce asexual as well as sexual spores. [p. 610]
   a. True
   b. False

4. All fungi are multicellular. [p. 606]
   a. True
   b. False

5. Which of the following is the mode of nutrition in fungi? [p. 607]
   a. They photosynthesize their nutrients.
   b. They ingest their food.
   c. They secrete enzymes to digest the nutrients in their substrate and then absorb them.

6. Which of the following phylum includes species that produce flagellated spores? [p. 612]
   a. Chytridiomycota
   b. Ascomycota
   c. Basidiomycota
   d. Zygomycota
   e. Glomeromycota

7. Which of the following phylum includes species that produce mushrooms? [p. 617]

a. Chytridiomycota

b. Ascomycota

c. Basidiomycota

d. Zygomycota

e. Glomeromycota

8. Which of the following phylum includes molds? [p. 615]

a. Chytridiomycota

b. Ascomycota

c. Basidiomycota

d. Zygomycota

e. Glomeromycota

9. Which of the following phyla includes species that produce 4–8 haploid sexual spores in a sac? [p. 614]

a. Chytridiomycota

b. Ascomycota

c. Basidiomycota

d. Zygomycota

e. Glomeromycota

10. Which of the following phyla includes species that live in the roots of the plants? [p. 614]

a. Chytridiomycota

b. Ascomycota

c. Basidiomycota

d. Zygomycota

e. Glomeromycota

11. Which of the following correctly describes the relationship between mycorrhizae and plant roots? [p. 614]

a. parasitic

b. decomposers

c. symbiotic

12. Imperfect fungi are given the name because [p. 619]

a. they do not have a definite shape.

b. they are parasitic fungi.

c. they do not reproduce.

d. their sexual reproduction has not been observed.

13. The fungal partner in a lichen is referred to as the [p. 620]

a. mycobiont.

b. photobiont.

c. mycobiont and photobiont.

14. Soredia are [p. 621]

a. clusters of fungal spores.

b. clusters of algal spores.

c. clusters of fungal and algal cells.

15. Lichens reproduce only asexually. [p. 621]

a. True

b. False

INTEGRATING AND APPLYING KEY CONCEPTS

1. Millions of plants and animals die each year. Discuss how fungi play an important role in the decay and recycling processes.

2. Fungi are considered as opportunistic organisms by virtue of their reproductive diversity. Explain how their reproduction helps them in adapting to their environment.

3. Fungi are now believed to be closer to animals than plants. Discuss the evidence.

29

INVERTEBRATE ANIMALS

CHAPTER HIGHLIGHTS

- An animal is defined with an overview of animal phylogeny and classification.
- Adaptations associated with animals are explained.
- Characteristics and species examples of Parazoans and Eumetazoans are described.
- The protostomes—Lophotrochozoans and Ecdysozoan groups are described.

STUDY STRATEGIES

- First, focus on the animal phylogeny and classifications schemes.
- Examine the adaptations associated with animals.
- Next, identify and understand the various characteristics of animal groups and why certain animals are placed in their respective classification groups.
- Next, make comparisons between the various animals classification groups, focusing on diversity and success of animal groups in their environments.

INTERACTIVE EXERCISES

Why It Matters [pp. 627–628]

Section Review (Fill-in-the-Blanks)

In the kingdom (1) _____, approximately 2 million living (2) _____ have been identified and described. The majority of theses animals are (3) _____, clearly a very successful group of animals without a(n) (4) _____. The (5) _____ species, animals with a(n) (6)_____, are fewer in number but still successful in a variety of different environments.

29.1 What Is an Animal? [pp. 628–629]

29.2 Key Innovations in Animal Evolution [pp. 629–633]

Section Review (Fill-in-the-Blanks)

Animals are (7) _____ organisms with more than (8) _____ cell or (9) _____. Unlike plants, animals cells (10) _____ cell (11) _____. Animals acquire (12) _____ and (13) _____ by eating plants and other (14) _____, thus they are (15) _____. Typically, animals are able to interact with their environment by being (16) _____, however, some groups are (17) _____, at least in a portion of their life cycle. Reproduction is either (18) _____ or (19) _____. Animals have a wide array of (20) _____ plans. Two major groups or branches of animals have been identified, those with (21) _____ or (22) _____, and those that are (23) _____ _____, the (24) _____. Animals with (25) _____ have either a(n) (26) _____ body plan with (27) _____ and (28) _____; or a(n) (29) _____ body plan with the additional tissue type (30) _____. Animal (31) _____ is typically (32) _____ or (33) _____. The sponges or (34) ____ are irregular in (35) _____ and are (36) _____. The presence or absence of a(n) (37) _____ cavity is used to classify bilateral animals into different phyla. (38) _____ lack a body cavity, while (39) _____ animals have a(n) (40) _____ body cavity or (41) _____.

Matching

Match each of the following terms with its correct definition.

42.	____	Front	A.	True fluid-filled, lined body cavity
43.	____	Lower	B.	Cleavage pattern of protostomes
44.	____	Head end	C.	Solid body, no body cavity
45.	____	Upper	D.	Cavity from the splitting of mesoderm
46.	____	Back	E.	Blocks of repeating tissue
47.	____	Coelom	F.	Developmental fate determined at cleavage
48.	____	Peritoneum	G.	Ventral
49.	____	Spiral cleavage	H.	Posterior
50.	____	Determinate cleavage	I.	Cleavage pattern of deuterostomes
51.	____	Acoelomates	J.	Space which pinching off from archenteron
52.	____	Indeterminate cleavage	K.	Anterior
53.	____	Mesenteries	L.	Surrounds internal organs that are suspended in the coelom
54.	____	Schizocoelom	M.	Cephalization
55.	____	Enterocoelom	N.	Dorsal
56.	____	Segmentation	O.	Tissue which lines a coelom
57.	____	Radial cleavage	P.	Development fate determined after cleavage

29.3 An Overview of Animal Phylogeny and Classification [pp. 633–635]

Section Review (Fill-in-the-Blanks)

Relationships between animals have previously been based on (58) _____ and (59) _____ evidence. With advances in technology, these relationships are now based on molecular analysis of (60) _____ sequences in (61) _____ RNA and (62) _____ DNA. It is important to note that molecular phylogenic relationships between animals in a new area of study and a true working (63) _____.

29.4 Animals without Tissues: Parazoa [pp. 635–636]

29.5 Eumetazoans with Radial Symmetry [pp. 636–639]

Section Review (Fill-in-the-Blanks)

There is only one animal group that are (64) _____, the (65) _____ or phylum (66) _____. The (67) _____ have two animal phyla, the (68) _____ and (69) _____. Sponges lack (70) _____, and the adult or mature form is (71) _____. Cnidarians and ctenophores have (72) _____ and undergo (73) _____ symmetry during development. These animals have a(n) (74) _____ cavity, with (75) _____ opening. During development of these animals, the (76) _____becomes both the (77) _____ and (78) _____. The body plan is (79) _____, however, most species have a gelatinous (80) _____ between the (81) _____ and (82) _____. In the life cycle of most cnidarians, there is a(n) (83) _____ stage and a(n) (84) _____ stage, while most ctenophores are primarily (85) _____.

Matching

Given the characteristic, match the most likely phyla.

86. _____	Nematocysts	A. Phylum Porifera
87. _____	Mesoglea	B. Phylum Cnidaria
88. _____	Oscula	C. Phylum Ctenophora
89. _____	Primarily motile	D. Phyla Cnidaria & Ctenophora
90. _____	Gastrovascular cavity	E. Phyla Porifera & Cnidaria
91. _____	Both polyp & medusa	F. Phyla Porifera, Cnidaria, & Ctenophora
92. _____	Have sessile stages	
93. _____	Use cilia for locomotion	
94. _____	Anthozoans	
95. _____	Comb jellies	
96. _____	Sea anemone	
97. _____	Venus flower basket	
98. _____	Corals	
99. _____	Asymmetrical	
100. _____	Transparent	

29.6 Lophotrochozoan Protostomes [pp. 641–653]

Section Review (Fill-in-the-Blanks)

The (101) _____ have two major lineages, the (102) _____ and the (103) _____. The Lophotrochozoan lineage has (104) _____ aquatic phyla with a true coelom which possess a(n) (105) _____. The other phyla in this group lack the lophophore. Phylum (106) _____ or (107) _____ are (108) _____ with a more complex (109) _____ body plan. There are (110) _____ flatworm groups or classes, which have adaptations for habitats that are (111) _____-_____ or (112) _____. Phylum (113) _____ are fresh water animals with a wheel-like structure, the (114) _____ used to obtain food and locomotion. Some of the rotifers undergo (115) _____, in which (116) _____ develop from (117) _____ _____ during unfavorable environmental conditions. Phylum (118) _____ has four major groups or classes, the (119) _____, (120) _____, (121) _____, and (122) _____. Mollusks have a body divided into (123) _____ regions, the (124) _____, which contains many of the major organ systems; the (125) _____, which is used for locomotion; and the (126) _____, which secretes a(n) (127) _____. The phylum (128) _____ are the (129) _____ worms with (130) _____ major groups or classes, the (131) _____, (132) _____, and (133) _____. The body segments are separated by (134) _____, and all annelids have (135) _____ except the (136) _____.

Matching

Match each of the following with its correct definition.

137. _____	platys	A. excretory system of flatworms
138. _____	endoparasites	B. cephalopods
139. _____	trochophore	C. few
140. _____	poly	D. leech
141. _____	helmis	E. folding door
142. _____	rota	F. bristles
143. _____	Parthenogenesis	G. ring
144. _____	gaster	H. excretory system of segmented worms
145. _____	radula	I. to carry
146. _____	pod	J. many
147. _____	oligos	K. unfertilized egg develops into a female
148. _____	flame cell	L. food scraping device of mollusks
149. _____	ferre	M. belly
150. _____	closed circulation	N. foot
151. _____	valva	O. worm
152. _____	anellus	P. larval form of mollusks
153. _____	chaite	Q. flat
154. _____	metanephridia	R. living within organ systems
155. _____	hirudo	S. wheel

29.7 Ecdysozoan Protostomes [pp. 653–663]

Section Review (Fill-in-the-Blanks)

The adaptation of shedding a(n) (156) _____ covering is one of the major characteristics which groups (157) _____ phyla in the lineage of (158) _____ protostomes. Two of these phyla have extremely (159) _____ numbers of species. Phylum (160) _____ are unsegmented worms or (161) _____. The nematode species occupy a diverse number and type of habitats and have a significant impact on humans. A small phylum, the velvet worms or phylum (162) _____ species are primarily in the (163) _____ hemisphere. Phylum (164) _____ are (165) _____ animals with a hard (166) _____, which sheds as the animal (167) _____. The body has (168) _____ regions, the (169) _____, (170) _____, and

(171) _____, in addition, the appendages of these animals are (172) _____. Like nematodes, the (173) _____ occupy a wide array of habitats and are extremely successful.

Short Answer

174. Identify and discuss one disadvantage of shedding (molting) the external covering or exoskeleton. __________
____.

175. Explain how parasitic nematodes are so successful. ____
____.

Matching

Match each of the following with its correct definition.

176. _____ onyx — A. claw
177. _____ lobos — B. two
178. _____ arthron — C. horn
179. _____ nema — D. three
180. _____ chela — E. jointed
181. _____ bi — F. thread
182. _____ tri — G. lobe
183. _____ keras

Choice

Given the animal characteristics, select the most appropriate phyla or subphyla.

184. ___Cephalothorax — A. Nematodes
185. ___Antennae — B. Arthropoda
186. ___Book lungs — C. Chelicerata
187. ___Compound eyes — D. Hexapoda
188. ___Malpighian tubules — E. Onchophora
189. ___Numerous unjointed legs
190. ___Metamorphosis

SELF-TEST

1. From the following, what is the most unique to animals? [pp. 628–629]
   a. sexual reproduction
   b. eukaryotic
   c. heterotrophic
   d. multicellular

2. Select the pair that are not appropriately matched. [pp. 629–633]
   a. Protostomes—spiral cleavage
   b. Deuterostomes—radial cleavage
   c. Bilateral symmetry—Cephalization
   d. Platyhelminthes—Coelomate

3. Identify the characteristic that is different between Protostomes and Deuterostomes. [pp. 629–633]
   a. Segmentation
   b. Origin of mesoderm
   c. Bilateral symmetry
   d. Triploblastic

4. If an animal has determinate cleavage and the nervous system located on the ventral side of the body, it is most likely a ________. [pp. 629–633]
   a. Protostome
   b. Deuterostome
   c. Coelomate
   d. Asymmetrical animal

5. If an animal has the ability to produce both eggs and sperm, they would be described as ______. [pp. 635–636]
   a. able to undergo parthenogenesis
   b. hermaphroditic
   c. able to undergo metamorphosis
   d. endoparasitic

6. Where would expect to find animals that have a radula? [pp. 647–650]
   a. bottom or rock dwellers
   b. in trees
   c. living in the respiratory system of other animals
   d. living on the surface of a fish

7. Cephalopods have a closed circulatory system, this is thought to be due to _______. [pp. 647–650]
   a. increased mobility
   b. slow movement along the bottom
   c. high oxygen requirement for movement
   d. both a and c are correct

8. Insects differ from spiders in that spiders have ________. [pp. 657–663]
   a. the first pair of appendages modified into pedipalps
   b. chelicerae
   c. malpighian tubules
   d. complete metamorphosis

9. If each segment has a pair of metanephridia and ganglia, the most likely animal would be [pp. 657–663]

a. a roundworm.

b. an annelid.

c. an insect.

d. a spider.

10. Insects that undergo complete metamorphosis are successful because the larvae and the adults _____. [pp. 657-663]

a. are both able to fly

b. are both able to undergo pupation

c. occupy different habitats and eat different food

d. none of these

INTEGRATING AND APPLYING KEY CONCEPTS

1. Discuss the major characteristics that make an organism an animal.
2. Address several reason why Nematodes and Arthropods are so successful.
3. Discuss the advantages and disadvantages of flight.

30

DEUTEROSTOMES: VERTEBRATES AND THEIR CLOSEST RELATIVES

CHAPTER HIGHLIGHTS

- Characteristics of deuterostomes; invertebrates through phylum Chordata are presented.
- Vertebrates' diversification and adaptation to habitat are described.

STUDY STRATEGIES

- First, focus on major characteristics of deuterostomes.
- Examine the various groups of vertebrates and how adaptations have enhanced their success in a given habitat.
- Next, make comparisons between vertebrate groups looking at the tremendous diversity of these animals.
- Evaluate the key characteristics which place each vertebrate group in the current classification scheme.

INTERACTIVE EXERCISES

Why It Matters [pp. 667–668]

Section Review (Fill-in-the-Blanks)

(1) _____are classified by (2) _____ development and (3) _____ _____ data. At present, there are (4) _____ phyla of deuterostomes. One of the unifying characteristics is that during development, the (5) _____ arises from a(n) (6) _____ opening, hence the term (7) _____.

30.1 Invertebrate Deuterostomes [pp. 668–671]

30.2 Overview of the Phylum Chordata [pp. 671–674]

Section Review (Fill-in-the-Blanks)

Phyla (8) _____, (9) _____, and (10) _____ all have a(n) (11) _____ body plan. Adult (12)_____ develop from a(n) (13) _____symmetrical larva. Interestingly, as the (14)_____ form develops, a(n) (15) _____ pattern develops often with (16) _____ rays. The phylum (17) _____ is a small group of (18) _____ worms. These animals are organized with an anterior (19) _____ surrounded with by a collar with tentacles, a pharynx with (20) _____ _____, and an elongated trunk. The phylum (21) _____ has (22) _____ subphyla. Two of the subphyla are (23)_____ and the most diverse, the (24) _____, contains a large number of groups or classes. (25) _____ are grouped by a set of distinct characteristics at sometime during their life cycle. These are (26) _____, (27) _____body, (28) _____ _____ _____ chord, and a(n) (29) _____ with slits or openings.

Matching

Match each of the following roots with its correct definition.

30. ____ chorda — A. lily
31. ____ ophioneos — B. half
32. ____ derma — C. head
33. ____ oura — D. string
34. ____ echinos — E. tail
35. ____ hemi — F. starlike
36. ____ holothourion — G. skin
37. ____ asteroeides — H. the back
38. ____ krinon — I. spiny
39. ____ kephale — J. snakelike
40. ____ noton — K. water polyp

Choice

For each of the following characteristics or structures, select the most appropriate animal group from the list below.

41. ____ Proboscis
42. ____ Dorsal hollow nerve chord
43. ____ Pedicellariae
44. ____ Vertebral column
45. ____ Oral hood
46. ____ Incurrent siphon
47. ____ Tube feet
48. ____ Bone
49. ____ Atriopore
50. ____ Sea daisies

A. Phylum Echinodermata
B. Subphylum Vertebrata
C. Phylum Chordata
D. Subphylum Urochordata
E. Phylum Hemichordata
F. Subphylum Cephalochordata

30.3 The Origin and Diversification of Vertebrates [pp. 674–675]

30.4 Agnathans: Hagfishes and Lampreys, Conodonts, and Ostracoderms [pp. 677–678]

Section Review (Fill-in-the-Blanks)

The (51) _____-_____ shape of animals appears to be determined by the (52) _____ genes. It appears that animal groups with a more (53) _____ structure have (54) _____ *Hox* genes than animal groups with a more (55) _____ structure. There is one group of vertebrates that lacks a(n) (56) _____, the (57) _____. All other vertebrates with (58) _____ are considered to be (59) _____, which include the (60) _____. Most (61) _____ use (62) _____ limbs for (63) _____. One lineage, the (64) _____, or animals with specialized (65) _____, have been very successful in a(n) (66) _____ habitat. Two groups of (67) _____ fish are living, the (68) _____ and (69) _____. Only (70) _____ of the other two groups, the (71) _____ and (72) _____, are available for study.

30.5 Jawed Fishes [pp. 678–683]

Matching

Match each of the following roots, terms or structures with its correct definition.

73. ____	Plax	A. oil in liver of sharks
74. ____	Chondros	B. Sarcopterygii
75. ____	Lateral-line system	C. chamber that covers gills
76. ____	Ichthys	D. increases surface area in digestive tract
77. ____	Sturgeons	E. hydrostatic organ
78. ____	Claspers	F. fish
79. ____	Akantha	G flat surface
80. ____	Squalene	H. cartilage
81. ____	Operculum	I. detects vibrations in water
82. ____	Spiral valve	J. Actinopteryglii
83. ____	Swim bladder	K. male reproductive specializations
84. ____	Lungfish	L. thorn

30.6 Early Tetrapods and Modern Amphibians [pp. 683–685]

Section Review (Fill-in-the-Blanks)

Life on land requires several important characteristics. These include a system for locomotion and adaptations for survival in a(n) (85) _____ environment. In addition, a means to detect (86) _____ waves instead of (87) _____ in water. Adult (88) _____ typically live on land, however, an adequate (89) _____ habitat is often necessary for the (90) _____ or developmental portion of their lifecycle. There are (91) _____ groups or classes, the (92) _____, or (93) ____ and (94) _____; the (95) _____ or (96) _____; and the (97) _____ or (98) _____.

Matching

Match each of the following with its correct definition.

99. ____	gymnos	A. tail
100. ____	bios	B. without
101. ____	oura	C. bone homologous to support structure of jaws
102. _____	stapes	D. both
103. _____	boney scales	E. membrane that vibrates with sound waves
104. ____	delos	F. snakelike
105. ____	ampho	G. naked
106. ____	an	H. visible
107. ____	ophioneos	I. life
108. ____	tympanum	J. caecelians

30.7 The Origin and Mesozoic Radiations of Amniotes [pp. 686–688]

Section Review (Fill-in-the-Blanks)

(109) _____ are characterized by a fluid-filled (110) _____ which surrounds the (111) _____, the (112) _____. These animals have several important characteristics which enhance their ability to survive in a(n) (113) _____ environment. First, their (114) _____ will not (115) _____ in air, due to skin cells filled with (116) _____ and (117) _____. Second, the (118) _____ egg with a hard, leathery (119) _____. This (120) _____ egg has (121) _____ for exchange between the egg and its environment. In addition, (122) _____ and (123) _____ provide the embryo with a source of (124) _____, (125) _____, and (126) _____. The third characteristics is a change in waste products, (127) _____ _____ instead of (128) _____, which requires a great deal of (129) _____ to reduce (130) _____. There are (131) _____ groups of amniotes, the (132) _____, the (133) _____, and the (134) _____.

30.8 Testudines: Turtles [pp. 688–689]

30.9 Living Nonfeathered Diapsids: Sphenodontids, Squamates, and Crocodilians [pp. 689–692]

30.10 Aves: Birds [pp. 692–694]

Section Review (Fill-in-the-Blanks)

The (135) _____ or (136) _____ are one of the amniotic groups, the (137) _____, which have (138) _____ _____ arches. (139) _____ are characterized by boxlike (140) _____. Another of the amniotic groups, the (141) _____, have (142) _____ _____ arches. Animal groups classified in this group are the (143) _____; (144) _____, or (145) _____ and (146) _____; the (147) _____, or (148) _____ and (149) _____; and (150) _____ or (151) _____.

Matching

Match each of the following with its correct definition.

152. _____ sphen — A. tooth
153. _____ apsis — B. lizard
154. _____ plastron — C. dorsal side of testudine shells
155. _____ keeled sternum — D. scale
156. _____ morphe — E. wedge
157. _____ odont — F. arch
158. _____ carapace — G. ventral side of testudine shells
159. _____ saurus — H. attachment bone of flight muscles
160. _____ syn — I. ruler
161. _____ archos — J. form
162. _____ lepis — K. with

30.11 Mammalia: Monotremes, Marsupials, and Placentals [pp. 695–697]

30.12 Nonhuman Primates [pp. 697–702]

30.13 The Evolution of Humans [pp. 702–707]

Section Review (Fill-in-the-Blanks)

(163) _____ belong to the last amniotic group, the (164) _____, which have (165) _____ _____ arch. There are four key characteristics that are associated with mammals, (166) _____ _____ rate and (167) _____ _____; specializations of (168) _____ and (169) _____; (170) _____ _____ and increased complexity of the (171) _____. One of the major differences between groups of mammals is their mode of (172) _____. Some mammals (173) _____ _____, these are the (174) _____ or (175) _____. The (176) _____ or (177) _____-_____ mammals are subdivided into (178) _____ and (179) _____ or (180) _____. There are several different groups or classes of placental mammals, including ungulates, herbivores, carnivores, insectivorous, and (181) _____ and (182) _____, which feed on fishes, other animals, and even plankton. Primates are mammals which include (183) _____, (184) _____, and (185) _____. Some of the primates are (186) _____ rather than live on the (187) _____. Most primates can also stand (188) _____ with more (189) _____ hip and shoulder joints. In addition, primates can (190) _____ objects with either their (191) _____ or (192) _____. In general, primates have a large (193) _____ region of the brain, which allows (194) _____ of information. Chimpanzees are sometimes (195) _____ for short distances, while humans have complete (196) _____ posture and (197) _____ locomotion. In addition, humans have both a(n) (198) _____ and (199) _____ grip, as well as (200) _____ brains.

Matching

Match each of the following with its correct definition.

201. _____ eu
202. _____ streptos
203. _____ incisors
204. _____ haploos
205. _____ marsupium
206. _____ therion
207. _____ protos
208. _____ canines
209. _____ meta
210. _____ ungula
211. _____ rhin

A. wild beast
B. hoof
C. between
D. pointed teeth for piercing
E. twisted
F. flattened teeth for cutting
G. single or simple
H. true
I. nose
J. first
K. pouch for additional development

SELF-TEST

1. If an animal has pedicellariae and a larval form with bilateral symmetry, it belongs to group [pp. 668–671]
   a. ophinuroidea.
   b. asteroidea.
   c. echinoidea.
   d. urochordata.

2. One of the major differences between animals of Echinodermata and Hemichordata is/are ___ [pp. 668–671]
   a. tube feet
   b. gill slits
   c. a dorsal hollow nerve chord
   d. both a and b are correct

3. Which of the following are retained and used in the adult jawed fishes? [pp. 671–674]
   a. perforated pharynx
   b. radial symmetry
   c. segmented nervous system
   d. none of these

4. Which of the following belong to the Gnathostomata lineage? [pp. 674–678]
   a. hagfishes
   b. lampreys
   c. Conodonts
   d. Acanthodian

5. If the swim bladder of a bony fish were destroyed, what would be the effect? [pp. 678–683]
   a. there would be no effect
   b. the fish would not be able to breath underwater
   c. the fish would be closer to the surface of the water
   d. the fish would be farther away from the surface of the water

6. One of the major differences between life in an aquatic vs a terrestrial environment is that terrestrial animals could _____. [pp. 683–685]
   a. become easily dehydrated
   b. become disoriented due to a smaller brain
   c. be less likely to find their mates
   d. die from starvation due to lack of food

7. In addition to the amniotic egg, which of the following is advantageous in a terrestrial environment? [p. 686]
   a. keratin and lipid in skin cells
   b. ammonium as a waste product
   c. perforated pharynx in the adult form
   d. all of these

8. Of the following, which is the most advantageous for flight? [pp. 692–694]

a. amniotic egg

b. bipedal locomotion

c. hollow limb bones

d. dorsal hollow nerve chord

9. Which of the following has a prototheria reproductive mode? [pp. 695–697]

a. monotremes

b. mammals which lay eggs

c. duck-billed platypus

d. All of these

10. Hominids are able to play the piano because of ________. [pp. 702–706]

a. a power grip

b. a precision grip

c. bipedal locomotion

d. both a and b are correct

INTEGRATING AND APPLYING KEY CONCEPTS

1. Discuss how the adult echinoderm—a sea star—is significantly different from a primate, yet both are considered to be deuterostomes.

2. Address the major characteristics or adaptation that enhanced or allowed movement from an aquatic to a terrestrial environment.

31

THE PLANT BODY

CHAPTER HIGHLIGHTS

- Meristems are responsible for the development of new tissue. They are, essentially, embryonic in nature.
- There are fundamental differences between the monocots and eudicots.
- Ground tissue, vascular tissue, and dermal tissue all have important adaptive functions in the life of plants.
- Among the functions of the primary shoot system of plants are support, growth, and photosynthesis.
- The primary root system is responsible for acquisition of water and minerals, support, and storage.
- Cambium is responsible for secondary growth in plants.

STUDY STRATEGIES

- Understand that plant tissues tend to be less specialized than animal tissues. Plant tissues often have multiple functions.
- Realize that the transition from aquatic to terrestrial life was possible only as a result of specific adaptations in form and function.

INTERACTIVE EXERCISES

Why It Matters [pp. 711–712]

Section Review (Fill-in-the-Blanks)

Humans began domesticating plants around (1) _____ years ago. Flowering plants, or angiosperms, belong to the phylum (2) _____.

31.1 Plant Structure and Growth: An Overview [pp. 712–715]

Section Review (Fill-in-the-Blanks)

The aboveground part of a plant is called the shoot system and the underground, nonphotosynthetic component is the root system. Each system is composed of (3) _____, structures that are composed of several tissues and that have a specific form and function. (4) _____ are groups of cells that function together to perform specific tasks. Botanists refer to the plant cell wall, membrane, and cytoplasm collectively as the (5) _____. The secondary cell wall of a plant cell is laid down inside the primary, cellulose cell wall. The process of (6) _____ involves depositing a waterproof substance over the cellulose fibers that also strengthens the cell wall. Stems, leaves, buds and, if present, flowers make up the (7) _____ of a plant. The underground parts of a plant that function in absorption of water and nutrients, support, and storage make up the (8) _____.

Animals usually grow to a certain size then stop. This is known as (9) _____. Plants, on the other hand, can continue to grow throughout their lifetime. This is known as (10) _____. New tissues in plants arise from (11) _____, an embryonic tissue found at the tip of roots and shoots. All vascular plants contain meristems at the tip of the shoot and root, known as (12) _____. These tissues give rise to the (13) _____ of a plant that make up the (14) _____, and they represent (15) _____. Other plants, especially ones that produce woody tissues, exhibit (16) _____ from cylinders of embryonic tissue known as (17) _____ that leads to growth in diameter. This type of growth is known as (18) _____ and forms the (19) _____.

The two major lineages of flowering plants have different body plans. These two lineages are known as (20) _____ and (21) _____. Plants can also be categorized according to their life spans. Plants that can complete the entire life cycle in one growing season are known as (22) _____. Other plants need two full growing seasons and may produce limited amounts of secondary tissue. There are known as (23) _____. In others, vegetative growth and reproduction continue throughout the life of the plant. These plants are known as (24) _____.

True/False

If the statement is true, write a "T" in the blank. If the statement is false, make it correct by changing the underlined word(s) and writing the correct word(s) in the answer blank.

25. ____________ The secondary cell wall is made strong and waterproof due to deposits of <u>pectin</u>.
26. ____________ Solutes move between adjacent plant cells through <u>plasmodesmata</u>.
27. ____________ Increases in height of a plant are due to cell divisions in the <u>root apical meristem</u>.
28. ____________ Lateral meristems give rise to <u>primary</u> tissues.
29. ____________ <u>Herbaceous</u> plants show little or no secondary growth.

Complete the Table

Complete the following table by filling in the appropriate description of a listed transport process or by naming the transport process based on the listed description.

Trait	*Monocots*	*Eudicots*
Flower parts	30.	Multiples of 4 or 5
Vascular bundles in stem	31.	Arranged in a ring
Leaf veins	Parallel	32.
Pollen pores or furrows	One	33.

31.2 The Three Plant Tissue Systems [pp. 715–721]

Section Review (Fill-in-the-Blanks)

Plants have three tissue systems that are the basis of plant organs. These tissues are (34) _____, which makes up most of the plant body; (35) _____, which makes up the plant's circulatory system; and (36) _____, which forms the "skin" of the plant.

There are three types of cells that make up ground tissue. (37) _____ cells are thin walled and irregularly shaped. They usually make up most primary growth of roots, stems, leaves, flowers, and fruits. (38) _____ cells are usually elongated and provide flexible support. (39) _____ cells typically have thick, lignified secondary cell walls that ultimately causes death of the cells. They function in rigid support and protection.

Vascular tissue is specialized for conduction of water and solutes. (40) _____ conducts water and minerals upwards from the roots. It is composed of two types of cells, (41) _____, which are elongated with overlapping tapered ends. (42) _____ are shorter cells with perforations in their ends. When arranged end to end, they create a tube called a(n) (43) _____ through which water flows. The other type of vascular cell is the (44) _____. It is composed of conducting cells called (45) _____. When laid end to end, they form a(n) (46) _____. The individual cells lose most of their cytoplasm and many organelles, including the nucleus. In many plants, adjacent, specialized parenchyma cells called (47) _____ communicate with sieve tube members via plasmodesmata and help regulate their function.

Dermal tissue is composed of several types of cells. (48) _____ covers the plant body in a continuous layer. It is covered with a(n) (49) _____, a waxy, waterproof layer. Crescent-shape cells of the epidermis that occur in pairs and are involved in gas exchange are known as (50) _____. The space between them is a(n) (51) _____. Other epidermal cells possess specializations such as (52) _____, which may contain sugars or irritants and (53) _____, which greatly increase surface area to maximize water and mineral uptake.

Choice

For each of the following plant structures, choose the most appropriate cell type from the list below.

a. collenchyma b. parenchyma c. sclerenchyma

54. ____ The hard outer layer of a cherry pit.

55. ____ The "string" in a stalk of celery.

56. ____ The cells in the flesh of an apple.

31.3 Primary Shoot Systems [pp. 721–727]

Section Review (Fill-in-the-Blanks)

The point on a stem where a leaf attaches is a(n) (57) _____ and the distance between two of them is a(n) (58) _____. The upper angle between a leaf and the stem is the (59) _____. The leaf surface that absorbs light for photosynthesis is called the (60) _____. Buds at the tip of the main shoot are (61) _____, while buds that produce branches are (62) _____. Terminal buds produce a hormone that inhibits the development of nearby lateral buds, a phenomenon known as (63) _____. When a cell in an apical meristem divides, one of the daughter cells becomes a(n) (64) _____, which remains as a part of the meristem, while the other becomes a(n) (65) _____, which will differentiate after several more divisions. As these latter cells differentiate, they give rise to three types of (66) _____, rather unspecialized tissues. One of these tissues, (67) _____, becomes the stem's epidermis. The second, (68) _____, ultimately forms the ground tissue of the plant. The primary vascular tissues are derived from the (69) _____. These vascular tissues are usually arranged into chords of xylem and phloem known as (70) _____. In the roots and stems of most eudicots, these chords are arranged into a cylinder, the (71) _____. The cells to the outside of this cylinder comprise the (72) _____, while those inside the cylinder are the (73) _____.

Along the sides of shoot apical meristems (74) _____ give rise to mature leaves. A mature leaf has an upper and lower epidermis, between which is the (75) _____ region, which can be divided into a palisades layer and a spongy layer. The vascular bundles of leaves are arranged in various patterns visible on the surface and are known as (76) _____.

Matching

Match each of the following plant parts with its correct name.

77.	____	ginger "root"	A. Bud
78.	____	onion "head"	B. Bulb
79.	____	potato "eye"	C. Rhizome

31.4 Root Systems [pp. 727–730]

31.5 Secondary Growth [pp. 730–734]

Section Review (Fill-in-the-Blanks)

The roots of most eudicots exhibit a(n) (80) _____ system, which consists of a main, large root with smaller branching roots called (81) _____. Most monocots have a highly branched root systems know as a(n) (82) _____ system. Roots that emanate from the stem or some other region of a plant, such as the prop roots of corn, are called (83) _____ roots. Root apical meristem is covered by a(n) (84) _____, which protects is as it grows through the soil. Some root apical meristems have a region of slowly dividing cells called the (85) _____, which can become active if it is needed to regenerate damaged parts. The more actively dividing cells behind it is the (86) _____. Behind this regions is the (87) _____, where most increase in length occurs. Above this region there may be a(n) (88) _____ where cells differentiate. The outer layer of root cortex cells may develop into a(n) (89) _____, which limits water loss and regulates ion uptake. The innermost layer of cortex is the (90) _____, which has similar functions. The cells immediately interior to the endodermis comprise the (91) _____, which, under the influence of chemical growth regulators, can give rise to (92) _____, which will develop into lateral roots.

Section Review (Fill-in-the-Blanks)

Secondary growth arises from two types of lateral meristems, (93) _____, which gives rise to secondary xylem and phloem, and (94) _____, which produces as secondary epidermis known as (95) _____. Lateral meristems in the stem give rise to two types of cells. (96) ____ originate from cambium within the vascular bundles and produces secondary xylem and phloem. (97) _____ are derived from parenchyma cells between the vascular bundles. As secondary xylem accumulates and ages, it hardens into the tissue known as (98) _____. Older wood, near the center of the stem, often accumulates material that clogs the vessels as it dries and hardens. These cells make up the (99) _____, while the younger, moister cells just interior to the vascular cambium comprises the (100) _____.

All the living tissue between the vascular cambium and the surface of the stem comprises the (101) _____. It consists of secondary phloem and the (102) _____, which is composed of cork, cork cambium, and secondary cortex.

True/False

If the statement is true, write a "T" in the blank. If the statement is false, make it correct by changing the underlined word(s) and writing the correct word(s) in the answer blank.

103. ______________ Bark consists of cork, cork cambium, and secondary cortex.

104. ______________ Heartwood consist of xylem vessels that are clogged and no longer conduct water.

105. ______________ Cambium within the vascular bundles gives rise to ray initials.

SELF-TEST

1. The layer between the primary cell wall of adjacent plant cells is the ________. [p. 712]

a. middle lamella

b. protoplast

c. secondary cell wall

d. vascular cambium

2. A ________ shoot produces flowers and, ultimately, fruits. [p. 713]

a. floral

b. nodal

c. reproductive

d. vegetative

3. The self-perpetuating embryonic cells of plants is called ________ tissue. [p. 713]

a. dermal

b. ground

c. meristematic

d. vascular

4. The tissue responsible for growth in diameter of a plant is ________. [p. 714]

a. lateral mersitem

b. root apical meristem

c. shoot apical meristem

d. b and c

5. Which cell type is most likely to be photosynthetic? [p. 716]

a. collenchyma

b. parenchyma

c. sclerenchyma

d. stone cell

6. Which type of cell assists in the functioning of sieve tube cells? [p. 719]

a. companion cells

b. endodermal cells

c. pith cells

d. stone cells

7. The distance between two leaves on a stem is known as a(n) ________. [p. 721]

a. axil

b. internode

c. leaf gap

d. node

8. Which of the following is NOT a primary meristem? [p. 722]

a. ground meristem

b. procambium

c. protoderm

d. vascular cambium

9. Most of the photosynthesis in a leave takes place in the ________. [p. 725]

a. guard cells

b. lower epidermis

c. palisades mesophyll

d. spongy mesophyll

10. The annual rings that form in trees in temperate climates are formed by alternating layers of ________. [p. 732]

a. fusiform initials and ray initials

b. heartwood and sapwood

c. pericycle and endodermis

d. spring wood and summer wood

INTEGRATING AND APPLYING KEY CONCEPTS

1. *Unanswered Questions:* Susan Lolle's research suggests that plants may have a backup copy of their genome in case the primary copy "crashes." What are some things that might cause the system to crash? Why does the sessile (stationary) lifestyle of plants make a backup copy more valuable to them than it might to motile animals?

2. *Insights from the Molecular Revolution:* Feather color in some bird species is not produced by pigments but is referred to as "structural color" because it is produced by light refracted off the feather surface. Relate this fact to the *mixta* mutant in snapdragons.

3. *Focus on Research:* Homeotic genes in many animal and plant species regulate developmental processes. Based on Sarah Hake's findings on the effect of inserting cloned homeotic genes in tobacco plants, what sorts of manipulations can you think of that might result in the improvement of food crops?

32

TRANSPORT IN PLANTS

CHAPTER HIGHLIGHTS

- Passive and active transport moves materials into and out of plant cells and these movements play a role in moving substances within a plant body.
- Water and minerals travel through roots via apoplast, symplast, and transmembrane routes.
- Cohesion, tension, adhesion, and transpiration are physical properties of water that assist in moving it through the vascular system of a plant.
- Solutes are moved through the phloem by the process of translocation which creates hydrostatic pressure differences in different parts of the plant body.

STUDY STRATEGIES

- Be sure you understand the principles of diffusion, osmosis, and active transport. Concentration gradients are critical aspects in the movement of water and materials through the plant body.
- Review the polar nature of water, its tendency to form hydrogen bonds, and how it relates to cohesion, adhesion, tension, and transpiration in order to move water over great vertical distances without the aid of a pump.

INTERACTIVE EXERCISES

Why It Matters [pp. 737–738]

Section Review (Fill-in-the-Blanks)

The cumulative effect of the rather weak forces such as (1) _____ and (2) _____ can be sufficient to move water in some plants over 100 m.

32.1 Principles of Water and Solute Movement in Plants [pp. 738–742]

Section Review (Fill-in-the-Blanks)

Two mechanisms may move water and solutes through plants. If the process requires no energy and involves movement of material down a concentration gradient, it is (3) _____. If energy is involved, it is (4) _____ process. Sometimes proteins embedded in the cell membrane called (5) _____ may assist in movement of material across the membrane. In plant cells, the cytoplasm contains more negatively charged ions than the extracellular fluid. This charge difference is known as a(n) (6) _____, which refers to the fact that it represents a source of potential energy. The fact that there are usually more H^+ ions outside the cell vs. inside allows for two kinds of cotransport to take place. As the H^+ ions move back into the cell, the energy released may power the simultaneous uptake of an ion against its gradient, a process known as (7) _____. If the cotransported ion moves in the opposite direction of the H^+, the process is known as (8) _____.

Within the vascular tissues of plants water and solutes move by (9) _____, the movement of molecules due to pressure differences between two regions. The water and dissolved materials that move as bulk in the xylem are known as (10) _____. Water moves passively across individual cell membranes by (11) _____. The potential energy that drives this process is called (12) _____ and is a function of solute concentration and physical pressure. This physical pressure may be exerted by the cell wall of plants and is known as (13) _____. If it is high enough, it can prevent the entry of water into a cell. The units of pressure are usually expressed in units called (14) _____. Most plant cells contain a large, water-filled (15) _____ surrounded by a membrane called a(n) (16) _____, which maintains the turgor pressure of the cell. If the cytoplasm loses water, it is replaced from the central vacuole through protein channels in the tonoplast membrane known as (17) _____. If a plant loses more water than it takes in over and extended period, (18) _____ occurs.

True/False

If the statement is true, write a "T" in the blank. If the statement is false, make it correct by changing the underlined word(s) and writing the correct word(s) in the answer blank.

19. _______________ In osmosis, water flows from an area of <u>higher water potential to an area of lower water potential</u>.

20. _______________ Antiport will move a substance across a cell membrane in the same direction as the flow of protons.

21. _______________ A slight charge difference on the two sides of a cell membrane is known as a <u>membrane potential</u>.

22. _______________ Facilitated diffusion is a type of <u>passive</u> transport that uses transmembrane protein channels.

23. _______________ Wilting is a result of loss of <u>turgor pressure</u>.

32.2 Transport in Roots [pp. 742–745]

32.3 Transport of Water and Minerals in the Xylem [pp. 745–750]

Section Review (Fill-in-the-Blanks)

Water enters the plant root in one of three ways. The first is the (24) _____, in which water travels through the spaces between cells of the root epidermis and cortex. The second is the (25) _____, in which the water moves through the cytoplasm of adjacent cells through plasmodesmata. Finally, water can pass from cell to cell not through plasmodesmata, but across cell membranes, a route called the (26) _____. Water continues to move through the root cortex until in encounters the cells of the endodermis, which are wrapped in a band of waterproof suberin called the (27) _____. This band forces the soil solution to pass through the membranes of living endodermal cells in order to reach the stele. The endodermal cells can now regulate passage of materials into the stele.

In 1914, Henry Dixon proposed a model of movement of sap in a plant based on the physical properties of water and that used (28) _____ as its driving force. This model is called the (29) _____. Another model to explain movement of sap in plants is the (30) _____ model that involves the active transport of ions from the soil into the stele creating a water potential difference across the endodermis. As more water enters the xylem significant pressure can build, enough to force water upwards in small plants and forcing it out of leaf pores, a phenomenon known as (31) _____. Some plants that live in hot, dry environments utilize (32) _____ photosynthesis, wherein CO_2 is fixed into an organic acid during the evening when humidity is higher and temperatures are lower, and stomates can be open without undue water loss. During the day, the stomates close, conserving water, while the organic acid donates its fixed CO_2 to the Calvin cycle for manufacture of sugar.

Matching

Match each of the following terms with its correct definition.

33. ____ abscissic acid (ABA)
34. ____ adhesion
35. ____ apoplast
36. ____ root pressure
37. ____ xerophyte

A. model that explains movement of water through xylem over short distances resulting in guttation
B. plant adapted to hot, dry conditions
C. plant hormone responsible for stomatal closure under stressful conditions
D. spaces between parenchyma cells in the root cortex
E. tendency of water molecules to stick to other types of molecules, including the cell walls of xylem

32.4 Primary Shoot Systems [pp. 750–754]

Section Review (Fill-in-the-Blanks)

The long-distance movement of sugars and many other substances through the sieve tubes of phloem is called (38) _____, and the mix of water and dissolved substances that flows within them is called (39) _____. The point of origin of these materials is called the (40) _____, and their destination is the (41) _____. The force that drives this movement is known as the (42) _____ mechanism. In some plants, companion cells are modified in order to actively transport substances into the sieve tube members. These cells are called (43) _____.

True/False

If the statement is true, write a "T" in the blank. If the statement is false, make it correct by changing the underlined word(s) and writing the correct word(s) in the answer blank.

44. ______________ <u>Fructose</u> is the most common type of sugar translocated in phloem.
45. ______________ Phloem transports sap <u>downward</u> from the source.
46. ______________ Translocation originates at the <u>source and ends at the sink</u>.
47. ______________ Movement of solutes into sieve tubes is accomplished by <u>active transport</u>.
48. ______________ <u>Water follows solutes</u> pumped into sieve tubes.

SELF-TEST

1. Symport and antiport are examples of ________. [p. 739]
   a. facilitated diffusion
   b. primary active transport
   c. pressure flow
   d. secondary active transport

2. ________ is the effect of dissolved material on water's tendency to move across a membrane. [p. 741]
   a. Membrane potential
   b. Pressure potential
   c. Solute potential
   d. Water potential

3. The membrane surrounding the central vacuole of a plant cell is the ________. [p. 741]
   a. apoplast
   b. protoplast
   c. symplast
   d. tonoplast

4. A plant cell that contains enough water to exert pressure on the cell wall is said to be ________. [p. 741]
   a. crenulated
   b. flacid
   c. plasmolyzed
   d. turgid

5. Water can enter a root and move toward the stele without entering the cytoplasm of a cell until it reaches the ________? [p. 744]
   a. cortical parenchyma
   b. endodermis
   c. epidermis
   d. pericycle

6. The "stretching" of hydrogen bonds that is involved in moving xylem sap up a tree is called ________. [p. 745]
   a. adhesion
   b. cohesion
   c. tension
   d. viscosity

7. The most important players in the "transpiration-photosynthesis compromise" are ________. [p. 721]
   a. companion cells and sieve tube members
   b. guard cells and stomates
   c. spongy and palisades mesophyll
   d. xylem and phloem

8. Which ions are most important in regulating stomate function? [p. 748]
   a. Ca^{++} and Mg^{++}
   b. H^{+} and Cl^{-}
   c. H^{+} and K^{+}
   d. K^{+} and Mg^{++}

9. Which plant would most likely utilize CAM photosynthesis? [p. 750]

a. maple tree

b. moss

c. *Sedum*

d. water lily

10. The sink for photosynthate ________. [p. 752]

a. can be anywhere on the plant

b. must be above the source

c. must be below the source

d. must be below the surface of the soil

INTEGRATING AND APPLYING KEY CONCEPTS

1. *Unanswered Questions:* Current research indicates that plasmodesmata are not simple, static channels between adjacent cells but dynamic, changing structures. Can you think of any advantages and/or disadvantages associated with this flexibility?

2. *Insights from the Molecular Revolution:* What might be some possible advantages to the central vacuole of a cell having aquaporins in its membrane vs. simply relying on osmosis to correct a cytoplasmic?

3. *Focus on Research:* Summarize Wright's and Fisher's work to measure pressure of phloem in sieve tubes.

33

PLANT NUTRITION

CHAPTER HIGHLIGHTS

- Plants require proper soil, water, minerals, gases, and light.
- Plants develop specific symptoms if they lack proper nutrients.
- The size of the particles, the amount of organic matter, air, water, minerals, and living flora in the soil is critical.
- The microenvironment around the roots affects the absorption of water and nutrients:
  - Bacteria and mycorrhizae are symbiotically associated with plant roots.
  - Bacterial metabolism is a major source of nutrition.
- Nitrogen supply:
  - Fertilizers.
  - Nitrogen-fixing bacteria.
  - Ammonifying bacteria.
  - Nitrifying bacteria.
  - Root nodules in legumes.
- Special adaptations: Carnivorous plants, parasitic plants, epiphytes.

STUDY STRATEGIES

- This chapter is comparatively simpler.
- Remember the goals of this chapter are to understand:
  - What plants need.
  - How plants obtain their nutrients from their environment.

TOPIC MAP

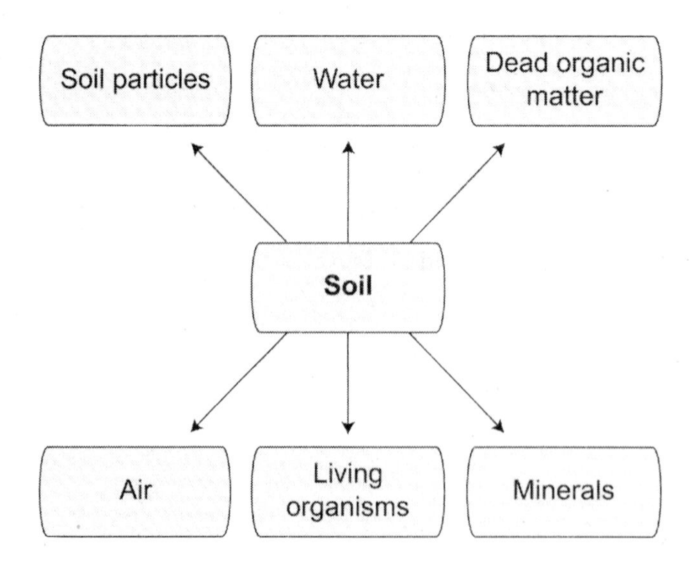

INTERACTIVE EXERCISES

Why It Matters [pp. 757–758]

Section Review (Fill-in-the-Blanks)

Plants need (1) ___________ and ___________ from the soil, and (2) __________ and __________ from the atmosphere. Plants have developed (3) ______________ and (4) ______________ features to match the environment in which they grow.

33.1 Plant Nutritional Requirements [pp. 758–762]

33.2 Soil [pp. 762–765]

Section Review (Fill-in-the-Blanks)

Plants are more than (5) _____% water. (6) __________________ is a technique where plants are grown in solutions containing measured amounts of (7) _______________.

(8) ______________ elements are necessary for plants to grow and reproduce. They acts as cofactors in (9) ______________ processes or help create osmotic (10) ______________ across the membrane for transportation of water. Out of 17 such elements, 9 are required in higher amounts called (11) __________________. These are (12) ______________, ______________, ______________, ______________, ______________, ______________, ______________, ______________, and ______________. Others are required in very small amounts called the (13) _________________. Some of these are (14) __________ _______, ________________, and ______________. Deficiency of these elements can lead to (15) ________________ growth and/or (16) ______________ of the leaves.

What soil is made of: A. Particles may be very fine such as clay that holds (17) ___________ tightly and lacks (18) _________ spaces, making it difficult for roots to penetrate and survive; or relatively coarse particles such as sand that has large (19) _________ spaces but drains (20) ______________ rapidly, making the soil dry. B. Decomposing dead plants and animals, called the (21) ___________, forms the organic component of the soil and helps in retaining (22) ___________ in the ground. C. Living organisms present in the soil such as saprophytic bacteria and fungi (23) _____________ organic matter; earthworms and insects that burrow in the soil add (24) ____________ to the soil. D. Minerals can be positively charged (25) ____________ or negatively charged (26) _____________. Their availability to the plant roots is affected by the (27) ______ of the soil.

Soil forms layers called the (28) __________________.

Matching

Match each of the following terms with its correct definition.

29. ____ Hydroponics
30. ____ Essential elements
31. ____ Macronutrients
32. ____ Micronutrients
33. ____ Chlorosis

A. Elements that are required in large amounts
B. Growing plants in a solution with known solutes
C. Yellowing of the plant tissue due to lack of chlorophyll
D. Elements that are necessary for normal growth and reproduction of the plants
E. Elements that are required in very small amounts

Complete the Table

34. Soil is a complex mix of important ingredients. Complete the following table by giving the role of each component of the soil.

Soil Ingredients	*Value in Soil*
Particles	
Humus	
Living organisms	
Minerals	

33.3 Obtaining and Absorbing Nutrients [pp. 765–772]

Section Review (Fill-in-the-Blanks)

Elements that are essential and often scarce in soil are (35) ______________, ______________, and ______________. Roots absorb ions just above the root tips where hundreds of (36) ______________ ______________ are present. The membranes have (37) ______________ proteins that help in absorption of ions. Symbiotic fungi, called the (38) ______________, and certain species of bacteria grow in and around the roots and help with the absorption of water and ions. In return, these organisms get (39) ______________ and (40) ______________ compounds from the plant. Once absorbed, water and minerals are stored in cell's (41) ______________, or used for the metabolic process in the cell's (42) ______________, or transported by (43) ______________ to other parts of the plant.

About (44) ____% of the atmosphere is nitrogen. Plants cannot use this gas form and lack (45) ______________ to convert it into nitrate and ammonium ions. Some bacteria, called the (46) ______________-______________ bacteria, live in the soil and convert the atmospheric nitrogen to ammonium ions. Other bacteria, called the (47) ______________ bacteria, decompose dead organic matter to produce ammonium ions. Plants use nitrogen in mostly (48) ______________ form but they absorb mostly in the form of nitrates. Bacteria called the (49) ______________ bacteria convert soil ammonium into nitrates for the plants. Once the plants absorb these nitrates, they convert it to (50) ______________ that is used to make (51) ______________ acids.

Certain (52) ______________-fixing bacteria live in the roots of the legumes and form swellings called the root (53) _______________. These bacteria are attracted by certain chemicals released by the roots, called the (54) ________________. Bacteria enter the root hair and release proteins expressed by their (55) ______ gene. These bacterial proteins stimulate root cells to multiply and form enlarged knots in the root called the root (56) ________________. Bacteria also become larger and immobile, now referred to as the (57) _______________. They have the enzyme (58) _______________ to convert nitrogen to ammonium. Root cells are also stimulated to produce a reddish, iron-containing protein called the (59) ____________________ that increases oxygen supply for the bacteria.

Other plant adaptations to get nitrogen:

A. (60) _______________ get their nitrogen by luring insects into an enzyme rich structure and digesting them. B. (61) _______________ develop special structures to penetrate other plants and tap their nitrogen, sugars, and water. C. (62) _______________ anchor on other plants to collect rainwater, dissolved nutrients, and light.

Matching

Match each of the following type of bacteria with their role in providing nitrogen to the plants.

63. ____ Nitrogen-fixing bacteria A. Bacteria that are able to decompose dead organic matter in the soil to ammonium

64. ____ Ammonifying bacteria B. Bacteria that are able to convert atmospheric nitrogen to ammonium

65. ____ Nitrifying bacteria C. Bacteria that are able to convert ammonium to nitrates for plants to absorb

66. ____ Bacteroids D. Enlarged, immobile bacteria in the legume root nodules

Match each of the following terms with their correct definition.

67. ____ Root nodule A. Bacterial gene that makes proteins to help nitrogen-fixing bacteria to enter plant roots

68. ____ Flavenoids B. Enzyme made by the bacteroids to convert atmospheric nitrogen to ammonium

69. ____ Nod gene C. A reddish, heme- and iron-containing protein that is produced by the root nodule cells to increase oxygen supply for the bacteria

70. ____ Leghemoglobin D. Chemicals released by plant roots to attract nitrogen-fixing bacteria

71. ____ Nitrogenase E. Swellings in the roots where nitrogen-fixing bacteria reside

Complete the Table

72. Plants have adapted in different ways to get their nutrients. Complete the following table with the definition of each adaptation.

Adaptations	*Description*
Carnivorous	
Parasite	
Epiphyte	
Symbiotic	

SELF-TEST

1. Plants absorb most of the nitrogen in the form of [p. 768]
   a. nitrogen gas
   b. ammonium
   c. nitrate
   d. amino acids

2. Plants use most of the nitrogen in the form of [p. 768]
   a. nitrogen gas
   b. ammonium
   c. nitrate
   d. amino acids

3. A fertilizer bag is labeled 15-30-25. Which minerals, in the correct order, do these numbers represent? [p. 762]
   a. carbon-hydrogen-oxygen
   b. carbon-nitrogen-oxygen
   c. nitrogen-phosphorus-potassium
   d. nitrogen-oxygen-phosphorus

4. Which of the following adds organic chemicals to the soil? [p. 763]
   a. soil particles
   b. humus
   c. live organisms
   d. minerals

5. Soil pH does not affect on the absorption of minerals by the plants. [p. 765]
   a. True
   b. False

6. Most of the absorption by the roots takes place in [p. 766]
   a. the root tip.
   b. above the root tip.
   c. older roots.
   d. dead cells of the root.

7. Orchids that live on the branches of trees are [p. 772]

a. parasites

b. saprobes

c. carnivores

d. epiphytes

8. Nitrifying bacteria [p. 769]

a. convert atmospheric nitrogen into ammonium.

b. decompose organic chemicals to make ammonium.

c. convert ammonium to nitrates.

d. convert nitrates into ammonium.

9. Bacteroids are [p. 769]

a. bacteria that live in the soil.

b. bacteria that live in root nodules.

c. bacteria that live on dead organic matter.

d. bacteria that infect plants and parasitize them.

10. Nod genes [p. 769]

a. allow plants to fix its own nitrogen.

b. stimulate plant cells to make leghemoglobin.

c. make the enzyme for converting nitrogen to nitrates.

11. Hydroponics is a technique that was developed to [p. 758]

a. study the role of different minerals in growth of a plant.

b. to replace agriculture industry.

c. to grow algae in water.

d. to multiply nitro-fixing bacteria.

12. Some of the minerals act as cofactors for the plant enzymes. [p. 761]

a. True

b. False

13. Macronutrients are more important than micronutrients. [p. 762]

a. True

b. False

14. Carbon, hydrogen, oxygen are [p. 761]

a. macronutrients.

b. micronutrients.

c. cofactors.

d. transport elements.

15. Where in the plant cells are water and minerals stored? [p. 767]

a. cytoplasm

b. central vacuole

c. nucleus

d. chloroplast

e. water

INTEGRATING AND APPLYING KEY CONCEPTS

1. Explain the role of nod genes in formation of the root nodule.

2. Explain the role of different types of bacteria involved in getting nitrogen to the plants.

3. List the different components of the soil and their role in supporting plant growth and development.

34

REPRODUCTION AND DEVELOPMENT IN FLOWERING PLANTS

CHAPTER HIGHLIGHTS

- Flowering plants are grouped as angiosperms.
- The plant is called a sporophyte because it bears male microspores and female megaspores.
- The floral shoot forms flowers that have:
    - Leaflike sepals that enclose unopened flowers.
    - Colorful petals to attract pollinators.
    - Stamens with bilobed anthers where the male gametophyte develops.
    - Carpals that have ovaries where the female gametophyte develops.
- Pollination is a transfer of pollen from an anther to the stigma of the flower.
- Fertilization is the fusion of male and female gametes.
- Embryos develop in the ovule and the ovule becomes the seed.
- The ovary, which becomes the fruit, houses the seeds.

STUDY STRATEGIES

- In this chapter you need to know the basic terminology.
- If you are already familiar with the reproduction process in animals, remember that plant reproduction follows a very similar process.
- Concentrate on the key concepts and on the sequence of events in each process.

TOPIC MAP

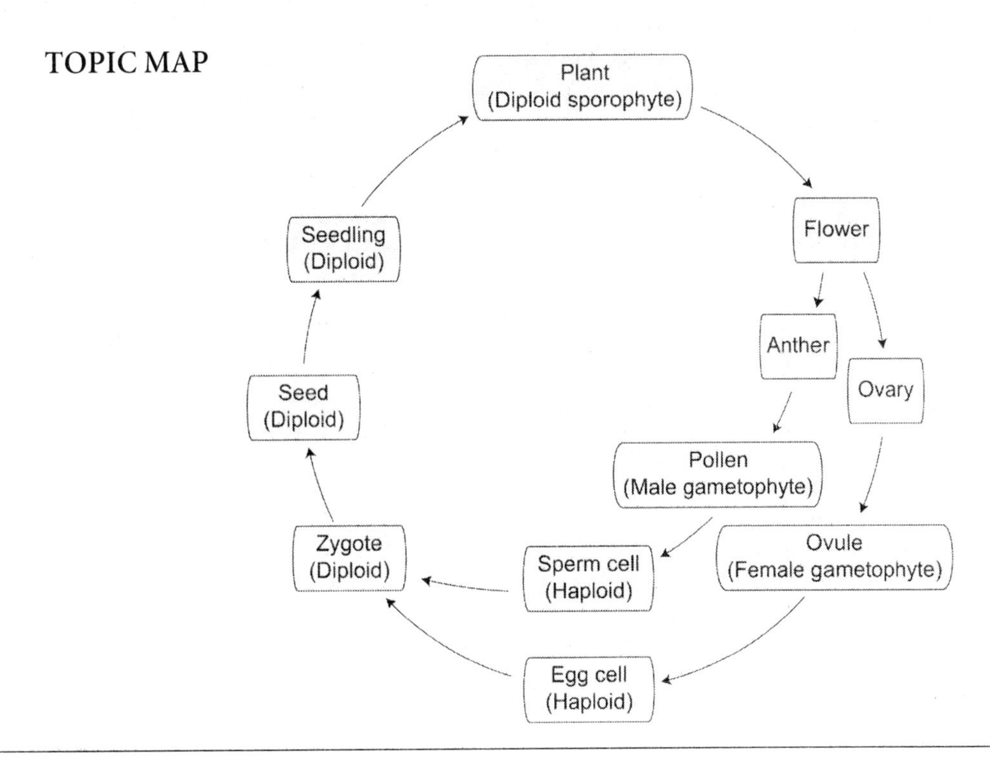

INTERACTIVE EXERCISES

Why It Matters [pp. 775–776]

Section Review (Fill-in-the-Blanks)

Angiosperms reproduce sexually by producing (1) ______________ that house male and female gametes. Male gametes are transferred to the female gamete, a process called (2) __________________. Once the gametes fuse, (3) ____________________, an embryo develops and (4) ____________ are formed in a fruit.

34.1 Overview of Flowering Plant Reproduction [pp. 776–777]

34.2 The Formation of Flowers and Gametes [pp. 778–780]

34.3 Pollination, Fertilization, and Germination [pp. 781–787]

Section Review (Fill-in-the-Blanks)

A plant is a diploid (5) _____________. It bears flowers where meiosis leads to the formation of male and female (6) ____________ spores. These spores divide by mitosis to form 3-celled male (7) ________________ called the pollen, and seven-celled female (8) _______________ called the embryo sac inside an ovule. The haploid male and female (9) ____________ fuse to form a(n) (10) ______________ zygote. The zygote divides by mitosis to form the new plant, the (11) _________________.

Once a vegetative shoot becomes a(n) (12) ____________ ______________, it bears a flower or a group of flowers called a(n) (13) ___________________. Each flower has four concentric whorls: A. The outermost, first whorl is called the (14) __________ that is made of leaflike (15)__________ to enclose the flower as a bud. B. The second whorl is called the (16) ___________ that is made of colorful (17) __________ to attract animals for pollination. C. The third whorl has the (18) ____________, each of which is made of a thin (19) __________ with a sac called the (20) ____________ containing the (21) __________ grains. D. The innermost whorl has one or more (22) __________. The lower part is called the (23) ___________ that contains the ovules. Its extension is called the (24) ____________ that ends in a flattened surface, the (25) ______________.

(26) ___________ flowers has all 4 whorls present. (27) ______________ flowers lack one or more whorls. (28) _____________ species have flowers of both sexes; (29) ________________ species have separate plants—plants with male flowers and plants with female flowers.

In a(n) (30) _____________, meiosis results into formation of many microspores, each one developing a thick (31) ________ ___________. Microspores divide by mitosis to form three-celled male (32) __________________ that contains two (33) ____________ cells and a pollen (34) __________ cell.

In a(n) (35) _____________, there may be one or more ovules formed. Inside each ovule, meiosis results into formation of (36) ______ mature megaspore while three others degenerate. A megaspore undergoes mitosis divisions to form an eight-celled female (37) _________________ that is referred to as the (38) _____________ sac. Each sac has (39) _______ egg cell, (40) __________ synergids, (41) _________ antipodal cells, and a single, large (42) _____________ cell. The egg cell and the synergids are located near the (43) _____________ of the ovule.

Air, water, birds, or other organisms transfer pollen grains from the (44) _____________ to the (45) _____________ of the ovary, a process called (46) _________________. Pollen grains germinate and the (47) ___________ tube goes through the (48) ___________ of the carpel to reach the ovules located inside an (49) _____________. After the fusion of the two gametes, (50) _________________, an embryo starts to develop. A(n) (51) _____________ is formed.

Incompatibility between two plants of the same species or two species is based on multiple (52) ______________ system of S gene. If pollen and the stigma have the same (53) _____________, pollen may not germinate or be able to grow inside the style.

As the pollen lands on the (54) ___________ of the carpel, a pollen tube cell grows through the (55) ___________ and reaches the micropyle of the (56) ___________. The two haploid (57) __________ cells enter the embryo sac, one fuses with the egg cell to form a diploid (58) _____________, and the other fuses with the diploid central nucleus to form a(n) (59) _____________ cell which eventually forms the (60) _______________, storing nutrients for the seedling. The zygote divides by mitosis to form a multicellular (61) _____________.

Eudicot seeds have two (62) _____________, which are usually thick and store nutrients. Monocots have only one (63) ______________ and a nutrient rich (64) _______________. The embryo in both has a(n) (65) ______________ that grows to become a root and a(n) (66) _______________ that becomes the leafy shoot. A seed also develops a protective seed (67) ____________. In monocots, a sheath of cells covers the embryo, (68) ______________ that covers the root and (69) _____________ that covers the shoot.

A fruit develops from the (70) __________ of the carpel, and its formation is stimulated by hormones released by the (71) ____________ grain. The wall of the fruit is called a(n) (72) _______________, and it develops from the wall of the (73) ____________.

(74) ___________ fruit develops from a single ovary (grains, nuts, peas, beans). (75) _____________ fruit develops from a flower that has more than one ovary (raspberries, strawberries). (76) _____________ fruit develops from ovaries of multiple flowers that fuse together (pineapple).

Once a seed is formed, it goes into a period of (77) ________________ during which it becomes metabolically inactive. Upon receiving proper environmental conditions, a seed absorbs water, a process called (78) ________________. During this period, the (79) ______________ produces hormones such as gibberellins that stimulate release of (80) ______________ enzymes to digest stored nutrients and provide them to the growing embryo. The (81) ______________ emerges from inside the seed coat to become the primary (82) ____________. The (83) ________________ grow to become the new foliage.

Matching

Match each of the following terms with its correct definition.

84.	____	Alternation of generation	A. A flower that lacks one of the four whorls
85.	____	Sporophyte	B. Where a plant bears the male and female flowers or flower parts on the different plants
86.	____	Gametophyte	C. The diploid plant which produces diploid microspores and megaspores
87.	____	Inflorescence	D. The haploid stage of the plant that bears the gametes
88.	____	Complete flower	E. Where the life cycle alternates between a diploid sporophyte and a haploid gametophyte
89.	____	Incomplete flower	F. A flower that has all four whorls—calyx, corolla, stamens, and carpels
90.	____	Monoecious	G. Flowers that are borne in a group
91.	____	Dioecious	H. Where a plant bears both the male and female flowers or flower parts on the same plant

Match each of the following flower part with its correct definition.

92.	____	Calyx	A. The second whorl of a flower that is made of brightly colored petals to attract pollinators
93.	____	Sepal	B. The third whorl of a flower that forms the male gametophytes
94.	____	Corolla	C. The innermost whorl that forms the female gametophytes
95.	____	Petal	D. The tip of the carpel where the pollen lands during pollination
96.	____	Stamen	E. The leafy parts of a flower—part of the outermost whorl
97.	____	Filament	F. The male gametophyte of a plant
98.	____	Anther	G. The brightly colored parts of a flower that attract pollinators
99.	____	Carpel	H. The outermost whorl that is made of leafy sepals
100.	____	Ovary	I. The thin stem of the stamens that bear anther at the tip
101.	____	Stigma	J. The lower part of a carpel that bears ovules
102.	____	Ovule	K. The round structure/s formed inside an ovary where the female gametophyte develops
103.	____	Pollen	L. The sac located at the tip of a stamen and bears the pollen grains

Match each cell of the male and female gametophytes with its correct definition.

104. _____ Sperm cells

105. _____ Pollen tube cell

106. _____ Egg cell

107. _____ Synergids

108. _____ Antipodal cells

109. _____ Central cell

A. The female gamete of the plant

B. The male gamete of the plant

C. The cell of the male gametophyte that grows as a tube inside the style of the carpel and transports the sperm cells to the ovule

D. Three cells located in the far end of the embryo sac that eventually degenerate

E. The large diploid cell that fuses with one of the sperm cells and forms the triploid endosperm

F. The two cells that flank the egg cell in the embryo sac and degenerate to guide the sperm cells to their receptive cells in the sac

Match each part of a seed with its correct definition.

110. _____ Cotyledon

111. _____ Endosperm

112. _____ Seed coat

113. _____ Radicle

114. _____ Plumule

115. _____ Coleorhiza

116. _____ Coleoptile

A. The fleshy structure that is formed by the triploid cells when the haploid sperm cell fuses with the diploid central cell

B. The protective covering of a seed

C. The part of the embryo in the seed that becomes the shoot of the seedling

D. A protective covering of the plumule in a monocot seed

E. One or two fleshy or thin structures derived from the embryo that often become the first leaves of the seedling

F. The part of the embryo in the seed that becomes the primary root of the seedling

G. A protective covering of the radicle in a monocot seed

Complete the Table

117. Complete the following table by description of each process.

Events	*Description*
A. Pollination	
B. Fertilization	
C. Germination	

118. Complete the following table by description of fruit types.

Fruits	*Description*
A. Simple fruits	
B. Aggregate fruits	
C. Multiple fruits	

34.4 Asexual Reproduction of Flowering Plants [pp. 787–789]

34.5 Early Development of Plant Form and Function [789–798]

Section Review (Fill-in-the-Blanks)

Multiplying from nonreproductive parts of the plant is the asexual reproduction referred to as the (119) ________________ reproduction. This is possible because many plant cells can dedifferentiate and divide to form a complete plant, a property called (120) ________________. (121) ________________ is when a part of the plant dedifferentiates and develops into a new plant. (122) ________________ is when an unfertilized egg or another diploid cell of the embryo sac divides and forms a seed without fertilization. Plant cuttings and grafting a quality stem on a root (123) __________ is a common practice in propagating fruits and flowers. Plant cells grown in culture often form an undifferentiated cell mass called the (124) ____________ that can be stimulated to redifferentiate by providing specific (125) ________________. Since mutations are common in cultured masses, new varieties of crops and their clones are often developed by this technique that is referred to as the (126) ________________ selection. The cell wall of the plant cells can be digested by enzymes to generate (127) ________________, which can then be fused with each other to combine varieties.

As soon as a plant zygote divides, a(n) (128) ____________-____________ axis is established where the (129) ______________ cell will eventually form shoots and the (130) ______________ cell will form the root system. After initial mitotic divisions, roots and shoots start forming, a process called (131) ________________. This process involves (132) ______________ cell division where the direction of the new cell influences the direction of the growth, and cell (133) ______________. Several (134) ____________ gene complexes have been identified that affect growth and differentiation in plants.

Matching

Match each of the following terms with its correct definition.

135. ____	Totipotency	A. When a piece of a plant detaches and form a new plant—a form of vegetative reproduction
136. ____	Dedifferentiation	B. A strong root used to graft a stem from a good variety of plant
137. ____	Fragmentation	C. When a cell has the potential to become a new organism
138. ____	Callus	D. When a specialized cell becomes undifferentiated
139. ____	Apomixis	E. A plant cell derived after enzymatic digestion of its cell wall
140. ____	Root stock	F. A set of genes that code for transcription factors that direct development of the organism
141. ____	Somaclonal selection	G. A mass of cells in culture grown from cells taken from a plant
142. ____	Protoplast	H. When a plant develops from unfertilized egg or a diploid cell of the embryo sac
143. ____	Root-shoot axis	I. Selecting new varieties from mutated cells of a callus
144. ____	Homeotic genes	J. Cells derived after the first division of the zygote that will eventually form the shoot and root respectively

145. ____	Morphogenesis	K. When early in the embryo development it is decided as to which cell will form root and shoot
146. ____	Apical and basal cells	L. Differentiation and development of roots and shoots in plants

SELF-TEST

1. A flowering plant is a [p. 776]

a. sporophyte.

b. gametophyte.

2. A three celled pollen grain is a [p. 776]

a. sporophyte.

b. gametophyte.

3. An embryo sac inside an ovule is a [p. 776]

a. sporophyte.

b. gametophyte.

4. Which of the following represents the cells in a pollen? [p. 779]

a. one sperm cell, two pollen tube cells

b. two sperm cells, one synergid, one central cell

c. three antipodal cells and one sperm cell

d. one pollen tube cell and two sperm cells

5. Which specific part of the flower transforms into a seed? [p. 784]

a. ovary

b. stamen

c. ovule

d. embryo sac

6. Which cell of the embryo sac fuses with the sperm cell to form endosperm? [p. 782]

a. egg cell

b. synergids

c. antipodal cells

d. central cell

7. Endosperm cells are [p. 782]

a. haploid

b. diploid

c. triploid

8. Eudicots store nutrients in their cotyledons and monocots store in endosperm. [p. 783]

a. True

b. False

9. A strawberry is a [p. 784]

a. simple fruit.

b. aggregate fruit.

c. multiple fruit.

10. A pineapple is a [p. 784]

a. simple fruit.

b. aggregate fruit.

c. multiple fruit.

11. A garden pea or a green bean is a [p. 784]

a. simple fruit.

b. aggregate fruit.

c. multiple fruit.

12. All the cells in a callus are genetically alike. [p. 789]

a. True

b. False

13. A radicle forms [p. 783]

a. stems.

b. leaves.

c. flowers.

d. roots.

14. Which of the following processes refers to the transfer of pollen grain from the anther to the stigma? [p. 781]

a. pollination

b. fertilization

c. imbibition

d. differentiation

15. Which of the following term refers to specialized cells becoming unspecialized? [p. 788]

a. differentiation

b. dedifferentiation

c. redifferentiation

INTEGRATING AND APPLYING KEY CONCEPTS

1. Totipotency is demonstrated in plants. Explain the concept by giving an example.
2. Describe the S gene concept to explain incompatibility between pollen and female tissues.
3. Which part/s of the flowering plant is a sporophyte and which part/s are gametophytes.

35

CONTROL OF PLANT GROWTH AND DEVELOPMENT

CHAPTER HIGHLIGHTS

- Plants produce hormones (Auxins, Gibberallins, Cytokinins, Ethylene, Brassinosteroids, Abscicic acid, Jasmonates) that regulate growth, development, and reproduction.
- Plants produce chemicals that protect them from infections and predators.
- Plants respond to light (phototropism), gravity (Gravitropism), physical stress (Thigmotropism) and exhibit nondirectional movement (Nastic movement).
- Plants maintain a 24-hour circadian rhythm that affects their movements, growth, flowering, seed dormancy, and germination.

STUDY STRATEGIES

- This chapter can be confusing because there are a number of overlapping functions of chemicals that plant produce.
- It helps to keep track of major functions first and then try to understand their interactions.

TOPIC MAP

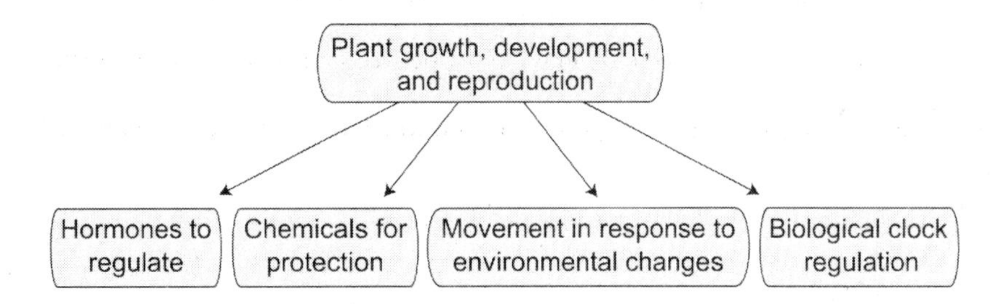

INTERACTIVE EXERCISES

Why It Matters [pp. 801–802]

Section Review (Fill-in-the-Blanks)

Plant growth and development are affected by environmental factors such as (1) ____________, ____________________, ________________, and nutrients. Plants also produce chemicals that affect their growth, development, flowering, and (2) _____________ production.

35.1 Plant Hormones [pp. 802–812]

Section Review (Fill-in-the-Blanks)

Plants respond to internal (3) ______________ changes and external (4) ______________ conditions by producing special chemicals called the (5) ______________. These chemicals are small, (6) ______________ chemicals that are required in very small amounts. They diffuse from one cell to another or they may be transported away by the (7) ______________ tissue. There are seven major groups of hormones:

A. Auxins—They are made by apical (8) ______________ and young leaves and stems. They (9) ______________ the growth of stems and lateral roots. They diffuse away from (10) ______________ and downwards from shoot to the roots. This causes the plant to grow differentially in response to (11) ______________ and (12) ______________. They also (13) ______________ fruit development.

B. Gibberellins—They are made mostly by root and shoot (14) ______________. They (15) ______________ cell division and hence the growth of the stem. They help in breaking seed and bud (16) ______________. They also stimulate elongation of floral stems in rosette plants, a process called (17) ______________.

C. Cytokinins—Produced mainly by the root (18) ______________ and transported by (19) ______________ vessels, cytokinins stimulate cell (20) ______________ rather than cell elongation.

D. Ethylene—Produced by most parts of the plant, ethylene stimulates aging and cell death, a process referred to as (21) ______________. Part of this process is dropping of leaves, flowers, and fruits, called (22) ______________. Ethylene also promotes ripening of the (23) ______________.

E. Brassinosteroids—These are steroid (24) ______________ that were first discovered in the mustard family. They (25) ______________ stem elongation, development of vascular tissue, and growth of pollen tubes.

F. Abscicic acid—This hormone is derived from (26) ______________ in the chloroplasts. It (27) ______________ stem growth and promotes abscission. It often accumulates in the seeds coats and must be slowly broken down before the seed can (28) ______________.

G. Jasmonates—These hormones are derived from (29) ______________ acids and protect plants from (30) ______________ and (31) ______________.

Matching

Match each of the following terms with its correct definition.

32. ____ Bolting — A. Aging process in plants that leads to breakdown and death of their cells
33. ____ Senescence — B. Dropping of leaves, flowers, and fruits in response to environmental changes
34. ____ Abscission — C. Period when seeds and buds go into growth-inhibiting phase
35. ____ Dormancy — D. Sudden elongation of the floral stem in rosette plants

Match each of the following scientists with their major contribution.

36. ____ Eiichi Kurosawa — A. Performed the first experiments to explain phototropism in grasses

37. ____ Francis and Charles Darwin — B. Demonstrated that the growth-promoting chemical travels from the shoot tip downwards

38. ____ Frits Went — C. Showed that auxins move laterally in response to light exposure

39. ____ Winslow Briggs — D. First one to discover the effect of gibberallins produced by a fungus on rice plants

Complete the Table

40. Complete the following table by describing the main actions of each hormone.

Hormone	*Action*
Auxins	
Gibberallins	
Cytokinins	
Ethylene	
Brassinosteroids	
Abscisic acid	
Jasmonates	

35.2 Plant Chemical Defenses [pp. 812–816]

35.3 Plant Responses to the Environment: Movements [817–821]

35.4 Plant Responses to the Environment: Biological Clocks [821–825]

35.5 Signal Responses at the Cellular Level [826–828]

Section Review (Fill-in-the-Blanks)

Plants are attacked by (41) ____________, ____________, ____________, and ____________. When a plant is attacked by an insect → a cascade of chemical and cellular reactions are stimulated by hormones such as (42) ____________ and ____________ → (43) ____________ acid and jasmonates are produced by the plant → plant cells make protease (44) ____________ → predators cannot use protease enzymes to digest plant (45) ____________ → cells around the wound produce hydrogen (46) ____________ → kills the cells around the wound → pathogen is physically blocked from healthy cells. This defense mechanism that separates the wounded tissue is referred to as the (47) ____________ response.

Plants also produce protective chemicals called the (48) ______________ metabolites such as (49) ________________________ that act like antibiotics or caffeine, cocaine, strychnine, tannins, and terpenes that ward off feeding herbivores.

Plants respond to certain environmental factors by growing away or towards the factor, a process referred to as (50) ________________: (51) ________________ is growth in response to unidirectional light, where a blue light absorbing pigment, (52) ________________, stimulates lateral movement of auxins that eventually leads to differential growth.

(53) ________________ is growth in response to gravity where amyloplasts, called the (54) ________________, move within the cells and stimulate roots to grow (55) ______________ and shoots to grow (56) ________________.

(57) ________________ is growth in response to physical contact with another object where auxins and ethylene stimulate thickening and bending of the plant.

(58) ____________ movement is reversible, temporary response to a stimulus, such as folding of the leaves, opening or closing of stomata, and movement of a flower or leaves with the movement of the sun.

Plants also exhibit a 24-hour cycle called the (59) ________________ rhythm that is often observed with opening and closing of stomata, flowers, or leaves. The rhythm may be divided into specific numbers of light and dark periods, and this phenomenon is referred to as (60) ____________________. A pigment called the (61) ____________________, which exists in two reversible forms, (62) ________ and (63)________, is responsible for signaling the light switch. Plants may be classified as (64) __________-______ plants or (65) __________-________ plants, depending upon the number of day-length needed.

Some plants do not flower until they go through hours of cold weather, called (66) ________________. Other plants have seeds and buds that go into a phase of no growth, referred to as the (67) ______________. Breaking this inhibition may require exposure to cold weather or production of hormones such as (68) ________________.

Chemical or environmental signals received by (69) ____________ proteins on a target cell → a cascade of (70) ____________ reactions begins inside the cell → causes production of second messenger cAMP, changes in membrane permeability, activation of enzymatic proteins or activation of genes that would lead to making of new (71) ____________.

Matching

Match each of the following terms with its correct definition.

72. ____	Salicylic acid	A. Chemicals produced by plants to prevent herbivores from consuming them
73. ____	Phytoalexins	B. Movement or growth of a plant in response to contact with an object
74. ____	Secondary metabolites	C. A chemical similar to aspirin that is produced by some wounded plants to protect themselves from infections
75. ____	Phototropism	D. A 24-hour cycle of response in plants
76. ____	Gravitropism	E. A blue-green pigment that exists in two forms and signals the light switch
77. ____	Thigmotropism	F. Refers to a low-temperature stimulation of flowering
78. ____	Nastic movement	G. Chemicals produced by plants infected with bacteria or fungi and act like antibiotics
79. ____	Circadian rhythm	H. Movement or growth of a plant in response to unidirectional light
80. ____	Phytochrome	I. Movement or growth of a plant in response to gravity
81. ____	Vernalization	J. Reversible or temporary movement in response to unidirectional stimulus
82. ____	Otoliths	K. Amyloplasts that move in the plant root and shoot tips to help in responding to gravity

---

SELF-TEST

1. Which of the following was identified as the first plant hormone? [p. 802]

a. Auxins

b. Gibberallins

c. Cytokinins

d. Ethylene

e. Abscisic acid

2. Which of the following hormones exhibits polar transport and moves away from unidirectional light? [p. 805]

a. Auxins

b. Gibberallins

c. Cytokinins

d. Ethylene

e. Abscisic acid

3. Which of the following hormones is involved in breaking of seed and bud dormancy? [p. 807]

a. Auxins

b. Gibberallins

c. Cytokinins

d. Ethylene

e. Abscisic acid

4. Which of the following hormones is involved in bolting seen in rosette plants? [p. 808]

a. Auxins

b. Gibberallins

c. Cytokinins

d. Ethylene

e. Abscisic acid

5. Which of the following hormones coordinates growth of roots and shoots in concert with the auxins? [p. 809]

a. Abscisic acid

b. Gibberallins

c. Cytokinins

d. Ethylene

6. Which of the following hormones is involved in ripening of fruits? [p. 810]

a. Auxins

b. Gibberallins

c. Cytokinins

d. Ethylene

e. Abscisic acid

7. Which of the following term is used for dropping of flowers, fruits, and leaves? [p. 810]

a. Senescence

b. Abscission

c. Tropism

d. Bolting

e. Vernalization

8. Which of the following terms is used for extension of the floral stem in rosette plants? [p. 808]

a. Senescence

b. Abscission

c. Tropism

d. Bolting

e. Vernalization

9. Which of the following refers to movement or growth of a plant in response to contact with an object? [p. 819]

a. Phototropism

b. Gravitropism

c. Thigmotropism

d. Nastic movement

e. Photoperiodism

10. Which of the following refers to temporary, reversible response to a unidirectional stimulus? [p. 819]

a. Phototropism

b. Gravitropism

c. Thigmotropism

d. Nastic movement

e. Photoperiodism

11. Which of the following refers to response of a plant due to changes in the length of light and dark periods during each 24-hour period? [p. 819]

a. Phototropism

b. Gravitropism

c. Thigmotropism

d. Nastic movement

e. Photoperiodism

12. Which of the following term is used for low temperature stimulation of flowering? [p. 824]

a. Senescence

b. Abscission

c. Tropism

d. Bolting

e. Vernalization

13. Which of the following terms is used for aging in plants? [p. 810]

a. Senescence

b. Abscission

c. Tropism

d. Bolting

e. Vernalization

14. In which of the following is phytochrome involved? [p. 821]

a. Phototropism

b. Gravitropism

c. Thigmotropism

d. Nastic movement

e. Photoperiodism

15. Gibberallins are made by plants and fungi. [p. 801]

a. True

b. False

INTEGRATING AND APPLYING KEY CONCEPTS

1. Knowing that ethylene is involved in fruit ripening, discuss how this information is used by fruit growers.
2. Discuss the hormones and the sequence of events involved in senescence.
3. Discuss the role of secondary metabolites produced by some plants.

36

INTRODUCTION TO ANIMAL ORGANIZATION AND PHYSIOLOGY

CHAPTER HIGHLIGHTS

- The organization of the animal body is presented—both structure and function.
- Function of tissues and organs is associated with structure.
- Homeostasis is a dynamic equilibrium to meet the changing demands of both the internal and external environments of animals.
- Explanation of feedback systems; the control mechanisms of homeostasis are introduced.

STUDY STRATEGIES

- First, focus on major levels of organization of animals—cells, tissues, organs, and organ systems.
- Examine and learn the structure of the various tissues, organs, and organ systems.
- Next, associate structure of tissues and organs with their function. Given the structure, be able to predict function.
- Evaluate the control mechanisms of feedback systems and maintenance of a constant, yet dynamic internal environment.

TOPIC MAPS

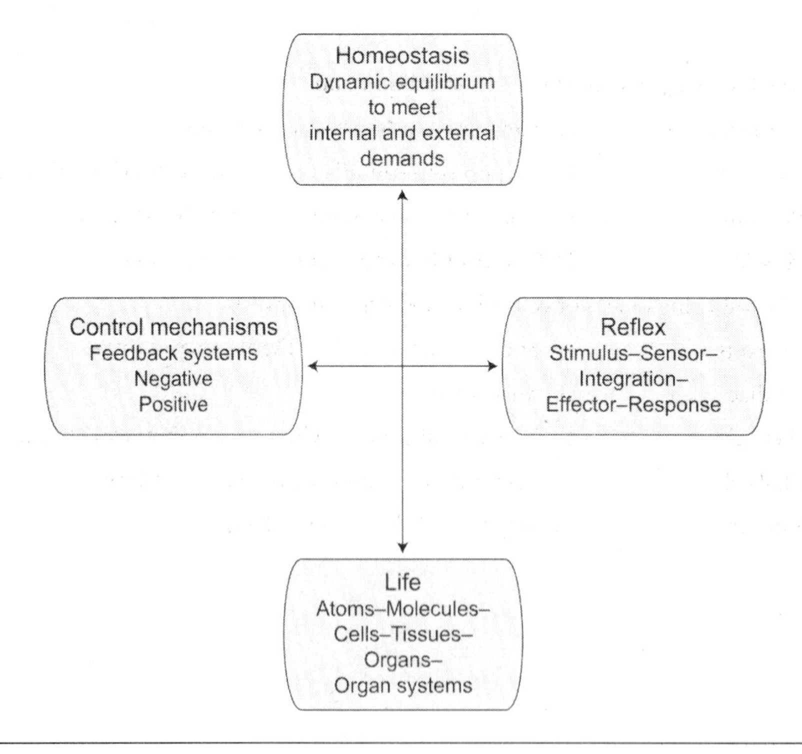

INTERACTIVE EXERCISES

Why It Matters [pp. 831–832]

Section Review (Fill-in-the-Blanks)

All living organisms are exposed to a variety of environmental factors. (1) _____ is the term to describe a stable, yet dynamic (2) ______ environment. Various control mechanisms are involved in (3) _____ of a constant internal environment. Animals are composed of cells, tissues, organs, and organ systems , the specific (4) _____ and (5) ______ of which play a significant role in homeostasis.

36.1 Organization of the Animal Body [p. 832]

Section Review (Fill-in-the-Blanks)

A cell has two environments, the (6) _____ and (7) _____ environments. From previous chapters, we have learned that living organisms are composed of a significant amount of (8) _____. An animal (9) _____ has the necessary components for life—the ability to uptake nutrients, convert the nutrients to usable energy, and function in a variety of demands, both internal and external. Cells with a similar structure and function are called (10) _____. A(n) (11) _____ is composed of two or more (12) _____ that carry out a specific (13) _____. A(n) (14) _____ _____ coordinates the activities of two or more (15) _____ to carry out a major body activity such as digestion or reproduction.

36.2 Animal Tissues [pp. 833–840]

Matching

Match each of the following tissues which its correct function.

16. ____ Epithelial
17. ____ Connective
18. ____ Nervous
19. ____ Muscle

A. Supports, binds, provides structure
B. Responds to stimuli and communicates with other cells and tissues
C. Contains actin and myosin—forcibly shortens
D. Covers or lines body surfaces and cavities

Match each of the following tissues or cells which its correct description.

20. ____ Neurons
21. ____ Cardiac
22. ____ Skeletal
23. ____ Glial cells
24. ____ Smooth

A. Short, branched cells with intercalated disks
B. Moves bones
C. Walls of tubes and cavities, primarily autorhythmic
D. Respond to stimuli with axon and dendrites
E. Support and electrical insulation

Choice

For each of the following statements, choose the most appropriate connective tissue components or type from the list below.

A. Collagen B. Tendons C. Loose CT D. Cartilage E. Bone
F. Blood G. Adipose

25. ____ Primary cell type is fibroblast, forms mesenteries

26. ____ Lipid storage

27. ____ Fibrous glycoprotein

28. ____ Function of oxygen delivery to tissues as well as immunity

29. ____ Attaches muscles to bones

30. ____ Primary cell type is the osteocyte

31. ____ Chondrocytes in collagen and chondroitin sulfate matrix

Section Review (Fill-in-the-Blanks)

Vertebrates have four major types of tissues, (32) ______, (33) ______, (34) ______, and (35) ______. (36) ______ tissue is classified by the number of (37) ______ ______ above the basement membrane and the (38) ______ of the outermost cell layer. The (39) ______ types of (40) ______ tissue are (41) ______, (42) ______, (43) ______, (44) ______, (45) ______, and (46) ______. Connective tissue is composed of (47) ______ and protein (48) ______ embedded in a(n) (49) ______ ______. The (50) ______ types of (51) ______ are (52) ______, (53) ______, and (54) ______. There are (55) ______ major cell types in (56) ______ tissue, (57) ______ and (58) ______ cells.

Complete the Table

59.

Tissue Type	Function	Location
Epithelial	A.	B.
C.	Support, Protection, Binds	Most body structures and systems
Muscle	D.	E.
Nervous	F.	Central and Peripheral Nervous system

36.3 Coordination of Tissues in Organs and Organ Systems [pp. 840–841]

Section Review (Fill-in-the-Blanks)

In multicellular organisms, the (60) ______ is the unit of life, yet it doesn't function in isolation. Similar cells from (61) ______ have specific functions. (62) ______ are composites of tissues that carry out a specific function, while (63) ______ ______ are composed of multiple organs that coordinate and integrate a series of events to accomplish vital tasks to ensure survival of the organism in its environment. Most animals have (64) ______ major organ systems. These organ systems work in a coordinated fashion to accomplish the qualities or characteristics associated with life, (65) ______, (66) ______, (67) ______, (68) ______, (69) ______, and reproduction.

Matching

Match each of the following organs with the most appropriate organ system

70. ____	Brain	A. Endocrine
71. ____	Heart	B. Muscular
72. ____	Skin	C. Nervous
73. ____	Thymus	D. Circulatory
74. ____	Bones	E. Lymphatic
75. ____	Thyroid	F. Integumentary
76. ____	Stomach	G. Reproductive
77. ____	Lungs	H. Excretory
78. ____	Cardiac and smooth	I. Digestive
79. ____	Spleen	J. Respiratory
80. ____	Testes	K. Skeletal
81. ____	Hair and nails	
82. ____	Pancreas	
83. ____	Liver	
84. ____	Spinal Cord	
85. ____	Kidneys	

36.4 Homeostasis [pp. 841–843]

Short Answer

86. Explain the importance of homeostasis with respect to a changing external environment. ________________

____.

87. Compare and contrast a negative and positive feedback mechanism with respect to output or product. ______

____.

Matching

Match each of the following with its correct definition.

88. ____	Response	A. coordinates multiple input
89. ____	Stimulus	B. detects change—temperature, pH, touch
90. ____	Effector	C. output
91. ____	Sensor	D. input—environmental change
92. ____	Integrator	E. responds to stimulus, produces the output

SELF-TEST

1. The demands placed on a cell are both _____. [p. 832]
   a. internal and external
   b. physical and anatomical
   c. functional and physiological
   d. none of these

2. Select the level that is encompassed by or within the other three. [p. 832]
   a. liver
   b. epithelium
   c. mitochondria
   d. hepatic (liver) cell

3. Given the characteristics of lines a body cavity with little or no extracellular matrix, select the major tissue type. [pp. 833–840]
   a. nervous
   b. muscle
   c. connective
   d. epithelial

4. This tissue has cells involved in immunity as well as oxygen delivery to cells. [pp. 833–840]
   a. nervous
   b. muscle
   c. connective
   d. epithelial

5. This tissue is found in every organ or organ system. It is the most varied and is characterized by various cell types, fibers, and an extracellular matrix. [pp. 833–840]
   a. nervous
   b. muscle
   c. connective
   d. epithelial

6. Maintenance of osmotic balance, electrolytes, and pH are important functions of this organ system. [pp. 840–841]
   a. Circulatory
   b. Excretory
   c. Respiratory
   d. Endocrine

7. This organ system is involved in homeostatic mechanisms and chemical regulation which is carried in circulation. [pp. 840–841]
   a. Circulatory
   b. Excretory
   c. Respiratory
   d. Endocrine

8. This organ system has sweat glands and provides protection to the organism. [pp. 840–841]

a. Muscular

b. Integumentary

c. Skeletal

d. Nervous

9. If the stimulus resulted in amplification of the response, this would be an example of ____. [pp. 841–843]

a. negative feedback

b. positive feedback

c. homeostasis

d. integration

10. If body temperature exceeded the set point of the hypothalamus, you would expect the stimulus to activate sweat glands would _____. [pp. 841–843]

a. increase

b. decrease

c. not change

d. Either b or c

INTEGRATING AND APPLYING KEY CONCEPTS

1. Explain how negative feedbacks are involved in normal physiological processes, while positive feedbacks are often (but not always) associated with abnormalities or disease states.

2. Compare and contrast the "division of labor" of a eukaryotic cell and the organ systems of a cat.

37

INFORMATION FLOW AND THE NEURON

CHAPTER HIGHLIGHTS

- Neurons are specialized cells that transmit information by electrical impulses. The impulses result from changes in ion distribution across the membrane of the neuron.
- Neuronal circuits are made up of neurons that 1) sense environmental information, 2) integrate that information, and 3) effect an appropriate response to the information.
- Neurons communicate with others neurons in a circuit or with other targets (muscles, glands) via synapses. Some synapses consist of direct contact of adjacent cells that communicate through gap junctions. Other synapses involve chemical signals (neurotransmitters) released by the communicating cell that binds to the target cell.
- Some neurotransmitters bind to ion channels (ligand-gated channels) to directly alter ion movement, whereas other neurotransmitters bind to G-protein receptors that, in turn, produce second messengers and alter cellular activity, including ion movement.
- Neurons integrate stimulatory and inhibitory information to initiate electrical impulses.

STUDY STRATEGIES

- The study of neurons and neural circuits is important for understanding how animal function is controlled.
- This chapter contains a lot of new and unfamiliar material. Go slow. Take one section at a time, then work through the companion section(s) in the study guide.
- Practice drawing neurons and labeling their component parts. Also practice drawing action potentials, being sure to label each phase and describing the bases for each phase.

INTERACTIVE EXERCISES

Why It Matters [p. 847]

Section Review

The (1) ____ integrates a variety of sensory inputs, then makes compensating adjustments to the activities of the body.

37.1 Neurons and their Organization in Nervous Systems: An Overview [pp. 848–851]

Section Review

(2) ____ is communication by (3) ____, which has four components. (4) ____ is the detection of a stimulus. (5) ____ is the sending of a message along a neuron, then relaying the message to another neuron or to a muscle or gland. (6) ____ is the sorting and interpretation of the message. (7) ____ is the output or action that results from the message. (8) ____, also known as (9) ____, transmit sensory stimuli to (10) ____, which integrate the information to formulate an appropriate response. (11) ____ carry response signals to (12) ____, such as muscle and glands. The (13) ____ of a neuron contains the nucleus. (14) ____ receive signals and transmit them toward the cell body,

whereas (15) _____, which arise from the (16) _____, conduct impulses away from the cell body. Axons end in buttonlike swellings called (17) _____. The association of neurons (connected from the axon of one to the dendrite of another) form a chain called a(n) (18) _____. (19) _____ provide nutrition and support to neurons. Star-shaped (20) _____ provide physical support and help maintain ion concentrations in the extracellular fluid. In vertebrates, (21) _____ in the central nervous system and (22) _____ in the peripheral nervous system wrap around axons to form sheaths. Gaps between Schwann cells along an axon are called (23) _____. A(n) (24) _____ is the junction between a neuron and another neuron or between and neuron and a muscle cell. On one side of the synapse is the (25) _____ and on the other side is the (26) _____. Communication across the synapse occurs in one of two ways: a) by direct flow of electrical current, making a(n) (27) _____, or b) by chemical transmission, making a(n) (28) _____. A(n) (29) _____ is the chemical released by a presynaptic cell at a chemical synapse, which crosses a narrow gap or (30) _____, and then binds to receptors on the postsynaptic cell.

Building Vocabulary

Prefixes	Meaning
pre-	before, prior to
post-	after, following
inter-	between, among

Prefixes	Meaning
-neuro(n)	of or relating to a nerve

Prefix	Suffix	Definition
_____	-synaptic	31. the neuron that transmits a signal to a specific synapse
_____	-synaptic	32. the neuron that receives the signal at a specific synapse
_____	_____	33. neurons that integrate information

Choice

For each of the following characteristics, choose the most appropriate type of synapse.

a. chemical synapse b. electrical synapse

34. _____ Plasma membranes of presynaptic cells and postsynaptic cells are in direct contact, and the electrical response flows between the two cells through gap junctions.

35. _____ Plasma membranes of presynaptic cells and postsynaptic cells are separated by a cleft, and the electrical impulse is conveyed by neurotransmitters.

For each of the following descriptions, choose the most appropriate type of cell.

a. presynaptic cell   b. postsynaptic cell

36. ____ neuron that transmits signal

37. ____ neuron that receives signal

For each of the following descriptions, choose the most appropriate type of cell.

a. afferent neuron   b. interneuron   c. efferent neuron

38. ____ transmits stimuli collected by sensory neurons

39. ____ integrates sensory information to formulate appropriate response

40. ____ carries impulses to effectors such as muscles and glands

Making Comparisons

41. Distinguish between motor neurons and efferent neurons.

____

____

Sequence

Arrange the following events in neural signaling in the correct order.

42. ____   a. integration

43. ____   b. transmission

44. ____   c. response

45. ____   d. reception

Complete the Table

Type of cell	Description
glial cells	46.
47.	Star-shaped cells that provide physical support and help maintain ion concentration
oligodendrocytes	48.
Schwann cells	49.
50.	Cells of the nervous system specialized to generate electrical impulses

Matching

Match each of the following with its correct definition.

51. ____ synapse   A. communication by neurons

52. ____ neural circuits   B. targets of efferent neurons such as muscles and glands

53. ____ neurotransmitter   C. a chain of neurons

54. ____ node of Ranvier   D. gaps between Schwann cells that expose axons

55. ____ effector   E. a special junction between a neuron and another neuron or between a neuron and other effector cell

56. ____ synaptic cleft   F. chemical released from an axon terminus into a synapse

57. ____ neural signaling   G. the narrow gap of a chemical synapse

Identify

Identify the various parts on the neuron.

58. _____

59. _____

60. _____

61. _____

62. _____

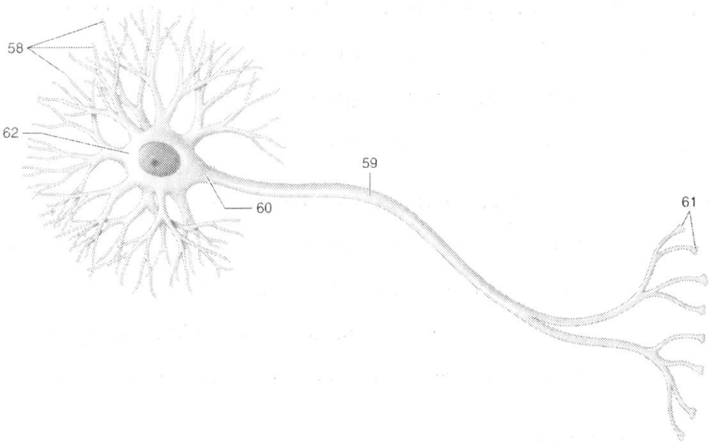

37.2 Signal Conduction by Neurons [pp. 851–857]

37.4 Conduction across Chemical Synapses [pp. 858–862]

37.5 Integration of Incoming Signals by Neurons [pp. 862–863]

Section Review

(63) ____ is the difference in charge across the membrane of all animal cells. The steady negative membrane potential of a neuron that is not conducting is called the (64) ____, which makes the cell (65) _____. When a neuron conducts an impulse, there is an abrupt transient change in membrane potential called the (66) ____. As the membrane becomes less negative, the membrane becomes (67) ____. Depolarization proceeds slowly at first until (68) ____ is reached, then the potential changes rapidly, rising to as much as +30mV inside with respect to the outside. The potential then falls, sometime dropping below the resting level and resulting in the membrane becoming (69) ____. An action potential only results if the stimulus is strong enough to cause the membrane potential to reach threshold; once triggered, the action potential results regardless of stimulus strength. This phenomenon is known as the (70) ____. At the peak of the action potential, the membrane enters a(n) (71) ____, which lasts until the resting potential is reestablished. During the refractory period, the neuron cannot be stimulated again, which keeps the impulses traveling only in one direction. The action potential is produced by movements of Na^+ and K^+ through (72) ____. (73) ____ of the electrical impulse occurs along the axon as a wave of depolarization. Impulses are sped when action potentials hop along the axon in a process called (74) ____.

In chemical synapses, neurotransmitters are released from the (75) ____, diffuse across the cleft, and alter ion conduction by activation of (76) ____ on the (77) ____. Neurotransmitters are stored in presynaptic cells in (78) ____ and are released by the process of (79) ____. (80) ____ directly bind ligand-gated channels to alter the flow of ions. (81) ____ bind to G-protein coupled receptors that trigger the generation of a second messenger (e.g., cAMP); the second messenger then affects ion channels to alter ion movement. Many different types of chemicals serve as neurotransmitters, including amines, amino acids, and peptides.

Neurotransmitters may stimulate or inhibit the generation of action potentials in the postsynaptic cell. Stimulation opens a ligand-gated sodium channel, depolarizing the cell and driving closer to threshold. Such a potential change is called a(n) (82) ____. An inhibitory neurotransmitter opens a ligand-gated channel that allows Cl^- to flow into the cell and K^+ to leave, causing the cell to become hyperpolarized and taking the cell further from threshold; such a potential change is called a(n) (83) ____. In contrast to all-or-nothing action potentials, EPSPs and IPSPs are (84) ____. (85) ____ is the accumulation of EPSPs from a single presynaptic neuron over a short period of time. (86) ____ is the accumulation of EPSPs produced from several different presynaptic neurons.

Building Vocabulary

Prefixes	Meaning
neuro-	of or relating to a neuron
hyper-	above, to a greater extent
de-	to undo

Prefix	Suffix	Definition
____	-transmitter	87. the chemical released at a chemical synapse
____	-polarized	88. a change in membrane potential that makes the cell less negative inside compared to outside
____	-polarized	89. a change in membrane potential that makes the cell more negative inside compared to outside

Choice

For each of the following characteristics, choose the most appropriate term from the list below.

a. presynaptic membrane b. postsynaptic membrane

90. ____ membrane from which exocytosis occurs to release neurotransmitters

91. ____ membrane on which neurotransmitters bind

For each of the following characteristics, choose the most appropriate neurotransmitter from the list below.

a. indirect neurotransmitter b. direct neurotransmitter

92. ____ neurotransmitter that binds to ligand-gated channel

93. ____ neurotransmitter that binds to a receptor to trigger production of a second messenger

For each of the following, choose the most appropriate potential from the list below.

a. excitatory postsynaptic potential b. inhibitory postsynaptic potential

94. ____ change in membrane potential of postsynaptic cells that brings it closer to threshold

95. ____ change in membrane potential of postsynaptic cells that brings it further from threshold

For each of the following characteristics, choose the most appropriate type of summation from the list below.

a. temporal summation b. spatial summation

96. ____ ____ change in membrane potential of postsynaptic neuron produced by the firing of several presynaptic neurons

97. ____ ____ change in membrane potential of postsynaptic neuron produced by successive firing of a single presynaptic neuron over a short period of time

For each of the following characteristics, choose the most appropriate type of channel from the list below.

a. ligand-gated channel   b. voltage-gated channel

98. ____ membrane proteins that control ion flow by opening and closing as membrane potential changes

99. ____ membrane proteins that control ion flow in response to binding a neurotransmitter.

Making Comparisons

100. Distinguish between membrane potential and resting potential.

____

____

Matching

Match each of the following aspects of neural signaling with its correct description.

101.	____	all-or-nothing principle	A. an abrupt, transient change in membrane potential
102.	____	synaptic vesicles	B. secretory vesicles that contain neurotransmitters
103.	____	action potential	C. rapid conduction of an impulse in which action potentials "hop" along an axon
104.	____	propagation	D. the progression of action potentials along an axon
105.	____	salutatory conduction	E. once triggered, the change in membrane potential that takes place regardless of stimulus strength
106.	____	graded potential	F. potential that moves up or down in response to stimulus without triggering an action potential; amplitude depends on strength of stimulus.

Identify

Identify the various parts of an action potential.

107. __________

108. __________

109. __________

110. __________

111. __________

112. __________

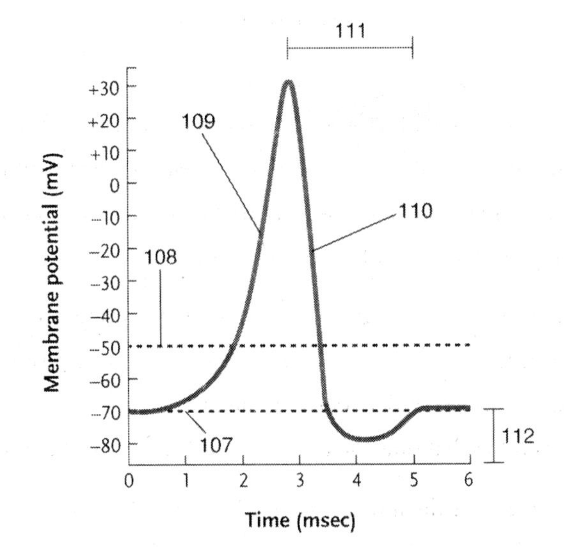

SELF-TEST

1. Neural signaling involves [p. 848]
   a. differentiation.
   b. integration.
   c. response.
   d. transmission.
   e. reception.

2. Efferent neurons [p. 848]
   a. innervate muscles and glands.
   b. integrate information.
   c. transmit electrical impulses away from interneurons.
   d. transmit electrical impulses from sensory receptors.

3. The nodes of Ranvier [p. 850]
   a. speed up conduction of electrical impulses.
   b. make up a myelin sheath.
   c. expose the axon.
   d. are gaps between adjacent Schwann cells.

4. Which of the following are glial cells? [p. 849–850]
   a. astrocyte
   b. interneuron
   c. Schwann cell
   d. neurosecretory cell
   e. oligodendrocyte

5. Electrical synapses [p. 850]
   a. are plasma membranes of pre- and postsynaptic cells that are in direct contact.
   b. have narrow gaps calls clefts between cells.
   c. are electrical impulses that flow between cells through gap junctions.
   d. involve release of neurotransmitters.

6. The resting potential results from [p. 851–852]
   a. differential permeability of membranes that results in accumulation of proteins and other anions on inside of cell.
   b. an opening of voltage-gated sodium channels.
   c. an imbalance of Na^+ and K^+ inside and outside of the cells created by the Na^+/K^+ pump.
   d. an opening of ligand-gated sodium channels.

7. The ion movement accomplished by the Na^+/K^+ pump. [p. 851]
   a. Na^+ in, K^+ out
   b. K^+ in, Na^+ out
   c. both Na^+ and K^+ in
   d. more K^+ out than Na^+ in
   e. more Na^+ out than K^+ in

8. The refractory period [p. 854]

a. is initiated with the closing of the inactivation gate of the Na^+ voltage-gated channel.

b. is initiated with the opening of the activation gate of the Na^+ voltage-gated channel.

c. ceases with the opening of the activation gate of the K^+ voltage-gated channel.

d. lasts from the peak of the action potential to the reestablishment of resting potential.

9. Nerve impulses [p. 854–855]

a. always travel in both directions along an axon.

b. sometimes travel in both directions along an axon.

c. travel in only one direction along an axon .

d. always travel with the same intensity along the length of an axon.

e. diminish in intensity as they travel along axon.

10. Neurotransmitters [p. 858]

a. travel through gap junctions.

b. are released from presynaptic cell through the process of exocytosis.

c. can bind to ligand-gated ion channels.

d. can bind to G-protein-coupled receptors.

11. Inhibitory postsynaptic potentials (IPSPs) [pp. 862]

a. result from K^+ influx into cell.

b. result from hyperpolarization of postsynaptic cell.

c. are all-or-nothing.

d. are graded.

12. Spatial summation [p. 863]

a. is the change in membrane potential of a postsynaptic cell brought on by the firing of different presynaptic neurons.

b. is the change in membrane potential of a postsynaptic cell brought on by the successive firing of a single presynaptic neuron over a short period of time.

c. is the total of the membrane potentials that occur along an axon as the electrical impulse is transmitted.

INTEGRATING AND APPLYING KEY CONCEPTS

1. The salivary gland of vertebrates are innervated with adrenergic neurons (neurons that release epinephrine from their axon termini). Epinephrine stimulates both fluid and amylase (a carbohydrate hydrolyzing enzyme) secretion form the salivary gland. The fluid secretion is calcium dependent (can be inhibited by calcium channel blockers), while amylase secretion is cAMP dependent. Explain.

38

NERVOUS SYSTEMS

CHAPTER HIGHLIGHTS

- Nervous systems consist of networks of neurons.
- The complexity of nervous systems increases with the complexity of the animal group. The nervous systems of invertebrates are generally simple, whereas those of vertebrates are generally elaborate and display pronounced cephalization.
- In vertebrates, the central nervous system (CNS) consists of the brain and spinal cord. The peripheral nervous system consists of all the neurons and ganglia that connect the brain and spinal cord to the rest of the body.
- The peripheral nervous system of vertebrates has two main components: the somatic nervous system, which connect the CNS to the body wall (including the skeletal musculature), and the autonomic nervous system, which connects the CNS to the internal organs and blood vessels.
- The autonomic nervous system has two antagonistic components: the sympathetic division, which predominates in stressful or strenuous situations, and the parasympathetic division, which predominates during low-stress and feeding situations.
- Memory, learning, and consciousness involve modifications of neuron behavior and connections between and among neurons in different parts of the brain.

STUDY STRATEGIES

- This chapter makes use of the concepts of cells structure and nerve function developed earlier. You should briefly review these concepts to make sure that you understand them.
- While many of the terms in this chapters may be familiar, others probably won't be, so don't be hasty in your study.
- This chapter is divided into sections. DO NOT try to go through them all in one sitting. Take one section at a time, then work through the companion section(s) in the study guide.
- Draw pictures of the various neural circuits, being sure to label of the various parts. You also should outline the various components of the nervous system of vertebrates, carefully noting the hierarchical organization.

INTERACTIVE EXERCISES

Why It Matters [pp. 867–868]

Section Review

A(n) (1) ____ consists of a network of neurons.

38.1 Invertebrate and Vertebrate Nervous Systems Compared [pp. 868–871]

Section Review

Radially symmetric animals such as cniderians (jellyfish) and echinoderms (starfish, sea urchins) possess loose meshes of neurons called (2) ___. In more complex invertebrates, some neurons are clustered together in (3) ___. There is an evolutionary trend toward cephalization in which a distinct head region forms that contains a central ganglion or (4) ____ connected to one or more (5) ____ that extend to the rest of the body. The brain and nerve cord constitute the (6) ____, and the nerves that extend from the CNS to the rest of the body make up the (7) ____. During embryonic development, the nervous system of vertebrates arises from a hollow (8) ____, the anterior portion of which develops into the brain, and the rest gives rise to the (9) ____. The central cavity of the neural tube persists in adults as the (10) ____ of the brain and the (11) ____ of the spinal cord. The (12) ____, (13) ____, and (14) ____ are the three distinct regions of the brain early in development that give rise to the adult brain.

Building Vocabulary

Prefixes	Meaning
hypo-	under, below
post-	behind, after
pre-	before, prior to

Prefix	Suffix	Definition
____	-thalamus	15. part of the brain located below the thalamus and primary integration center for regulation of the viscera
____	-ganglionic	16. relates to a neuron leading away from a ganglion
____	-ganglionic	17. relates to a neuron leading into a ganglion

Choice

For each of the following statements, choose the most appropriate term from list below.

a. ventricles b. central canal

18. ____ spaces within the brain

19. ____ space within the spinal cord

Complete the Table

Structure	Feature/Description
Nerve net	20.
21.	Functional clusters of neurons
Nerve cord	22.

38.2 The Peripheral Nervous System [pp. 871–872]

38.3 The Central Nervous System and Its Functions [pp. 872–879]

Section Review

The (23) ____ control body movements that are primarily voluntary. The (24) ____ controls largely involuntary processes such as digestion, secretion by sweat glands, and circulation of blood. Situations involving adaptation to stress or strenuous activity are mediate by the (25) ____, whereas the (26) ____ predominates during digestion and periods of quiet and relaxation. The (27) ____ surrounds and protects the brain and spinal cord. (28) ____ circulates in the central canal and in the ventricles. The butterfly-shaped core of the spinal cord is (29) ____, which is surrounded by (30) ____. Interneurons in the gray matter are involved in (31) ____. The (32) ____ connects the forebrain with the spinal cord. The surface layer of the forebrain is the (33) ____. Tight junctions between cells of brain capillaries form a(n) (34) ____ and prevent the movement of many substances.

Incoming sensory information is filtered by the (35) ____ before going to other CNS centers. The (36) ____ helps fine tune balance and body movements. The (37) ____ relays sensory information to regions of the cerebral cortex concerned with motor responses. The (38) ____ helps coordinate temperature and osmotic homeostasis. The (39) ____ surrounds the thalamus and moderates voluntary movements directed by motor centers in the cerebrum. The (40) ____ is made up of the thalamus, hypothalamus, and basal nuclei, as well as the (41) ____, (42) ____, and (43) ____. The two cerebral hemispheres are connected via the (44) ____. The (45) ____ in each hemisphere of the of the cerebral cortex registers information on touch, pain, temperature, and pressure. (46) ____ of the cerebral cortex integrate information from the sensory areas, formulate responses, and pass them on to the (47) ____. The localization of some brain functions in one of the two hemispheres is called (48) ____.

Choice

For each of the following descriptions, choose the most appropriate structure from the list below.

a. spinal nerves b. cranial nerves

49. ____ connect with the CNS at the level of the brain

50. ____ connect with the CNS at the level of the spinal cord

For each of the following descriptions, choose the most appropriate characteristic from the list below.

a. efferent b. afferent

51. ____ neuron that conducts information away from the CNS

52. ____ neuron that conducts information away toward the CNS

For each of the following descriptions, choose the most appropriate characteristic from the list below.

a. gray matter b. white matter

53. ____ consists of nerve cell bodies and dendrite

54. ____ consists of axons, many of which are surrounded my myelin sheaths

Complete the Tables

Component	Subcomponent	Function
somatic	-----------------	55.
56.	-----------------	Control internal organs and blood vessels (most involuntary)
-----------------	57.	Predominates in situations involving stress, excitement, etc.
-----------------	58.	Predominates in low-stress situations

Structure	Function
Corpus callosum	59.
60.	Registers information about touch, pain, temperature, and pressure
61.	Causes movement of specific part of the body (each hemisphere controls the opposite side)
	62.

Matching

Match each of the following type of junction with its correct definition.

63. ____ meninges
64. ____ basal nuclei
65. ____ amygdala
66. ____ lateralization
67. ____ cerebral cortex
68. ____ reflex
69. ____ blood brain barrier
70. ____ thalamus
71. ____ hippocampus
72. ____ hypothalamus
73. ____ reticular formation
74. ____ olfactory bulb
75. ____ cerebrospinal fluid
76. ____ brainstem

A. Connective tissue layers that surround and protect brain and spinal cord
B. Fluid that circulates in central canal and the ventricles of the brain
C. Programmed movement that takes place without conscious effort
D. Results from tight junction between capillary cells in the brain
E. Phenomenon in which some brain functions are localized to one hemisphere or the other
F. Connects brain with spinal cord
G. Surface layer of cerebrum
H. Part of the brainstem that connects thalamus to spinal cord
I. Receives sensory information and relays it to high CNS centers
J. Moderates voluntary movements directed by the cerebrum
K. Sends information to frontal lobes
L. Relays information about experience and emotions
M. Relays olfactory information to cerebral cortex
N. Regulates basis homeostatic functions

38.4 Memory, Learning, and Consciousness [pp. 879–883]

Section Review

(77) ____ is the storage of an experience. (78) ____ involves a change in the response to a stimulus based on information or experience stored in memory. (79) ____ is awareness of ones self and his or her surroundings. (80) ____ stores information for up to an hour or so, whereas (81) ______ stores information for days, years, or even for life. (82) ____ is a lasting increase in the strength of synaptic connections in neural pathways. (83) ____ is an increased responsiveness to mild stimuli after experiencing a strong stimulus. Changes in neural activity can be recorded by a(n) (84) ____. During (85) ____, an individual's heart rate and respiration decrease, their limbs twitch, and their eyes move rapidly behind closed eyelids.

Choice

For each of the following descriptions, choose the most appropriate structure from the list below.

a. short-term memory b. long-term memory

86. ____ lasts up to an hour or so
87. ____ lasts for days, years, or even life.

Matching

Match each of the following types of junction with its correct definition.

88.	____	memory	A. Storage/retrieval of a sensory or motor experience
89.	____	sensitization	B. Changes in response to a stimulus
90.	____	long-term potentiation	C. Awareness of self and surroundings
91.	____	rapid eye-movement (REM) sleep	D. Long-lasting increase in strength of response
92.	____	consciousness	E. Increased response to a mild stimulus after experience a strong stimulus
93.	____	learning	F. A phase of sleep in which brainwaves similar to those in the waking state are observed

SELF-TEST

1. The principal integration center of homeostatic regulation and leads to the release of hormones. [p. 870]

a. cerebellum

b. cerebrum

c. thalamus

d. hypothalamus

e. association area

2. The regulation of blood pressure is primarily under the control of ___. [p. 872]

a. somatic nervous system

b. autonomic nervous system

c. parasympathetic division

d. sympathetic division

3. The regulation of feeding and digestion are primarily under the control of ___. [p. 872]

a. somatic nervous system

b. autonomic nervous system

c. parasympathetic division

d. sympathetic division

4. Cerebrospinal fluid [pp. 873]

a. circulates in the central canal.

b. circulates in the ventricles of the brain.

c. is within layer of th meninges.

d. mixes with blood.

e. cushions brain and spinal cord from jarring movements.

5. The part of the mammalian brain that integrates information about posture and muscle tone. [p. 875]

a. cerebrum

b. cerebellum

c. myencephalon

d. medulla oblongata

6. The part of the brain that coordinates muscular activity. [p. 875]

a. cerebrum

b. cerebellum

c. myencephalon

d. medulla oblongata

7. The brain and the spinal cord are wrapped in a connective tissue layer called the ____. [p. 873]

a. gray matter

b. meninges

c. dura matter

d. sclera

8. In non-REM sleep compared to REM sleep there is ____. [p. 882]

a. more delta waves

b. faster breathing

c. lower blood pressure

d. more dream consciousness

9. If only one temporal lobe is damaged, you would expect ____. [pp. 877–878]
   a. blindness in one eye
   b. total blindness
   c. loss of hearing in one ear
   d. partial loss of hearing in both ears

10. The cerebrum typically is ____. [p. 876–878]
   a. divided into two hemispheres
   b. mostly gray matter
   c. mainly cell bodies and some sensory neurons
   d. mostly white matter
   e. mainly axons connecting various other parts of the brain

INTEGRATING AND APPLYING KEY CONCEPTS

1. *Insights from the molecular revolution*: What would you predict to be the effects of genetic defect on CaMKII or in the NMDA receptor?

2. Discuss the selective pressure for cephalization in animals. Would there be a difference between those with sessile or motile life histories?

39

SENSORY SYSTEMS

CHAPTER HIGHLIGHTS

- Sensory receptors convey information about the external and internal environments to the nervous system.
- Mechanoreceptors detect pressure and vibration and provide information about movement and position.
- Photoreceptors detect radiant energy and are involved with perception of visual images.
- Chemoreceptors respond to different kinds of chemicals and are involved with the perception of taste and smell.
- Thermoreceptors detect heat and are involved with temperature regulation. Nociceptors detect pain and protect animals from dangerous stimuli.
- Electroreceptors detect electrical fields and are used for communication and to detect prey. Magnoreceptors detect magnetic fields and are used for navigation.

STUDY STRATEGIES

- This chapter makes use of the concepts of cell structure, cell communication, and nerve function developed earlier. You should briefly review these concepts to make sure that you understand them.
- While many of the terms in this chapters may be familiar, others probably won't be, so don't be hasty in your study.
- This chapter is extremely long. DO NOT try to go through it all in one sitting. Take one section at a time, then work through the companion section(s) in the study guide.
- Draw pictures of the various sensory systems, being sure to label of the various parts and noting their linkage to the nervous system.

INTERACTIVE EXERCISES

Why It Matters [pp. 885–886]

Section Review

Information about the internal and external environment is conveyed by the (1) ____ to the central nervous system.

39.1 Overview of Sensory Receptors and Pathways [pp. 886–888]

39.2 Mechanoreceptors and the Tactile and Spatial Senses [pp. 888–891]

39.3 Mechanoreceptors and Hearing [pp. 891–894]

Section Review

(2) ____ are formed by dendrites of a neuron or are specialized cells that synapse with afferent neurons. The conversion of a stimulus into a change in membrane potential is called (3) ____. The five types of sensory receptors are (4) ____, (5) ____, (6) ____, (7) ____, and (8) ____. The intensity and extent of a stimulus is registered by the (9) ____ and the (10) ____. (11) ____ is the effect of reducing the response to a stimulus at a constant level.

Some mechanoreceptors that detect touch and pressure are free nerve endings, while others, such as (12) ____, have structures that surround the nerve ending to help detect stimuli. (13) ____ detect stimuli that provide information about the position of limbs and is used to maintain balance. (14) ____ are fluid-filled chambers that contain (15) ____ and (16) ____ to detect position in some invertebrates. Fish and some amphibians detect vibrations and water current with a(n) (17) ____. Some fish also have domed-shaped (18) ____ that contain sensory hair cells covered with (19) ____ that extend into a gelatinous (20) ____ and is used to detect orientation and velocity.

The (21) ____ perceives position and motion of the head and consists of three (22) ____ and two chambers, the (23) ____ and the (24) ____, which contain small crystals called (25) ____. The two types of (26) ____ that detect position and movement of limbs are called (27) ____ in muscles and (28) ____ in tendons.

Some invertebrates have auditory organs that consist of a thinned region of exoskeleton or (29) ____ stretched over a hollow chamber. The (30) ____ of the (31) ____ focuses sound waves into the auditory canal where they strike the (32) ____. The (33) ____ is an air-filled cavity containing three interconnected bones, the (34) ____, (35) ____ and (36) ____, the later of which is attached to an elastic (37) ____. The (38) ____ contains several fluid-filled compartments, (39) ____, (40) ____, and (41) ____, as well as a spiraled (42) ____. Within the cochlea is the (43) ____, which contains sensory hair cells that detect sound vibrations that dissipate when they reach the (44) ____. Many vertebrates locate prey and avoid obstacles by (45) ____, a process that involves generating sounds and listening for the echoes that bounce back.

Building Vocabulary

Prefixes	Meaning
chemo-	chemical
oto-	ear
proprio-	one's own
thermo-	heat, warm
photo-	light
noci-	pain, injury

Suffix	Meaning
-lith	stone

Prefix	Suffix	Definition
____	-receptor	46. specialized to detect chemical stimuli
____	-receptor	47. specialized to detect heat/temperature
____	____	48. calcium carbonate crystals in inner ear of vertebrates
____	-ceptor	49. sensory receptors in muscles, tendons, and joints
____	-receptor	50. specialized for the detection of radiant energy at particular wavelengths
____	-ceptor	51. specialized for the detection of tissue damage/noxious chemicals

Choice

For each of the following statements, choose the most appropriate term from list below.

a. sensory transduction b. sensory adaptation

52. ____ The conversion of a stimulus into a change in membrane potential of a sensory cell
53. ____ The reduction in the response of a sensory cell to a stimulus of constant intensity

Complete the table

Structure	Description
54.	Detects vibrations and currents in water
statocysts	55.
vestibular apparatus	56.
57.	detects stretch of muscle
58.	contains sensory hairs that detect sound waves
Golgi tendon organ	59.
60.	provides information about orientation and velocity of fish

Matching

Match each of the following structure with its correct definition.

61. ____ proprioceptors
62. ____ echolocation
63. ____ tympanum
64. ____ cupula
65. ____ otoliths
66. ____ stereocilia
67. ____ stretch receptor
68. ____ statoliths
69. ____ sensory hair cell

A. Type of mechanoreceptor that detects position and movement of body parts
B. Type of proprioceptor that detects position and movement of limbs
C. Calcium carbonate crystal inside vestibular apparatus
D. Sensory cell with long hairlike projection of plasma membrane
E. Gelatinus matrix inside neuromasts
F. Stonelike body inside statocysts
G. Microvilli on hair cells of neuromasts
H. Process of generating sound waves and detecting echos
I. Thinned region of exoskeleton on some invertebrates specialized for detecting vibrations

Labeling

Identify each numbered part of the following illustration.

70. ______________

71. ______________

72. ______________

73. ______________

74. ______________

75. ______________

76. ______________

77. ______________

78. ______________

79. ______________

80. ______________

81. ______________

82. ______________

83. ______________

39.4 Photoreceptors and Vision [pp. 894–899]

39.5 Chemoreceptors [pp. 899–902]

39.6 Thermoreceptors and Nociceptors[pp. 902–904]

39.7 Electroreceptors and Magnoreceptors [pp. 904–906]

Section Review

The simplest eye in invertebrates, a(n) (84) ____, detects light but does not form an image. One type of image-forming eye in invertebrates is the (85) ____, which have hundreds to thousands of faceted visual units called (86) ____, which in insects are focused by a(n) (87) ____ onto a bundle of photoreceptive cells containing the (88) ____, rhodopsin. The other type of image-forming eye in invertebrates is the (89) ____, which resembles the eye of vertebrates in its camera-like operation. Light is concentrated by the (90) ____ onto the (91) ____, a layer of photoreceptors at the back of the eye. Muscle of the (92) ____ adjusts the size of the (93) ____ to regulate the amount of light entering the eye. (94) ____ is a process by which the lens changes to enable the eye to focus on objects at different distances. The structures of the vertebrate eye are similar to those of the invertebrate single-lens eye. (95) ____ fills the space between the cornea and the lens, and the jellylike (96) ____ fills the main chamber of the eye between the lens and the retina. The lens of many vertebrates is focused by changing its shape by contraction of muscles of the (97) ____ an adjusting the tension of the ligaments that anchor the lens to the muscles. The two types photoreceptors in the vertebrate eye are (98) ____, which are specialized for detection of light at low intensities, and (99) ____, which are specialized for detection of different wavelengths (colors). In mammals and birds, cones are concentrated in and around the (100) ____. Images focused there can be seen distinctly, while the surround image is termed (101) ____. (102) ____ are comprised of the light-absorbing molecule, (103) ____, covalently bound to proteins called (104) ____. (105) ____ is the photopigment in rods. Photoreceptors are linked to a network of neurons that integrate and process initial visual information. (106) ____ synapse with rods and cones on one end and with (107) ____ on their other end. In addition, (108) ____ connect photoreceptor cells and bipolar cells. (109) ____ connect bipolar with ganglion cells. In (110) ____, horizontal cells inhibit bipolar cells that are outside a spot of light striking the retina; this visual processing sharpens the edges of the image and enhances contrast. Many animals have color vision, which depends on the number and types of cones. Humans and other primates have three types of cones, based on the form of (111) ____ that they possess. Just behind the eye, optic nerves converge and a portion of each optic nerve crosses over to the opposite side of the brain, forming the (112) ____. Axons enter the (113) ____ of the thalamus, where they synapse with neurons leading to the visual cortex.

Insects have taste receptors inside hollow sensory bristles called (114) ____. (115) ____ of vertebrates have distinct receptors that respond to sweet, sour, salty, bitter, and umami (savory). Olfactory receptor cells possess (116) ____; the density of these receptors determines olfactory sensitivity.

The (117) ____-gated calcium channel family acts as heat receptors; different channels have different temperature thresholds. Pain receptors do not exhibit adaptation.

Electroreceptors are specialized for detecting electric fields. (118) ____ detect the earth's magnetic field and help provide directional information important for navigation.

Distinguish between Members of the Following Sets of Terms

119. Compound eye and single-lens eye ____________________

____

120. Opsins, rhodopsins, and photopsins ____________________

____

Describe

121. What is accommodation? ____________________

____

Complete the Tables

Cell type	Function
122.	neurons that synapse with rods/cones on one end and ganglion cells on the other end
ganglion cell	123.
124.	connect with bipolar cells and ganglion cells
horizontal cell	125.

Type of receptor	Stimuli	Function
126.	127.	visual image formation
taste bud	chemicals	128.
129.	130.	perception of smell
131.	Electric field	132.
magnoreceptor	133.	directional movement and navigation
134.	135.	perception of pain

Matching

Match each of the following type of junction with its correct definition.

136. ____	ocellus	A. A simple photoreceptor that does not form a visual image
137. ____	photopigment	B. The individual visual unit of an invertebrate compound eye
138. ____	optic chiasm	C. Specialized for detecting light of low intensity
139. ____	rod cell	D. Specialized for detecting light of different wavelengths
140. ____	cone cells	E. Type of visual processing that sharpens edges and enhances contrast
141. ____	ommatidia	F. Region where portions of each optic nerve cross over
142. ____	lateral inhibition	G. Region in the thalamus where optic nerve axons terminate
143. ____	lateral geniculate nuclei	H. Hollow sensory bristle that contains taste receptors in insects
144. ____	sensilla	I. Light-absorbing complex of retinal and protein

Labeling

Identify each numbered part of the following illustration.

145. ______________

146. ______________

147. ______________

148. ______________

149. ______________

150. ______________

151. ______________

152. ______________

153. ______________

154. ______________

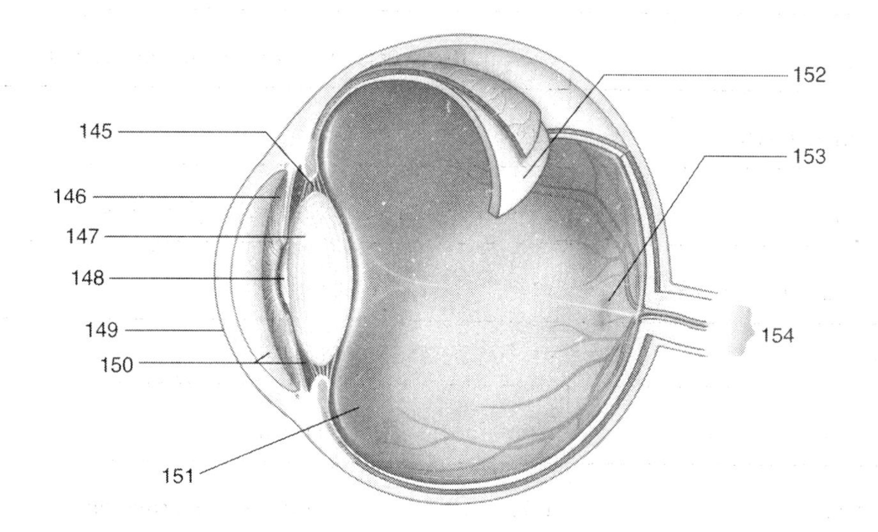

SELF-TEST

1. Sensory transduction involves ____. [pp. 887]
   a. one sensory stimulus being converted to another
   b. an increase in the amplitude of an action potential
   c. stimulus causing a change in membrane potential of sensory cell
   d. a reduced response of a sensory cell in the face of constant intensity of stimulus

2. Sensory structures that help provide information about the position/orientation of a body. [pp. 888–892]
   a. statocysts
   b. tympanum
   c. vestibular apparatus
   d. organ of Corti
   e. neuromasts

3. Variations in the quality of sound are recognized by ____. [p. 894]

a. number of hair cells stimulated

b. pattern of hair cells stimulated

c. amplitude of action potential

d. frequency of action potential

4. Movement of limbs is detected by ____. [p. 901]

a. muscle spindles

b. Golgi tendon organ

c. joint receptors

d. barrow receptors

e. carotid bodies

5. The membrane in contact with the stapes that transmits sound waves to the inner ear. [p. 893]

a. oval window

b. round window

c. tympanic membrane

d. basilar membrane

6. Visual images can be focused by ____. [p. 897]

a. lateral inhibition

b. moving the lens back and forth relative to the retina

c. altering the number of ommatidia

d. changing the shape of the lens

7. Responsible for differences in absorption characteristics that underlie color vision. [p. 899]

a. rod cells

b. retinal

c. carotine

d. opsin

e. cone cells

8. Photoreceptive cells that are specialized for detection of light of low intensity. [p. 897]

a. rod cells

b. ganglion cells

c. horizontal cells

d. cone cells

e. bipolar cells

9. Chemicals that act as "natural painkillers" [p. 905]

a. capsaicin

b. substance P

c. insulin

d. endorphins

10. Used by animals for the location and capture of food. [pp. 895; 904–905]

a. echolocation

b. electroreceptors

c. pit organs

d. cochlea

e. Pacinian corpuscle

INTEGRATING AND APPLYING KEY CONCEPTS

1. Despite the independence of sensory quality and information at the receptor level, animals perceive a unified representation of their environment within which information from the entire complement of sensory channels is seamlessly integrated. Why is such integration important? How is the integration accomplished?

2. *Insights from the Molecular Revolution*: Explain the body's responses to eating spicy food in light of capsaicin. Discuss the adaptive significance of temperature-sensitive calcium channels.

40

THE ENDOCRINE SYSTEM

CHAPTER HIGHLIGHTS

- The endocrine system produces chemical signals, or hormones, that work with the nervous system to regulate the various processes of animals, including growth, development, reproduction, and metabolism.
- Hormones operate at various levels, including through the blood and locally on neighboring cells, or even on the cell that secreted it.
- Although hormones are chemically unique so that they can be distinguished from other chemicals, they can be classified into four families: amine, peptide, steroid, and fatty acid derivative.
- Many organs in vertebrates produce hormones and serve as endocrine glands, including the brain, pituitary, thyroid, pancreas, adrenal, gonads, kidney, liver, and the gastrointestinal tract.

STUDY STRATEGIES

- This chapter makes use of the concepts of biomolecules and cell communication developed earlier. You should briefly review these concepts to make sure that you understand them.
- This chapter is extremely long and has many terms. DO NOT try to go through the chapter all in one sitting. Take one section at a time, then work through the companion section(s) in the study guide.
- Draw diagrams of the various endocrine pathways (source organ, target organ, feedbacks), being sure to label the various parts and noting the chemical nature of the hormone as well as its actions on the target organ. Some of these pathways are complex and involve intermediate or multiple targets as well as multiple hormonal players.

INTERACTIVE EXERCISES

Why It Matters [pp. 909–910]

Section Review

(1) ____ are chemicals released by one cell that affect the activity of another cell. The network of cells that produces hormones is the (2) ____.

40.1 Hormones and their Secretions [pp. 910–912]

40.2 Mechanisms of Hormone Action [pp. 912–919]

Section Review

(3) ____ are ductless glands that secrete hormones. (4) ____ are specialized neurons that also secrete hormones. Hormones can be classified into four chemical types: (5) ____, (6) ____, (7) ____, and (8) ____. (9) ____ are examples of peptide hormones and (10) ____ are examples of a fatty acid derivative.

Building Vocabulary

Prefixes	Meaning
neuro-	of or relating to a nerve
endo-	within, to the inside
hyper-	over
hypo-	under

Suffix	Meaning
-crine	secretion

Prefix	Suffix	Definition
_____	-hormone	11. a hormone secreted by a neuorsecretory cell
_____	-secretion	12. a diminished/under secretion
____	-secretion	13. an excessive/over secretion
_____	-glycemia	14. an abnormally high level of glucose in the blood
____	____	15. a mode of secretion from cells in a ductless gland

Complete the Tables

Mode of secretion	Description
16.	The release of a hormone from a neuron
Endocrine	17.
Paracrine	18.
19.	The release of a chemical into the extracellular fluid that regulates the activity of the cell that secreted it

Chemical type	Hormonal example
amine	20.
peptide	21.
steroid	22.
23.	prostaglandins

Matching

Match each of the following structures with its correct definition.

24. ____ growth factor
25. ____ neurosecretory neurons
26. ____ endocrine system
27. ____ neurosecretory neuron

A. A chemical signal released from one cell that affect another cell usually at a distance from its site of secretion
B. Network of cells/glands that secrete hormones
C. A group of peptide hormones that regulate growth and differentiation
D. A neuron specialized to release hormones

40.3 The Hypothalamus and Pituitary [pp. 919–922]

Section Review

The (28) ____ consists of two lobes: the (29) ____, which contains axons and nerve ending that originate in the hypothalamus, and the (30) ____. Some of the axons that terminate in the posterior pituitary release (31) ____ that travel through the portal veins and affect the secretions of the anterior pituitary, while others release nontropic hormones that enter the general blood circulation. The two types of tropic hormones are (32) ____ and (33) ____. The two nontropic hormones that enter the general circulation are (34) ____, which helps regulate water balance, and (35) ____, which stimulates milk ejection from the mammary glands of mammals. The anterior pituitary produces and secretes (36) ____, (37) ____, (38) ____, (39) ____, (40) ____, and (41) ____. FSH and LH are referred to as (42) ____ because of their action on gonads. In some species the anterior pituitary has a distinct intermediate lobe that produces and releases (43) ____ and (44) ____; in those species without an intermediate lobe, these hormones are produced by cells dispersed in the other regions of the anterior pituitary.

Distinguish between members of the following pair of terms

45. Anterior pituitary and posterior pituitary ____

____

Complete the Table

Hormone	Chemical type	Site of secretion	Function
Growth hormone	peptide	46.	47.
48.	49.	50.	Influence reproductive activity, stimulates milk synthesis
51.	52.	Anterior pit	Stimulates thyroid gland to produce T4
ACTH	peptide	Anterior pit	53.
FSH	54.	55.	56.
LH	57.	58.	59.
60.	61.	62.	Controls pigmentation/coloration

endorphin	63.	64.	65.
ADH	66.	67.	68.
69.	70.	71.	Stimulates smooth muscle contraction (includes milk "letdown" or secretion)

Choice

For each of the following descriptions , choose the most appropriate structure from the list below.

a. releasing hormone b. inhibiting hormone

72. ____ tropic hormone that stimulates release of hormones from anterior pituitary

73. ____ tropic hormone that inhibits the release of hormones from the anterior pituitary

Labeling

Identify each numbered part of the following illustration.

74. ______________

75. ______________

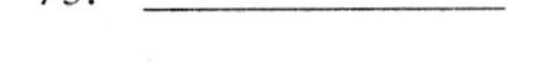

76. ______________

77. ______________

78. ______________

79. ______________

80. ______________

81. ______________

82. ______________

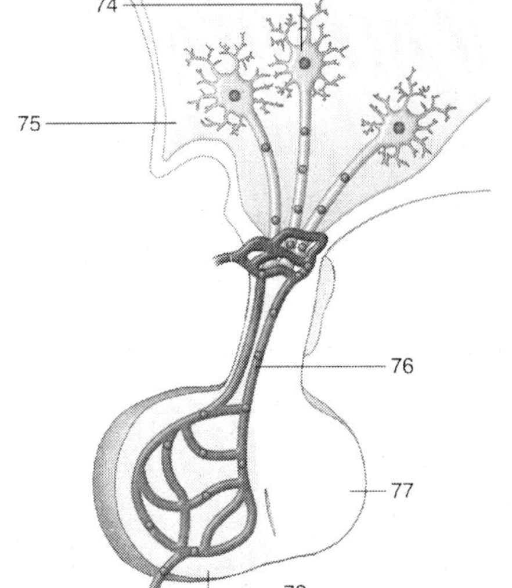

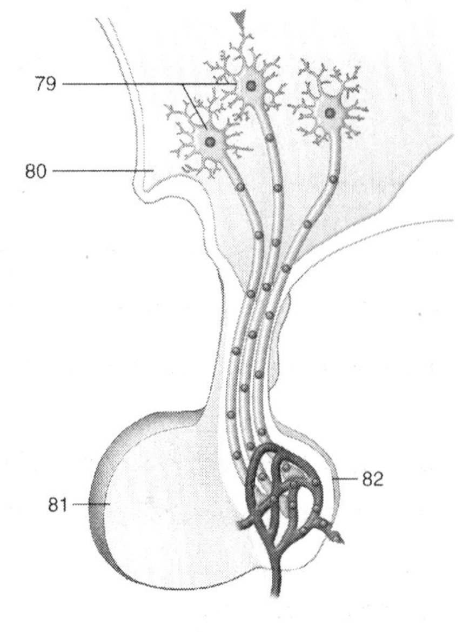

Matching

Match each of the following type of junction with its correct definition.

83. ____ tropic hormone A. A hormone that stimulates the release of a second hormone

84. ____ gonadotropin B. Modulates the growth-promoting actions of growth hormone

85. ____ insulin-like growth factor C. A class of hormone that stimulates the gonads to produce other hormones

40.4 Other Major Endocrine Glands of Vertebrates [pp. 922–928]

Section Review

The (86) ____ produces (87) ____, which is converted to its biologically active form, (88) ____, prior to binding to intracellular receptors. In amphibians, T4 triggers (89) ____. In mammals, the thyroid gland also produces (90) ____, which helps lower blood calcium levels. (91) ____ is produced by the (92) ____ and increases blood calcium. PTH also promotes the conversion of (93) ____ to its active form; the latter of which increases calcium absorption by the gut and synergizes with PTH to release calcium from bone. The adrenal gland has two parts: a centrally located (94) ____, and a peripherally located (95) ____. The medulla releases two (96) ____, hormones derived from the amino acid tyrosine: (97) ____ and (98) ____. The cortex produces two classes of steroid hormones: (99) ____, which regulate carbohydrate metabolism, and (100) ____, which help regulate ion and water balance. (101) ____ is a specific glucocorticoid in mammals and (102) ____ is a specific mineralocorticoid in mammals. The gonads, (103) ____ and (104) ____, are a primary sources of three classes of sex steroids: (105) ____, (106) ____, and (107) ____. (108) ____ is a specific androgen, whereas (109) ____ is a specific estrogen, and (110) ____ is a specific progestin. The (111) ____ of the (112) ____ produces several hormones that regulate metabolism. (113) ____ is generally anabolic and stimulates the uptake of nutrients into cells (e.g., glucose) as well as the synthesis of large macromolecules (e.g., glycogen). (114) ____ is generally catabolic and stimulates the breakdown of large macromolecules in tissues and the release of monomers into the blood. (115) ____ can result from several defects, including the inability to produce insulin or the failure of insulin to function properly in targets cells; conditions that generally all lead to abnormally high glucose levels in the blood. The (116) ____ produces (117) ______, which helps maintain daily biorhythms.

Matching

Match each of the following types of junction with its correct definition.

118. ____ metamorphosis
119. ____ catecholamines
120. ____ pancreas
121. ____ gonad
122. ____ estrogen
123. ____ progestin
124. ____ diabetes mellitus
125. ____ androgen
126. ____ vitamin D
127. ____ anabolic steroid

A. change in body form (e.g., tadpole to adult)
B. steroidlike molecule that when activated stimulates Ca^{++} absorption
C. class of amines derived from tyrosine
D. steroid hormones that stimulate muscle development
E. typically has both exocrine and endocrine (islets of Langerhans) components
F. a disease that typically displays abnormally high glucose levels in the blood
G. organ that produces gametes and sex steroids
H. class of sex steroid primarily produced in males
I. an example is estradiol
J. an example is progesterone

Distinguish between Members of the Following Pairs of Terms

128. Glucocorticoid and mineralocorticoid ____

____

129. Thyroxine (T4) and Triiodothronine (T3) ____

____

Complete the Table

Hormone	Chemical type	Site of secretion	Function
thyroxin	130.	Thyroid gland	131.
132.	peptide	133.	Lowers blood calcium
parathyroid hormone	peptide	Parathyroid gland	134.
135.	amine	136.	Increase blood flow to muscles; stimulates breakdown of macromolecules
137.	138.	Adrenal cortex	Promotes fat and protein breakdown; stimulates gluconeogenesis
aldosterone	139.	140.	141.
142.	steroid	testes	Stimulates sperm maturation/maintenance of 2nd sex characteristics
143.	144.	ovary	145.
progesterone	steroid	ovary	146.
147.	peptide	148.	Stimulates secretion of FSH and LH
insulin	149.	150.	151.
152.	153.	154.	Stimulates glycogen breakdown
155.	156.	Pineal gland	157.

40.5 Endocrine Systems in Incertebrates [pp. 928–930]

Section Review

The three hormones that regulate molting and metamorphosis in insects are (158) ____, (159) ____, and (160) ____.

In crustaceans, (161) ____ helps regulate molting by inhibiting ecdysone secretion.

Matching

Match each of the following type of junction with its correct definition.

162. ____ brain hormone
163. ____ molt-inhibiting hormone
164. ____ ecdysone
165. ____ juvenile hormone

A. peptide hormone that stimulates prothoracic glands
B. steroid hormone secreted by prothoracic glands
C. peptide hormone secreted by corpora allata
D. peptide hormone secreted by gland in eye stalk

SELF-TEST

1. A chemical released from an epithelial cell in a gland that enters the blood to affect the activity of another cell some distance away. [p. 910]
   a. hormone
   b. neurohormone
   c. pheromone
   d. exocrine gland secretion

2. Control of hormone secretion by a negative feedback mechanism generally includes [p. 912]
   a. no change in hormone secretion.
   b. decrease in hormone secretion.
   c. change that maintains homeostasis.
   d. increase in hormone secretion.

3. Small hydrophobic hormones that hormone-receptor complexes in the cytoplasm of target cells. [pp. 913-914]
   a. steroids
   b. cAMP
   c. thyroid hormone
   d. prolactin
   e. often result in altered gene express

4. The hormone and target involved in elevating blood glucose by glycogenolysis and gluconeogenesis. [p. 927]
   a. insulin and pancreas
   b. glucagon and liver
   c. TSH and thyroid
   d. ACTH and adrenal cortex
   e. T4 and skeletal muscle

5. A person that exhibits nervous behavior, irratability, rapid/irregular heartbeat, and inflamed/protruding eyes may suffer from [p. 923]
   a. low metabolic rate.
   b. Cushing's disease.
   c. Graves's disease.
   d. low plasma T4 levels.
   e. high plasma T4 levels.

6. Increased skeletal growth results directly and/or indirectly from the activity of [p. 921]
   a. aldosterone.
   b. growth hormone.
   c. hormones produced in the hypothalamus.
   d. progestins.
   e. epinephrine.

7. The activity of the anterior pituitary is controlled by [p. 919]

a. the hypothalamus.

b. ADH.

c. epinephrine.

d. releasing hormones.

e. inhibiting hormones.

8. Normal individuals have a fasting glucose level of 100 mg/100 ml blood. A person with a fasting level of 300 mg/100 ml blood ____. [p. 927]

a. is hypoglycemic

b. is hyperglycemic

c. has too much insulin

d. probably has cells not using enough glucose

9. Diabetes mellitus is an adult with ample numbers of functioning beta cells probably ____. [p. 927]

a. suffers from type I DM

b. suffers from type II DM

c. indicates that not enough insulin is being secreted

d. indicates that insulin is not active on target cells

e. is unusual since this condition normally occurs in children

10. Removing the eye stalk of a lobster would ____. [p. 929]

a. increase secretion of ecdysone

b. decrease secretion of ecdysone

c. accelerate molt cycle

d. decelerate molt cycle

e. have no effect on ecdysone or molt cycle

INTEGRATING AND APPLYING KEY CONCEPTS

1. Discuss the validity of the following statement: The endocrine system and nervous system are separate and distinct systems that serve to coordinate the function of animals.

2. The integration of genetic and various environmental factors (e.g., nutritional status) leads to the coordination of animal growth via the interplay of numerous hormones. A central component of growth coordination is the growth hormone–insulin-like growth factors "axis." Describe the various components of this axis and how regulation at each level would affect growth.

41

MUSCLES, BONES, AND BODY MOVEMENTS

CHAPTER HIGHLIGHTS

- Skeletal muscles move the joints of the body.
- Skeletal muscle is organized in a hierarchical manner; the functional unit is the sarcomere, which shortens as a result of the relative movement of myosin and actin.
- There are three kinds of skeletons in animals: a hydrostatic skeleton, an exoskeleton, and an endoskeleton.
- The bones of an endoskeleton are connected by three types of joints: a synovial joint, a cartilaginous joint, and a fibrous joint.
- Most bones are moved by antagonistic pairs of muscles.

STUDY STRATEGIES

- This chapter makes use of the concepts of cell structure and nerve function developed earlier. You should briefly review these concepts to make sure that you understand them.
- While many of the terms in this chapters may be familiar, others probably won't be, so don't be hasty in your study.
- This chapter is divided into sections. DO NOT try to go through them all in one sitting. Take one section at a time, then work through the companion section(s) in the study guide.
- Draw pictures of the functional unit of muscle (sarcomere), being sure to label the various parts and note their role in contraction.

INTERACTIVE EXERCISES

Why It Matters [pp. 933–934]

Section Review

There are three types of muscle tissue: (1) ____, (2) ____, and (3) ____. Most (4) ____ are attached by tendons to the skeleton of vertebrates

44.1 Vertebrate Skeletal Muscle: Structure and Function [pp. 934–941]

Section Review

Skeletal muscle consists of elongated, cylindrical cells called (5) ____. Inside the cell, there are contractile elements or (6) ____ that consist of regular arrangements of (7) ______, containing the protein myosin, and (8) ____, containing the protein actin. The functional unit of the myofibril is the (9) ____, which extends Z-line to Z-line. The plasma membrane of the muscle cell has deep indentations called (10) ______, which are in close association with the cell's complex network of endoplasmic reticulum called (11) ____. A neuron comes in contact with a muscle cell to form a(n) (12) ____. The release of (13) ____ from the neuron causes the muscle cell to depolarize and release calcium from the lumen of the sarcoplasmic reticulum and raise cytosolic calcium concentration. Calcium enables the interaction between myosin and actin so that the filaments move relative to one another in a process called the (14) ____; this movement shortens the sarcomere and results in muscle contraction. A single action potential arriving at the neuromuscular junction usually causes a single weak contraction of the muscle cell called a(n) (15) ____. (16) ____ results when muscle fibers can't relax between rapidly arriving stimuli. (17) ____ contract relatively slowly and the intensity of the contraction is low, whereas (18) ____ contract relatively quickly and powerfully. Controlled contraction of the overall muscle results from organized activation of (19) ____.

Building Vocabulary

Prefixes	Meaning
sarco-	of or relating to flesh/muscle
myo-	of or relating to muscle

Suffix	Meaning
-mere	unit, part

Prefix	Suffix	Definition
____	____	20. The functional unit of a contractile element in a muscle cell
____	-fibril	21. The contractile element inside of a muscle cell
____	-globin	22. Oxygen-storing molecule in some muscle cells

Distinguish between Members of the Following Pair of Terms

23. Slow muscle fibers and fast muscle fibers ____

____

Choice

For each of the following statements, choose the most appropriate type of filament from the list below.

a. thick filament b. thin filament

24. ____ Parallel bundles of myosin molecules

25. ____ Mostly composed of two linear chains of actin arranged in a double helix

Matching

Match each of the following structures with its correct definition.

26. ____ skeletal muscle

27. ____ myoglobin

28. ____ neuromuscular junction

29. ____ myofibrils

____

30. ____ muscle fibers

31. ____ muscle twitch

A. Connect to bones of the skeleton by tendons

B. Muscle cell

C. Cylindrical contractile elements in a muscle cell

D. The weak contraction of a muscle cell in response to a single action potential

E. Process in which there is relative movement of actin and myosin

F. Collection of muscle fibers controlled by branch of same neuron

32 ____ tetanus

33. ____ motor units

34. ____ sliding filament mechanism

G. Specialized junction between a neuron and a muscle cell

H. Oxygen-storing protein in some muscle

I. Continuous contraction of a muscle fiber

Labeling

Identify each numbered part of the following illustration.

35. ____________

36. ____________

37. ____________

38. ____________

39. ____________

41.2 Skeletal Systems [pp. 941–943]

41.3 Vertebrate Movement: The Interactions between Muscles and Bones [pp. 943–946]

Section Review

A(n) (40) ____ consists of muscles and fluid, with the fluid either within the muscle or in compartments. A(n) (41) ____ is a rigid external covering that provides support, whereas (42) ____ is made up of internal body structures that provide support such as bone. The skeleton of vertebrates is organized into the (43) ____ and the (44) ____.

The bones of vertebrates are connected by three types of joints: (45) ____, (46) ____, and (47) ____. A muscle that causes movement at a joint is a(n) (48) ____. Most bones are moved by muscles in (49) ____: (50) ____, which extend the joint, and (51) ____, which retract the joint.

Choice

For each of the following descriptions, choose the most appropriate type of skeleton from the list below.

a. hydrostatic skeleton b. exoskeleton c. endoskeleton

52. ____ rigid external body covering

53. ____ supportive internal body structures

54. ____ consists of muscles and fluid

For each of the following descriptions, choose the most appropriate type of skeleton from the list below.

a. axial skeleton b. appendicular skeleton

55. ____ contains skull and vertebral column

56. ____ contains shoulders, hips, and limbs

For each of the following descriptions, choose the most appropriate type of muscle from the list below.

a. extensor muscle

b. flexor muscle

57. ____ retracts a joint

58. ____ extends a joint

Matching

Match each of the following types of junction with its correct definition.

59. ____ agonist

60. ____ cartilaginous joint

61. ____ fibrous joint

62. ____ ligaments

63. ____ antagonistic pair

64. ____ synovial joint

A. A muscle that causes movement in a joint when it contracts

B. Arrangement of muscle groups in which one muscle has the opposite effect of the other

C. Moveable joint enclosed by a fluid-filled capsule

D. Moveable joint without a fluid-filled capsule

E. Joint connected by stiff connective tissue

F. Connective tissue that joins bones on either side of a joint

---

SELF-TEST

1. The sliding filament mechanism states that ____. [p. 936]

a. nebulin and titin filaments move relative to one another

b. actin and myosin filaments become arranged perpendicular to one another

c. actin and myosin filaments move relative to one another

d. The two actin chains unwind and dissociate into G-actin

2. Myofilaments are composed of ____. [pp. 935–936]

a. fibers

b. myofibrils

c. actin

d. myosin

e. sarcoplasmic reticulum

3. The human skull is part of the ____. [p. 944]

a. girdle

b. axial skeleton

c. atlas

d. appendicular skeleton

e. hydrostatic skeleton

4. The human axial skeleton includes: [p. 944]
   a. ulna.
   b. shoulder blades.
   c. centrum.
   d. femur.
   e. sternum.

5. Vertebrate appendages are connected to ____. [p. 944]
   a. the cervical complex
   b. the atlas
   c. the appendicular skeleton
   d. the axial skeleton
   e. girdles

6. Animal group in which the primary means of support is a hydrostatic skeleton. [pp. 942–943]
   a. annelids
   b. cnidarians
   c. flatworms
   d. echinoderms
   e. lobsters and cratfish

7. Internal skeletons are found in ____. [p. 944]
   a. annelids
   b. cnidarians
   c. echinoderms
   d. lobsters and crayfish
   e. vertebrates

8. The primary neurotransmitter released at neuromuscular junctions in vertebrates. [p. 936]
   a. substance P
   b. inositol triphosphate
   c. epinephrine
   d. acetylcholine
   e. endorphin

9. Facilitates the diffusion of oxygen into tissues from blood and stores oxygen in tissues. [p. 940]
   a. myoglobin
   b. hemoglobin
   c. opsin
   d. bilirubin
   e. dystrophin

10. Connective tissue that joins bones together on either side of a joint. [p. 945]
   a. periosteum
   b. stratum corneum
   c. tendons
   d. ligaments
   e. meninges

INTEGRATING AND APPLYING KEY CONCEPTS

1. *Insights from the molecular revolution*: Describe the factors that regulate gene expression and identify the possible targets of action for drugs aimed at increasing utropin content in muscle cells of DMD patients.

2. Coordinated muscle contraction requires neuronal stimulation of motor units. Discuss the roles of calcium in neuromuscular function and the potential consequences if calcium deficiency. (Be sure to consider both the muscle cell and the nerve cell.)

42

THE CIRCULATORY SYSTEM

CHAPTER HIGHLIGHTS

- Circulatory systems distribute nutrients and gases to tissues and collects waste products.
- The heart serves as a pump to distribute fluid through the circulatory system.
- Some circulatory systems are open, in which fluid bathes tissues directly. Other systems are closed, in which blood is distributed through a system of vessels and is kept separate from interstitial fluid.
- The blood of vertebrates is a connective tissue that consists of numerous cell types (e.g., red blood cells, white blood cells) and a fluid matrix (plasma).
- The lymphatic system collects excess interstitial fluid and participates in the immune response of vertebrates (covered in more detail in the next chapter).

STUDY STRATEGIES

- This chapter makes use of the concept of diffusion developed earlier. You should briefly review this concept to make sure that you understand it.
- While many of the terms in this chapters may be familiar, others probably won't be, so don't be hasty in your study.
- This chapter is divided into sections. DO NOT try to go through them all in one sitting. Take one section at a time, then work through the companion section(s) in the study guide.
- Draw pictures of the various circulatory systems, being careful to detail the flow of blood, especially the path through the heart.

INTERACTIVE EXERCISES

Why It Matters [pp. 949–950]

Section Review

A(n) (1) ____ consists of fluid, a heart, and vessels for conducting molecules (e.g., nutrient, wastes, gases) through the organism.

42.1 Animal Circulatory Systems: An Introduction [pp. 950–953]

42.2 Blood and Its Components [pp. 953–956]

Section Review

In a(n) (2) ____, vessels leaving the heart release (3) ____ directly into body spaces called (4) ____. In a(n) (5) ____, the blood is confined to a network of vessels and is separated from the interstitial fluid. In closed circulatory systems, (6) ____ conduct blood away from the heart and break into highly branched (7) ____ that are specialized for exchange of material between the blood and interstitial fluid. Blood then flows into (8) ____, which return it to the heart. The heart of vertebrates, depending on species, may consist of one or two (9) ____, which receives blood returning to the heart, as well as one or two (10) _____, which pumps blood from the heart. In amphibians, which have a single ventricle, most of the oxygenated blood enters the (11) _____, while deoxygenated blood is directed into the (12) _____. In reptiles, which also have a single ventricle, oxygenated blood enters the systemic circuit, while deoxygenated blood is directed into a(n) (13) _____. Mammals and birds have two atria and two ventricles so that there are two separate circuits: a pulmonary circuit and a systemic circuit.

Blood is a connective tissue that consists of several cells types and a liquid matrix called (14) _____. There are three classes of protein in the liquid matrix: (15) _____, (16) _____, and (17) _____. (18) _____, commonly called (19) _____, are the cells that carry O_2 to tissues. (20) _____ is a hormone produced by the kidney and stimulates RBC production. White cells, or (21) _____, play a role in the body's defense against invading organisms. (22) _____ are cell fragments that contain factors important for blood clotting, including enzymes that convert fibrinogen into (23) _____.

Building Vocabulary

Prefixes	Meaning
erythro-	red
leuko-	white (without color)
hemo-	blood

Suffix	Meaning
-cyte	a cell

Prefix	Suffix	Definition
_____	_____	24. red blood cell
_____	_____	25. white blood cell
_____	-globin	26. protein in red blood cells specialized for binding O_2

Choice

For each of the following statements, choose the most appropriate circuit of blood from the list below.

a. systemic circuit b. pulmonocutaneous circuit c. pulmonary circuit

27. _____ The circuit that distributes oxygenated blood to the body
28. _____ The circuit in amphibians that distributes deoxygenated blood to the lungs and skin
29. _____ The circuit in reptiles, birds, and mammals that distributes deoxygenated blood to the lungs

For each of the following statements, choose the most appropriate type of circulatory system from the list below.

a. open circulatory system

b. closed circulatory system

30. ____ Fluid exits heart and bathes tissues directory

31. ____ Fluid is confined to a system of vessels

For each of the following statements, choose the most appropriate chamber(s) of the heart from the list below.

a. atrium/atria

b. ventricle(s)

32. ____ receives blood returning to the heart

33. ____ pumps blood from the heart

Complete the Table

Cell/Component	Function
34.	Eliminates dead/dying cells; removes cellular debris
erythrocytes	35.
36.	Cell fragments produced in bone marrow; play a role in blood clotting

Distinguish between the Following Pair of Terms

37. Fibrinogen and fibrin ______________________________

Matching

Match each of the following structures with its correct definition.

38. ____ veins
39. ____ hemolymph
40. ____ sinuses
41. ____ ____ erythropoietin
42. ____ albumins
43. ____ arteries
44. ____ globulins
45. ____ plasma
46. ____ capillaries

A. Fluid distributed to tissues in an open circulatory system
B. Body spaces that surround organs
C. Vessels that conduct blood away from the heart
D. Vessels specialized for exchange of material between blood and interstitial fluid
E. Vessels that return blood to heart
F. Fluid component of blood that contains nutrients, ions, gases, etc.
G. Proteins important for osmotic balance, pH buffering, and transport
H. Proteins important in transport and as antibodies
I. Hormone that stimulates red blood cell production

42.3 The Heart [pp. 956–961]

42.4 Blood Vessels of the Circulatory System [pp. 961–965]

42.5 Maintaining Blood Flow and Pressure [pp. 965–966]

42.6 The Lymphatic System [pp. 966–967]

Section Review

Oxygenated blood leaves the heart and enters the (47) ____. The cardiac cycle consists of a period of heart muscle contraction or (48) ____ and a period of relaxation or (49) ____. As the atria contract, blood flows through the (50) ____ and into the ventricle. When the ventricle contract, the AV valves close and blood flows through the semilunar valves in the pulmonary and systemic circuits. Some animals such as crabs and lobsters have a(n) (51) ____ in which the heart beat is under the control of the nervous system. Other animals, including all vertebrates, have (52) ____ in which the heart has its own endogenous rhythm that does not require outside signals. In mammals, contraction is initiated in the right atrium by the (53) ____, which contains specialized (54) ____. The electrical impulse travels from the atria to the ventricles through the (55) ____. The electrical activity of the heart can be recorded externally to produce a(n) (56) ____. (57) ____ is a condition of chronically elevated blood pressure.

As blood flows from the heart, it enters progressively smaller arteries until finally reaching the (58) ____. Blood flow into individual capillary beds is controlled by a(n) (59) ____. Capillaries join together to form (60) ____, which merge to form larger veins that return blood to the heart.

Blood pressure is influenced by (61) ____ (the product of stroke volume x heart rate), the degree of contraction of blood vessels (primarily arterioles), and blood volume.

The (62) ____ is a network of vessels that collects excess interstitial fluid or (63) ____, and returns it to the systemic circulation (venous side). (64) ____ act a filters and participate in immune responses.

Choice

For each of the following descriptions, choose the most appropriate type of heart from the list below.

a. myogenic heart b. neurogenic heart

65. ____ possess endogenous electrical activity that initiates contract

66. ____ requires neural signals to initiate contraction

For each of the following descriptions, choose the most appropriate region of the heart from the list below.

a. sinoatrial node b. atrioventricular node

67. ____ in the right atrium and possesses pacemaker cells; initiates contraction of atria

68. ____ conducts electrical impulses from atria to ventricles

For each of the following descriptions, choose the most appropriate phase of the cardiac cycle from the list below.

a. systole b. diastole

69. ____ period of contraction and emptying of heart chambers

70. ____ period of relaxation and filling of heart chambers

For each of the following descriptions, choose the most appropriate valve from the list below.

a. atrioventricular valves
b. semilunar valves

71. _____ valves between atria and ventricles

72. _____ valves between ventricles and the arteries that leave the heart

Complete the Table

Component	Function
Lymphatic system	73.
74.	Filters blood and participates in immune responses
75.	Interstitial fluid that enters lymph vessels

Describe

76. What is an electrocardiogram?_______________

____

____

Matching

Match each of the following types of junction with its correct definition.

77. _____ pacemaker cells
78. _____ cardiac cycle
79. _____ venules
80. _____ aorta
81. _____ cardiac output
82. _____ hypertension
83. _____ precapillary sphincter
84. _____ arterioles

A. Cells that initiate the endogenous rhythms of myogenic hearts
B. Chronic elevation in blood pressure above normal levels
C. Small vessels that supply blood to capillaries
D. Small vessels that drain blood from capillarieis
E. Control blood flow into capillary
F. Is a function of force of contract and heart rate
G. The sequence of heart contraction-relaxation
H. The large vessels into which blood enters after leaving the heart

SELF-TEST

1. The fluid-filled space that surrounds organs of animals with an open circulatory system. [p. 951]
   a. coelom
   b. lymphocoel
   c. atracoel
   d. sinus

2. Erythrocytes are _____. [pp. 954-955]
   a. also called red blood cells
   b. spherical shaped
   c. produced in bone marrow
   d. a kind of white blood cell
   e. specialized to transport O_2

3. In mammals, blood enters the right atrium from the ____. [p. 957]

a. right ventricle
b. superior vena cava
c. inferior vena cave
d. pulmonary artery
e. jugular vein

4. Pulmonary arteries carry blood that is [p. 958]

a. low in O_2.
b. low in CO_2.
c. high in O_2.
d. high in CO_2.
e. on its way to the lungs.

5. Very small vessels that supply blood to capillary beds. [p. 963]

a. arterioles
b. venules
c. arteries
d. veins
e. capillaries

6. Fibrinogen is ____. [p. 954]

a. a plasma lipids
b. a gamma globulin
c. a protein
d. a precursor to fibrin
e. involved in the clotting mechanism

7. The ventricular muscle cells of the mammalian heart receive stimuli to contract from the ____. [p. 960]

a. SA node
b. AV node
c. atria
d. Purkinje fibers

8. In mammals, a semilunar valve is found between a ventricle and ____. [p. 958]

a. an atrium
b. another ventricle
c. the aorta
d. a pulmonary artery
e. a pulmonary vein

9. Three-chambered hearts generally consist of the following numbers of atria/ventricles: [pp. 952-954]

a. 2/1.
b. 1/2.
c. 1/1.
d. 0/3.
e. 3/0.

10. In general, the path of blood in the systemic circulation in a vertebrate occurs in the following order: [pp. 963–964]

a. veins, venules, capillaries, arterioles, arteries.
b. arterioles, capillaries, arteries, venules, veins.
c. venules, veins, capillaries, arteries, arterioles.
d. veins, arteries, arterioles, venules, capillaries.

INTEGRATING AND APPLYING KEY CONCEPTS

1. *Evolution link*: What is the adaptive significance of a closed circulatory system?

2. *Insights from the molecular revolution*: Explain why some arterioles vasoconstrict in response to epinephrine while others vasodilate. What is the functional significance of the different responses?

43

DEFENSES AGAINST DISEASES

CHAPTER HIGHLIGHTS

- Humans and other vertebrates have three lines of defense against pathogens: physical barriers, innate immunity, and adaptive immunity.
- In innate immunity, molecules on the surface of pathogens are recognized by various types of host cells to initiate inflammation and the complement systems that lead to the destruction of the pathogen.
- Adaptive immunity is carried out by B cells and T cells and has two components: an antibody-mediated response and a cell-mediated response.
- In the antibody-mediated response, B cells produce antibodies that bind to the pathogen and mark it for elimination. T cells stimulate B cells in this process.
- In the cell-mediated response, T cells act as killer cells to eliminate infected cells from the body.

STUDY STRATEGIES

- The concepts of this chapter build on elements of cell biology (phagocytosis, endomembrane system), transcription/translation/protein synthesis, and cell communication developed earlier. You should briefly review these concepts to make sure that you understand them.
- This chapter contains a lot of new information and new terminology. DO NOT try to go through it all in one sitting. Take one section at a time. Start by skimming the section and writing down the bold face terms and studying the figures, then go back and read the text followed by working through the companion section(s) in the study guide. Repeat this process for each section.
- Draw diagrams of the various response pathways, carefully labeling each component and noting its function.

TOPIC MAP

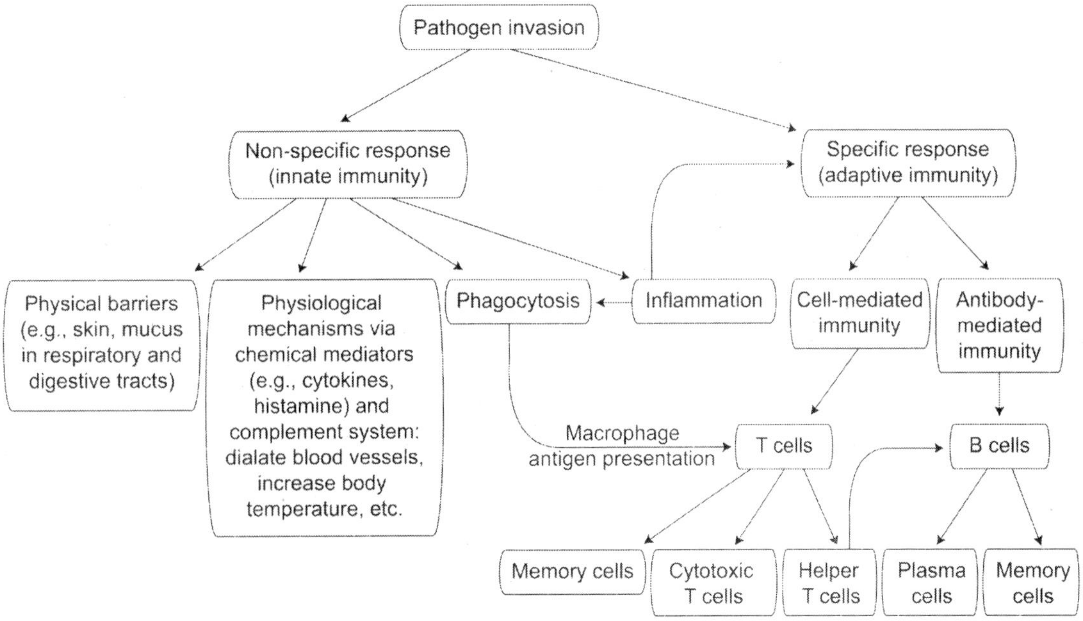

Pathogen invasion

Non-specific response (innate immunity)

Specific response (adaptive immunity)

Physical barriers (e.g., skin, mucus in respiratory and digestive tracts)

Physiological mechanisms via chemical mediators (e.g., cytokines, histamine) and complement system: dialate blood vessels, increase body temperature, etc.

Phagocytosis

Inflammation

Cell-mediated immunity

Antibody-mediated immunity

Macrophage antigen presentation

T cells

B cells

Memory cells

Cytotoxic T cells

Helper T cells

Plasma cells

Memory cells

INTERACTIVE EXERCISES

Why It Matters [pp. 971–972]

Section Review

The (1) ____ is the main defense against disease.

43.1 The Function of Gas Exchange [pp. 972–973]

43.2 Adaptations for Respiration [pp. 973–976]

Section Review

The three lines of defense against invasion by foreign agents in humans and other vertebrates is physical barriers, the (2) ____ and the (3) ____. The (4) ____ is the defensive reactions of the immune system.

(5) ____ is a rapid response to injury that involves heat, pain, redness, and swelling. (6) ____ are phagocytes that are usually first to recognize pathogens, engulf them, and become activated to secrete (7) ____. (8) ___ also become activated by foreign substances and release histamine. (9) ____ are attracted to an infection site by (10) ____, as are (11) ____ to help kill large pathogens. The (12) ____, a group of over 30 proteins, also become activated, and some of these form (13) ____ on the surfaces of pathogens. (14) ____ are produced by viral-infected cells and initiate events that result in RNS degradation and inhibition of protein synthesis in the infected cell. (15) ____ destroy viral-infected cells often by triggering programmed cell death or (16) ____. NK cells are a type of (17) ____, a type of leukocyte that carries out most of its actions in the lymphatic system.

Building Vocabulary

Prefixes	Meaning
anti-	against, opposite
mono-	alone, single, one
lyso-	loosening, decomposing

Suffix	Meaning
-cyte	cell

Prefix	Suffix	Definition
_____	-body	18. A specific protein that binds to a pathogen and marks it for elimination
_____	-histamine	19. A drug that blocks the effects of histamine
____	____	20. A white blood cell located in lymphoid tissue
____	-zyme	21. An enzyme that attaches to and degrades pathogens

Choice

For each of the following statements, choose the most appropriate type of immunity from the list below.

a. innate immunity b. adaptive immunity

22. ____ immediate, nonspecific response to an invading pathogen
23. ____ specific response to a particular pathogen

For each of the following statements, choose the most appropriate type of chemical mediator from the list below.

a. cytokines b. chemokines c. interferons

24. ____ proteins produced by infected cells that initiate degradation of RNA and cessation of protein synthesis
25. ____ released by activated macrophages and initiates heat, redness, and swelling of inflammation
26. ____ released by activated macrophages and attracts neutrophils to infected site

Complete the Table

Type of cell	Function
macrophage	27.
28.	releases histamine upon activation
neutrophil	29.
30.	type of leukocyte that helps kill large pathogens by secreting lysozmes and defensins
31.	destroys viral-infected cells by perforating their plasma membrane

Matching

Match each of the following structures with its correct definition.

32. ____ apoptosis
33. ____ histamine
34. ____ inflammation
35. ____ complement system
36. ____ lymphocyte
37. ____ immune response
38. ____ membrane attack complex

A. The collection of defensive actions of the immune system
B. Rapid response that involves swelling
C. Proteins that circulate in the blood and become activated by molecules on the surface of pathogens
D. Activated complement proteins that perforate cell membranes
E. Type of leukocytes that carries out its actions in the lymphatic system
F. Programmed cell death
G. Causes dilation of blood vessels leading to the infected site

43.3 Specific Defenses: Adaptive Immunity [pp. 976–989]

Section Review

A foreign molecule that triggers an adaptive immunity response is a(n) (39) ____. These molecules are recognized by (40) ___, which are derived and mature in bone marrow, and (41) ____, which are derived in bone marrow and mature in the (42) ____. The two types of adaptive immunity are (43) ____ and (44) ____. In antibody-mediated immunity, B cells differentiate into (45) ____ that secrete (46) ____. In cell-mediated immunity, a subclass of T cells become activated and, with other cells of the immune system, attach to foreign cells and destroy them. Some activated lymphocytes differentiate into (47) ____ that circulate and can initiate a rapid response upon reexposure to the same antigen. (48) ____ and (49) ____ bind to specific regions or (50) ____ of an antigen molecule. Antibodies are a large class of proteins known as (51) ____; an individual protein of this class consists of four subunits: two identical (52) ____ and two identical (53) ____. A(n) (54) ____ engulfs invading cells (e.g., bacteria) and becomes activated. Proteins of a foreign cell are broken down, combine with (55) ____ proteins made in the dendritic cell, and the foreign protein fragments (serving as antigens) are presented on the surface of the cell, thus making the dendritic cell a(n) (56) ____. The (57) ____ are derived from a large cluster of genes expressed in a few immune cell types (e.g., dendritic cells, macrophages, B cells). The antigen on the APC binds to a CD4 receptor on a(n) (58) ____ and initiates (59) ____, or the proliferation of activated (60) ____. These clonal cells differentiate into (61) ____, which is one type of (62) ____. When a helper T cell encounters a B cell displaying the same antigen, they link together; this linkage activates the B cells and stimulates proliferation to form a clone of these activated B cells. Some of those cloned cells differentiate into short-lived (63) ____, while others differentiate into (64) ____. (65) ____ is the process by which a lymphocyte is specifically selected for cloning. The persistence of memory B cells and (66) ____ in an inactive state in the lymphatic system provides (67) ____ of foreign antigens. When exposed to a foreign antigen for the first time, a(n) (68) ____ results, whereas when a foreign antigen enters the body for a second or subsequent time, a(n) (69) ____ results. (70) ____ is the production of antibodies in response to a foreign antigen, whereas (71) ____ is the acquisition of antibodies by direct transfer from another individual. In (72) ____, cytotoxic T cells kill host cells infected with pathogens. In this process, the infected cell, displaying foreign protein fragments on its surface (making it an APC), binds to (73) ____ on the surface of (74) ____. The activated T cell proliferates to form a clone, and some of these cells differentiate into (75) ____. (76) ____ react only against the same epitope of a single antigen. (77) ____ are composite cells that arose from the fusion of B cells and cancerous lymphocytes called myeloma cells.

Describe

78. What is the major histocompatibility complex (MHC)? ____________________

____

____

Choice

For each of the following descriptions, choose the most appropriate type of adaptive immunity from the list below.

a. cell-mediated immunity   b. antibody-mediated immunity

79. ____ involves activation of T cell derivatives that attach to foreign cells and kill them

80. ____ involves production of specific proteins by B cell derivatives that bind to foreign molecules

For each of the following descriptions, choose the most appropriate type of response from the list below.

a. primary immune response b. secondary immune response

81. ____ response mounted by exposure to an antigen for the first time

82. ____ response mounted by exposure to an antigen for a second or subsequent times

For each of the following descriptions, choose the most appropriate type of immunity from the list below.

a. passive immunity b. active immunity

83. ____ The production of antibodies in the body in response to exposure to a foreign antigen

84. ____ The acquisition of antibodies by direct transfer from another individual

Complete the Table

Type of cell	Function
B cell	85.
86.	Lymphocyte that differentiates from stem cells in bone marrow but matures in thymus gland
Plasma cell	87.
Memory cell	88.
Dendritic cell	89.
90.	Derived from activated CD4+ cells and help stimulate B cells
Helper T cell	91.
92.	A type of T cell that binds to an antigen-presenting cell via a CD* receptor and leads to cell-mediated immunity
CD8 cytotoxic T cell	93.

Matching

Match each of the following structures with its correct definition.

94. ____ thymus
95. ____ antigen
96. ____ monoclonal antibody
97. ____ clonal expression
98. ____ clonal selection
99. ____ epitopes
100. ____ immunoglobins
101. ____ antigen-presenting cell
102. ____ immunological memory
103. ____ hybridomas
104. ____ antibodies

A. Organ of lymphatic system involved in maturation of T cells
B. Protein that binds to antigens and marks them for elimination
C. Specific region of an antigen molecule
D. Family of proteins that serve as antibodies
E. A cell that presents antigens on its surface
F. The proliferation of a particular clone of cells
G. The process by which a lymphocyte is specifically selected for cloning
H. Ability to recognize previous antigens and foreign cells
I. React with a particular epitope of an antigen
J. A composite cell
K. A foreign molecule that triggers an adaptive immunity response

44.4 Malfunctions and Failures in the Immune System [pp. 989–992]

44.5 Defenses in Other Animals [pp. 992–994]

Section Review

(105) ____ protects that body's own molecules from attack by the immune system; failure of this process can lead to (106) ____. (107) _____ are a distinct class of antigen that initiate an allergic reaction. A severe reaction can bring on (108) ____, an inflammation that blocks airways and interferes with breathing.

Although invertebrates do not have antibodies, they do produce proteins such as (109) ____ that are members of the immunoglobin family.

Matching

Match each of the following types of junction with its correct definition.

110 . ____	allergen	A. protection of a body's own molecules from immune system attack
111. ____	hemolin	B. production of antibodies against molecules of one's own body
112. ____	immune tolerance	C. substance responsible for initiating an allergic reaction
113. ____	anaphylactic shock	D. extreme inflammatory response that can restrict air passages
114. ____	autoimmune reaction	E. an Ig-family protein in moths

---

SELF-TEST

1. T cells ___. [p. 977]
   a. are lymphocytes
   b. are called LGLs
   c. mature in the thymus gland
   d. are platelets
   e. are involved in adaptive immunity

2. An allergic reaction involves ___. [p. 992]
   a. killer T cells
   b. mast cells
   c. over production of histamine
   d. production of IgE
   e. histocompatibility complex

3. The histocompatibility complex is ___. [pp. 982–984]

a. found in cell nuclei

b. different in each individual

c. the same on individuals comprising a species

d. derived from a group of closely linked genes

4. The millions of different types of antibodies produced by the immune system are most likely due to ____. [pp. 980–981]

a. a one gene-one antibody ratio

b. many C regions

c. gene rearrangement involving many combinations of V and J segments

d. spaces between V and C regions

5. Complement ____. [p. 974]

a. is a system of many proteins

b. is an antibody

c. is highly antigen specific

d. helps destroy pathogens

6. Active immunity can be artificially induced by ____. [p. 986]

a. transfusions

b. injecting vaccines

c. passing maternal antibodies to a fetus

d. injecting gamma globulin

7. Nonspecific defense mechanisms in vertebrates include: [pp. 972–973]

a. skin.

b. acid secretion.

c. inflammation.

d. phagocytosis by phagocytes.

e. antibody production.

8. B cells ___. [pp. 977–978]

a. are granular

b. are lymphocytes

c. clone after contacting its targeted antigen

d. are derived from plasma cells

e. include many antigen-binding forms

9. T cell receptors ____. [p. 977]

a. bind antigens

b. have no known function

c. are identical in all T cells

d. are found on killer cells

e. stimulate antibody production

10. The secondary immune response is due to ____. [pp. 984–985]

a. killer T cells

b. memory cells

c. plasma cells

d. macrophages

e. helper T cells

INTEGRATING AND APPLYING KEY CONCEPTS

1. *Focus on Research*: How can knockout mice be used to study asthma?

2. *Insights from Molecular Revolution*: Devise a treatment strategy for killing melanoma cells based on the *Fas-Fas*L system.

44

GAS EXCHANGE: THE RESPIRATORY SYSTEM

CHAPTER HIGHLIGHTS

- All animals exhibit physiological respiration by exchanging O_2 and CO_2 with the environment.
- Gases move by simple diffusion between the respiratory medium of either air or water and the animal across their respiratory surface.
- In small animals, the body serves as the respiratory surface. In larger animals, specialized structures such as the tracheal system of insects, gills, and lungs serve as respiratory structures.
- The carrying capacity of blood/body fluids is increased by respiratory pigments such as hemoglobin.

STUDY STRATEGIES

- The concepts of this chapter build on elements of diffusion developed earlier. You should briefly review these concepts to make sure that you understand them.
- This chapter contains a lot of new information and new terminology. DO NOT try to go through it all in one sitting. Take one section at a time. Start by skimming the section and writing down the bold face terms and studying the figures, then go back and read the text followed by working through the companion section(s) in the study guide. Repeat this process for each section of the text.
- Draw diagrams of the various respiratory systems, carefully labeling each component and noting its function.

INTERACTIVE EXERCISES

Why It Matters [pp. 997–998]

Section Review

The (1) ____ is the organ system that allows animals to exchange CO_2 produced in the body for O_2 from the surroundings.

44.1 The Function of Gas Exchange [pp. 998–1000]

44.2 Adaptations for Respiration [pp. 1000–1003]

Section Review

(2) ____ is the process animals use to exchange gas with their surroundings. The source of O_2 and the "sink" for CO_2 in the environment is called the (3) ____, which can be either air or water. (4) ____ is the exchange of gases with the respiratory medium. Gases move by simple diffusion across a(n) (5) ____. The entire body serves as a respiratory surface in many small animals. Insects possess an extensive (6) ____ to distribute air to internal organs. In larger animals, (7) ____ and (8) ____ are used for gas exchange with water and air, respectively. Increasing (9) ____, the flow of the respiratory medium over the respiratory surface, and (10)____, the flow of blood/body fluid on the

internal side of the respiratory surface, maintains the diffusion gradient for gases and maximizes the rate of exchange.

Gills are evaginations of the body surface; (11) ____ extend from the body and do not have protective coverings, whereas (12) ____ are located within chambers of the body and have a protective cover. Many aquatic vertebrates maximize gas exchange with (13) ____, a process whereby water flows over the gills in the opposite direction of the flow of blood beneath the respiratory surface. Air enters and exits the tubes or (14) ____ of the unique respiratory system of insects through (15) ____. In some air-breathing fish (e.g., lungfish) and amphibians, air is gulped and forced into the lungs by a process called (16) ____. Other vertebrates, such as reptiles, birds, and mammals, use (17) ____, in which the air pressure in the lung is lowered by expanding the lung cavity through muscular contraction, resulting in air being pulled inward. The mammalian lung consists of millions of tiny air pockets or (18) ____. Birds possess nonrespiratory air sacs that assure all air in the lungs comes in contact with respiratory surfaces and which creates a(n) (19) ____ flow of air that maximizes gas exchange via cross-current exchange (i.e., air flows perpendicular to blood flow).

Building Vocabulary

Prefixes	Meaning
hyper-	over, above
hypo-	under, below
ventila-	to fan

Suffix	Meaning
-tion	the process of

Prefix	Suffix	Definition
____	____	20. The process of moving air or water across a respiratory surface
____	____	21. Excessively rapid breathing
____	____	22. Excessively slow breathing

Describe

23. Describe the process of counter-current exchange in the gills of fish and its adaptive significance. ____________

____

____

Choice

For each of the following statements, choose the most appropriate type of gill from the list below.

a. external gills b. internal gills

24. _____ gills that extend from the body and are in direct contact with water

25. _____ gills that are contained within a body cavity and have a protective cover.

For each of the following statements, choose the most appropriate form of breathing from the list below.

a. positive pressure breathing b. negative pressure breathing

26. _____ inflation of lungs by forcing air into them

27. _____ inflation of lungs by reducing pressure inside of them to below atmospheric pressure

For each of the following statements, choose the most appropriate aspect of the respiratory system from the list below.

a. respiratory medium b. respiratory surface

28. _____ the type of environment with which the animal exchanges gases

29. _____ the epithelial cell layer across which gases diffuse

For each of the following statements, choose the most appropriate process from the list below.

a. ventilation b. perfusion

30. _____ the flow of blood/body fluid on the internal side of the respiratory surface

31. _____ the flow of the respiratory medium over the respiratory surface

Complete the table

Animal group	Principal respiratory medium	Principal respiratory structure
insects	32.	33.
most fish	34.	35.
reptiles	36.	37.
mammals	38.	39.

Matching

Match each of the following structures with its correct definition.

40.	_____	breathing	A.	The exchange of gases with the respiratory medium
41.	_____	lungs	B.	The air tubes of the unique respiratory system of insects
42.	_____	alveoli	C.	Air hole on insect exoskeleton that controls air flow
43.	_____	tracheae	D.	Small air pocket of mammalian lung
44.	_____	tracheal system	E.	Process by which animals exchange gases with their environment
45.	_____	spiracles	F.	Respiratory structures that represent evaginations of the body
46.	_____	physiological respiration	G.	Respiratory structures that represent invaginations of the body
47.	_____	gills	H.	Extensive system of air tubes in insects

44.3 The Mammalian Respiratory System [pp. 1004–1007]

Section Review

Air enters the body through nasal passages and the mouth and enters the (48) ____, a common pathway leading to the digestive tract and lungs. For respiration, air travels from the pharynx first into the (49) ____ then into the (50) ____, which branches into two (51) ____, one leading to each lung. The terminal airways or (52) ____ in the lungs lead into cup-shaped pockets called alveoli. The lungs are covered with (53) ____, a double layer of epithelial tissue. Inhalation of the lung occurs by contraction of the (54) ____ and the (55) ____, actions which expand the thoracic cavity and reduce air pressure in the lung below atmospheric pressure, thereby enabling the lungs to fill passively. Exhalation occurs by relaxation of these muscle groups. With increased activity, air can be forcibly expelled from the lungs by contraction of the (56) ____. The volume of air entering and leaving the lung is the (57) ____; the maximum such volume is the (58) ____. A(n) (59)____ of air is left in the lungs after exhalation. Respiration is controlled by the medulla oblongata and the pons of the brainstem, which integrate chemosensory information about O_2 and CO_2 levels centrally in the medulla and peripherally from (60) ____ and (61) ____.

Choice

For each of the following descriptions, choose the most appropriate type of adaptive immunity from the list below.

a. carotid bodies   b. aortic bodies

62. ____ chemoreceptors located in the aortic arch

63. ____ chemoreceptors located in the carotid arteries

For each of the following descriptions, choose the most appropriate type of response from the list below.

a. tidal volume   b. vital capacity   c. residual volume

64. ____ the maximum tidal volume of an individual

65. ____ the volume of air entering and leaving the lungs during inhalation and exhalation

66. ____ the volume of air left in the lungs after exhalation

Labeling

Identify each numbered part of the following illustration.

67. ______________

68. ______________

69. ______________

70. ______________

71. ______________

72. ______________

73. ______________

74. ______________

75. ______________

76. ______________

77. ______________

78. ______________

79. ______________

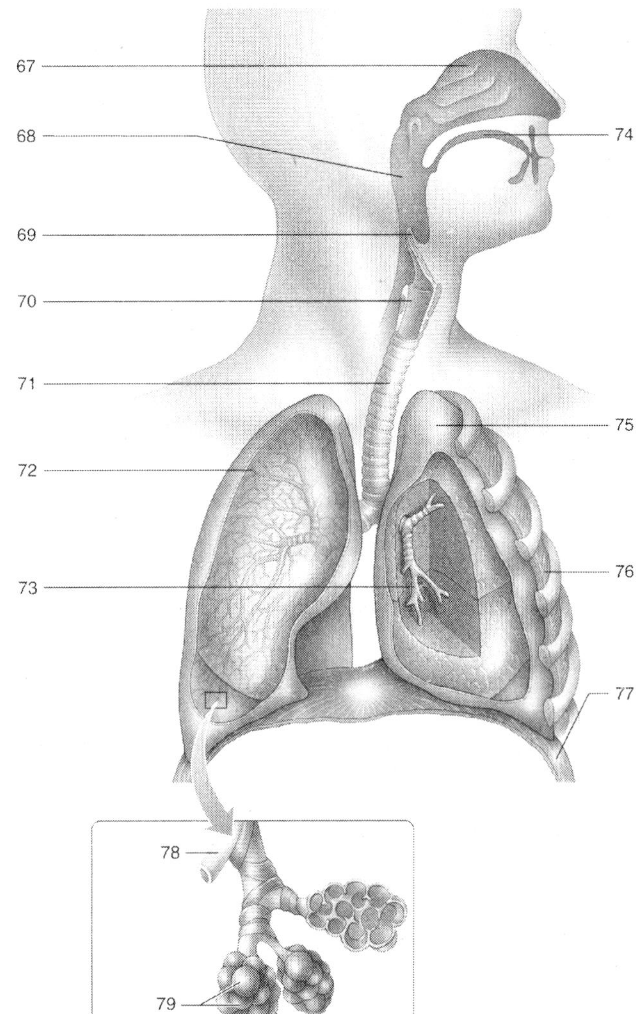

44.4 Mechanisms of Gas Exchange and Transport [pp. 1007–1010]

44.5 Respiration at High Altitudes and in Ocean Depths [pp. 1010–1012]

Section Review

The (80) ____ of O_2 in clean, dry air at sea level is about 160 mmHg. The carrying capacity of O_2 in the blood of vertebrates is increased by the presence of the respiratory pigment (81) ____. The binding of O_2 to Hb, a protein with four subunits, can be described as a sigmoid(S)-shaped (82) ____. The majority of CO_2 is transported in the plasma in the form of bicarbonate (HCO_3^-) produced by the enzyme (83) ____ in red blood cells and represents an important (84) ___ to control blood pH.

Humans that visit high altitudes adapt to lower atmospheric pO_2 levels by increasing the number of red blood cells (RBCs); RBC production is stimulated by the hormone (85) _____. Long-term residents of high altitudes display more permanent changes such as increased number of alveoli. High altitude-adapted animals such as the llama possess Hb that exhibits increased affinity for O_2. Diving animals such as seals use (86)_____ to store O_2 in their tissues.

Discuss

87. What are the reasons for the shape of the hemoglobin oxygen dissociation curve? ______________________________

____

Choice

For each of the following descriptions, choose the most appropriate type of adaptive immunity from the list below.

a. hemoglobin b. myoglobin

88. _____ Iron-containing respiratory pigment with four subunits

89. _____ Iron-containing respiratory pigment with one subunit and a very high affinity for O_2

Matching

Match each of the following types of junction with its correct definition.

90. _____ buffers

91. _____ erythropoietin

92. _____ partial pressure

93. _____ carbonic anhydrase

A. The pressure of an individual gas in a mixture

B. Resists changes in pH by accepting or donating electrons

C. Protein hormone that stimulates RBC production

D. Enzyme that catalyzes the reaction $CO_2 + H_2O$ Ξ $[H_2CO_3]$ Ξ $H^+ + HCO_3^-$

SELF-TEST

1. The alternating cycle of contraction and relaxation of the diaphragm and external intercostal muscles underlie _____. [p. 1005]

a. inspiration and expiration

b. inhalation and exhalation

c. expiration and inspiration

d. exhalation and inhalation

e. oxygen intake and CO_2 output

2. If your pet dog lived at atmospheric pO_2=150 mmHg and exhibited arterial pO_2=100 mmHg and tissues pO_2=10 mmHg, you would expect the dog to _____. [pp. 1007–1008]

a. die

b. accumulate CO_2

c. have a serious but nonlethal O_2 deficit

d. become dizzy from too much O_2

e. function normally

3. All respiratory surfaces share the following characteristics: [p. 998]

a. moist

b. thin

c. large surface area

d. alveoli

e. spiracles

4. A cockroach obtains O_2 for tissues by means of ____. [p. 1002]

a. internal gills

b. psuedolungs

c. counter-current exchange

d. a tracheal system

5. The normal path of airflow through a mammalian respiratory system is ____. [p. 1004]

a. trachea, larynx, pharynx, bronchus, bronchiole

b. larynx, pharynx, trachea, bronchus, bronciole

c. pharynx, larynx, trachea, bronchus, bronciole

d. pharynx, trachea, larynx, bronchus, bronchiole

6. Counter-current exchange ____. [p. 101]

a. is used by bony fish

b. maximizes gas exchange by maintaining diffusion gradient

c. respiratory medium moves in same direction as blood flow

d. respiratory medium moves in opposite direction of blood flow

7. As RBCs circulate into the region of actively metabolizing tissues (e.g., muscle), you would expect [pp. 1008–1009]

a. increase binding of CO_2 to Hb.

b. increase in reduced Hb.

c. increase in bicarbonate ion concentration in plasma.

d. movement of chloride out of RBC.

8. Detect O_2 levels in blood and provide information to the medulla oblongata. [p. 1006]

a. carotid bodies

b. islets of Langerhans

c. aortic bodies

d. nodes of Ranvier

9. The ability of oxygen to bind to Hb is affected by ____. [pp. 1007–1008]

a. pO_2 in respiratory medium

b. pH

c. temperature

d. pCO_2 in tissues

INTEGRATING AND APPLYING KEY CONCEPTS

1. What is(are) the disadvantage(s) of the adaptation(s) of a short-term visitor to high altitude?

2. Carbon monoxide is dangerous because it out competes O_2 for binding sites on Hb. The reaction between CO and Hb is reversible (like that between O_2 and Hb), whereas other pollutants, such as NO_x compounds, bind irreversibly to Hb. Discuss the differences in treatment strategies for individuals that you suspect to be suffering from CO and NO_x poisoning.

45

ANIMAL NUTRITION

CHAPTER HIGHLIGHTS

- Understand the importance of nutrition in animals.
- Various types of feeding and digestive processes in the animal kingdom are described.
- Digestive specialization in vertebrates are presented.
- Detailed explanation of the digestive process, including control mechanisms in humans and other mammals.

STUDY STRATEGIES

- First, focus on the various feeding mechanisms in animals.
- Examine the importance of nutrition, the digestive process, and why waste products are produced.
- Next, make comparisons between the various types of feeding, specializations, and nutritional requirements among animals.
- Understand and evaluate the digestive process in humans and other mammals.

INTERACTIVE EXERCISES

Why It Matters [pp. 1015–1016]

Section Review (Fill-in-the-Blanks)

Animal survival requires a compatible environment and a good water and adequate (1) _____ supply. (2) _____ is not just intake of food, it includes the processes of (3) _____ of materials and (4) _____ into cells. Understanding the various types of (5) _____ methods in animals as well as the (6) _____ processes or breakdown of the biological molecules (7) _____ (8) _____ (9) _____ and (10) _______ is the focus of nutrition in the animal kingdom.

45.1 Feeding and Nutrition [pp. 1016–1018]

Section Review (Fill-in-the-Blanks)

Among animals, the primary organic molecules used for food are (11) _____ and (12) _____. If the intake of nutrients is limited or the processing of nutrients is altered, a state of (13) _____ may develop. In undernourished animals, fats and glycogen stores are utilized first. However, if the condition persists, (14) _____, typically from (15) _____ and even (16) _____ _____, are metabolized for (17) _____ production. Death eventually occurs if nutrient intake is not reestablished. Given adequate nutrition, animals are able to (18) _____ many organic molecules by breaking down food and using the building blocks to produce necessary organic molecules not found in their diet. However, there are a group of (19) _____ building blocks that must be ingested and processed. These include essential (20) _____ _____, (21) _____ _____, (22) _____, and (23) _____.

Matching

Match each of the following feeding types with its correct description.

24. ____ Fluid Feeder — A. Consume whole or large chunks
25. ____ Suspension Feeder — B. Consume liquids containing organic molecules
26. ____ Deposit Feeder — C. Consume small organisms suspended in water
27. ____ Bulk Feeder — D. Consume small particles by scraping from sold material

45.2 Digestive Processes [pp. 1018–1020]

Section Review (Fill-in-the-Blanks)

Digestion is the (28) ______ larger particles into (29) _____ particles that can be (30) ______ into body fluids and cells. Many of the chemical bonds of larger particles are broken by (31) _____ ______. These enzymatic reactions are often very (32) _____. There is a separate enzyme for each type of (33) _____ or (34) _____. These digestive processes can be (35) _____ or (36) _____.

Choice

For each of the following statements, choose the most appropriate enzyme from the list below.

A. lipases B. amylases C. nucleases D. proteases

37. ____ catalyzes the bonds between carbohydrates such as starch
38. ____ catalyzes the bonds between amino acids
39. ____ catalyzes the bonds between glycerol and fatty acids
40. ____ catalyzes the bonds between DNA or RNA

Short Answer

41. Explain how endocytosis and exocytosis are involved in intracellular digestion. ______________________________

____.

42. Explain how extracellular digestion is advantageous with respect to food source. ______________________________

____

Matching

Match each of the following anatomical locations with the appropriate stage of digestion. More than one stages may occur in the various locations

43. ____ Crop of an Annelid — A. Mechanical processing
44. ____ Anus — B. Secretion of digestive enzymes and lubricates
45. ____ Gastric ceca of insects — C. Enzymatic hydrolysis
46. ____ Gizzard of Annelid — D. Absorption
47. ____ Intestine — E. Elimination
48. ____ Crop of insect or bird
49. ____ Proventriculus of bird

45.3 Digestion in Humans and Other Mammals [pp. 1020–1031]

Section Review (Fill-in-the-Blanks)

The digestive process in mammals is very similar to other animals. The five steps of digestion, mechanical processing, secretion of digestive aids, enzymatic hydrolysis, absorption, and elimination are carried out in specialized areas of the digestive system. As we previously learned, certain molecules can't be synthesized in mammalian cells; these are (50) _____ amino acids, fatty acids, vitamins and minerals. There are (51) _____ essential amino acids and (52) _____ essential fatty acids, which must be obtained from the (53) _____. There are two classes of vitamins, water-soluble or (54) _____ and fat-soluble or (55) _____. Two of the vitamins are not necessary in the diet: vitamin (56) _____ is synthesized in the (57) _____ by exposure to ultraviolet light and vitamin (58) _____ is synthesized by (59) _____, which inhabit the large intestine.

Matching

Match each of the following structures/components of the digestive system with its correct description.

60. ____	Mucosa	A.	Found in the stomach, fibers are oriented in a diagonal direction
61. ____	Submucosa	B.	Wave of contraction from the circular and longitudinal muscles
62. ____	Muscularis	C.	Outermost layer of the gut, secretes a thin, slippery lubricating fluid
63. ____	Serosa	D.	Smooth muscle rings that act as valves to ensure one-way flow
64. ____	Sphincter	E.	Muscle oriented perpendicular to the axis of the gut
65. ____	Peristalsis	F.	Innermost layer of the gut, absorption occurs in this layer
66. ____	Circular muscle	G.	Muscle oriented parallel to the axis of the gut
67. ____	Longitudinal muscle	H.	Major muscle coat of the gut—composed of 2–3 layers
68. ____	Oblique muscle	I.	Thick, connective tissue layer of the gut

Consecutive Order

Put the following structures or functions in the proper order—starting with #1.

69. ____ gastroesophageal sphincter
70. ____ esophagus
71. ____ anal sphincter
72. ____ large intestine
73. ____ ileocecal sphincter
74. ____ stomach
75. ____ food passes the epiglottis
76. ____ secretions of pancreas & liver added
77. ____ pyloric sphincter
78. ____ rectum
79. __1__ mechanical processing in mouth
80. ____ secretions of HCl and pepsinogen added
81. ____ secretions from salivary glands added
82. ____ anus
83. ____ chyme enters small intestine

Section Review (Fill-in-the-Blanks)

Digestion begins in the (84) _____. The mechanical processing begins with the action of the (85) _____ and the secretions from the (86) _____ _____. The processed food now quickly passes through the (87) _____ and through the (88) _____ sphincter into the (89) _____. Additional mixing occurs in the stomach due to the (90) _____ layers of smooth muscle. (91) _____ secretions are added that contain (92) _____ and (93) _____, which is inactive and becomes activated due to the (94) _____ of the stomach. In addition, there is a large amount of (95) _____ added for lubrication and protection. The chyme now passes through the (96) _____ sphincter and enters the (97) _____ _____. Here secretions from the (98) _____ and (99) _____ are added. In addition, the intestinal (100) _____ secretes both mucus and enzymes. The majority of (101) _____ and (102) _____ occurs in the small intestine. The remaining chyme will pass through the (103) _____ sphincter and enter the (104) _____. The primary function in this area is absorption of (105) _____ and bile salts. The final portion of the digestive system is the (106) _____. (107) _____, residual undigested/absorbed material, will exit the body via the (108) _____.

45.4 Regulation of the Digestive Process [pp. 1032–1033]

Matching

Match each of the following hormones with its correct action in digestion.

109. _____	Gastrin	A. Increases insulin release from the pancreas
110. _____	Cholecystokinin	B. Inhibits additional HCl release from gastric mucosa, stimulates HCO_3^- release
111. _____	Secretin	C. Stimulates secretion of HCl and pepsinogen and GI tract motility
112. _____	Leptin	D. Under control of hypothalamic neurons, thought to inhibit appetite
113. _____	Alpha-MSH	E. Presence of fat stimulates release, inhibits gastric activity, and increases pancreatic enzyme release
114. _____	GIP	F. Increases when fat deposition increases, reduces appetite

True/False

If the statement is true, write a "T" in the blank. If the statement is false, make it correct by changing the underlined word(s) and writing the correct word(s) in the answer blank.

115. _______________ Regulation and control of the digestive process is under local control as well as the <u>peripheral or voluntary</u> nervous system.

116. _______________ The presence of food in the GI tract <u>initiates</u> secretion of mucus and digestive enzymes as well as motility.

117. _______________ The neurons which make up the appetite center are located in the <u>mucosa of the stomach</u>.

118. _______________ A/An <u>decrease</u> in the blood level of leptin will stimulate or increase appetite.

45.5 Digestive Specializations in Vertebrates [pp. 1033–1036]

Section Review (Fill-in-the-Blanks)

The length of the digestive system is well correlated with the type of (119) _____. In general animals such as rabbits or cows, (120) _____, have (121) _____ intestinal tracts often with several areas for (122) _____. Animals such as dogs or tigers, (123) _____, have (124) _____ intestinal tracts. Herbivores often have (125) _____ microorganisms which often produce enzymes that participate in the digestive process. The classic example are the (126) _____ (cattle, deer, goats), which have (127) _____ chambered stomachs as well as symbiotic microorganisms which hydrolyze (128) _____ in plant material.

SELF-TEST

1. Digestion is the utilization of organic molecules for which fundamental processes associated with life? [pp. 1016–1017]
   a. ATP production
   b. synthesis of biological molecules
   c. reproduction
   d. Both A and B

2. Which of the following is incorrectly matched? [pp. 1016,1035]
   a. Giraffes—long intestinal tract
   b. Tigers—short intestinal tract
   c. Horse—long intestinal tract
   d. Kangaroo—short intestinal tract

3. Which of the following biological molecules has the most potential kcal per gram? [pp. 1017–1018]
   a. fats
   b. proteins
   c. carbohydrates
   d. nucleic acids

4. Essential amino acids and vitamins must be ______. [pp.1018–1019]
   a. synthesized before any of the others
   b. obtained in the diet
   c. metabolized before others can be synthesized
   d. obtained from symbiotic relationship

5. An animal with teeth and claws would most likely be a _____ feeder. [pp. 1018–1019]
   a. fluid
   b. suspension
   c. deposit
   d. bulk

6. Animals that intake food and remove wastes through the same opening have a _____. [pp. 1019–1022]
   a. saclike digestive system
   b. tubelike digestive system
   c. filtering digestive system
   d. gizzard

7. Which of the following structures are not typically involved with mechanical processing? [pp. 1022–1024]
   a. gizzard
   b. stomach
   c. pancreas
   d. crop

8. Joe has bleeding problems; he had been on antibiotics for 6 weeks. You know what is wrong. The long-term antibiotics __. [pp. 1022–1024]
   a. decreased the bacteria which produce vitamin D
   b. decreased the bacteria which produce vitamin K
   c. increased the level of water-soluble vitamins
   d. have nothing to do with the bleeding problems

9. If the pyloric sphincter were blocked, chyme could not move from the ______. [pp. 1022–1028]

a. esophagus to the stomach

b. large intestine to the rectum

c. stomach to the small intestine

d. oral cavity to the esophagus

10. One of the hormones produced by the intestine, secretin, was inactive. What effect would this have on the pH of the digestive secretions?. [pp. 1027–1031]

a. secretions would be more acidic

b. secretions would be more basic

c. secretions would be neutralized

d. secretin has no effect on pH of digestive juices

INTEGRATING AND APPLYING KEY CONCEPTS

1. Compare and contrast intracellular and extracellular digestion with respect to nutrient sources.

2. Explain how the thought of food, as well as the presence of food in the GI tract, stimulate the digestive process.

3. Explain why horses and cows are large compared to humans, especially since they basically eat "salad" with no dressing, which is considered diet food for humans.

46

REGULATING THE INTERNAL ENVIRONMENT

CHAPTER HIGHLIGHTS

- Osmoregulation is defined and comparison of osmoregulatory processes among animals is explored.
- Excretion processes and regulation in mammals is presented.
- Excretion between mammals and other vertebrates is compared.
- Thermoregulation among animals is explained with a focus on ectothermy and endothermy.

STUDY STRATEGIES

- First, focus on concepts of osmoregulation and thermoregulation.
- Examine the various types of osmoregulatory mechanisms in the animal kingdom.
- Next, make comparisons between invertebrate and vertebrate excretory mechanisms.
- Focus on the mammalian excretory system, both structure and function, making comparisons between mammals and other vertebrates.
- Evaluate thermoregulatory processes in the animal kingdom, with a focus on ectothermy and endothermy.

INTERACTIVE EXERCISES

Why It Matters [pp. 1039–1040]

Section Review (Fill-in-the-Blanks)

Water and heat movement is dependent on (1) _____ gradients. (2) _____ is the set of mechanisms of water and ion balance. In animals, the processes associated with (3) _____ function to maintain balance of (4) _____ and (5) _____, while removing (6) _____ _____.

46.1 Introduction to Osmoregulation and Excretion [pp. 1040–1043]

Section Review (Fill-in-the-Blanks)

(7) _____ is the process of water molecules moving from an area of (8) _____ concentration of (9) _____ molecules to an area of (10) _____ concentration of (11) _____ molecules through a(n) (12) _____ _____. This type of movement is (13) _____, in that a(n) (14) _____ gradient is required. The concentration of solutes in a solution is the (15) _____. When the osmolarity of two solutions are compared, specific terms are used: (16) _____ when the number of solute particles are the same, (17) _____ when one solution has more solute particles, and (18) _____ when the other solution has fewer particles. It is important to note that these terms are only comparing the number of particles between two solutions, not if the particle can cross the semipermeable membrane following their concentration gradient.

If an animal has control mechanisms which can maintain an internal osmolarity of its body fluids that is significantly different than the external environment, the animal is considered a(n) (19) ______. However, if osmoregulatory processes are lacking or adjust the internal environment to closely match the external environment, the animals is known as a(n) (20) ______. All animal cells produce (21) ______ wastes during the metabolism of organic molecules. Many of the wastes are toxic to cells, especially when levels are elevated. The process of (22) ______ is the (23) ______ of metabolic wastes while maintaining water and ion (24) ______, thus allowing the animal to survive in its environment.

Matching

Match each of the following components of excretion or type of metabolic waste with its correct description.

25. ____ Filtration
26. ____ Urea
27. ____ Reabsorption
28. ____ Secretion
29. ____ Ammonia
30. ____ Removal
31. ____ Uric acid

A. Elimination of waste into the external environment
B. Nontoxic waste that forms crystals in water
C. Highly toxic waste from protein breakdown
D. Selective removal of molecules from the excretory tubules
E. Soluble waste that is relatively nontoxic
F. Selective transport of molecules into excretory tubules
G. Nonselective removal of small molecules from body fluids at the proximal end of excretory tubules

46.2 Osmoregulation and Excretion in Invertebrates [pp. 1043–1045]

Section Review (Fill-in-the-Blanks)

(32) ______ invertebrates are (33) ______, while freshwater and (34) ______ invertebrates are (35) ______. Since marine invertebrates have an intracellular osmolarity that is similar to their (36) ______ environment, very little (37) ______ is utilized on osmoregulation. (38) ______ wastes, typically in the form of (39) ______, are directly released into the external environment. Freshwater and terrestrial invertebrates do expend (40) ______ to maintain an internal environment, which is (41) ______ to their external environment. The types of habitats these animals can inhabit is more (42) ______ than their (43) ______ counterparts.

Matching

Match each of the following types of excretory types with its correct definition.

44. ____ Protonephridia
45. ____ Metanephridia
46. ____ Malpighian tubules

A. closed tubule immersed in hemolymph
B. blind ended tubule with a flame cell
C. funnel-shaped tubule with cilia

Match each of the following animals with the appropriate excretory system type.

47. ____ Insects

48. ____ Flatworms

49. ____ Earthworm

50. ____ Larval clam

51. ____ Adult clam

A. Protonephridia

B. Metanephridia

C. Malpighian Tubules

46.3 Osmoregulation and Excretion in Mammals [pp. 1045–1052]

46.4 Regulation of Mammalian Kidney Function [pp. 1052–1054]

Section Review (Fill-in-the-Blanks)

The (52) _____ is the primary organ of excretion in mammals. The (53) _____ is the structural and functional unit of the kidney. The outermost region of the kidney is the (54) _____, while the innermost region is called the (55) _____. (56) _____ is the fluid produced in the nephron that contains water, electrolytes, and metabolic (57) _____ products. The urine is carried from the (58) _____ _____ of the kidney to the (59) _____ _____ through a tube, the (60) _____. Mammalian kidneys produce (61) _____ urine compared to body fluids. This is one of the mechanisms that conserves (62) _____. The structure and function of the (63) _____, (64) _____ _____ that surround the nephrons, and the increasing osmolarity of the kidney cortex to medulla, all play significant roles in production of a(n) (65) _____ urine.

The kidney has a(n) (66) _____ system to ensure adequate glomerular (67) _____. Regulation of kidney function is dependent on the (68) _____ apparatus, which is located at a region where the (69) _____ convoluted tubules comes into contact with the (70) _____ arteriole. In addition, (71) _____ hormone systems are involved in water and electrolyte balance. These are the (72) _____-_____-_____ system or (73) ____ and the (74) _____ hormone or (75) _____ mechanism.

Matching

Match each of the following regions of the nephron and associated area in the kidney with its primary characteristic. Multiple answers may occur.

76. ____ Glomerulus

77. ____ Proximal convoluted tubule

78. ____ Bowman's capsule

79. ____ Ascending limb—Loop of Henle

80. ____ Distal convoluted tubule

81. ____ Juxtaglomerular apparatus

82. ____ Descending limb—Loop of Henle

83. ____ Afferent arteriole

A. Last portion of the nephrons, permeable to water

B. Capillaries where filtration occurs

C. Blood vessel that supplies the glomerulus

D. Portion of loop which takes filtrate toward the medullary region of the kidney

E. Blood vessel that exits the glomerulus

F. Nephron portion of the filtration area

G. Distal convoluted tubule comes into contact with the afferent arteriole

H. Region of the loop that is impermeable to water

84. ____ Efferent arteriole

85. ____ Collecting Ducts

I. Region of the nephrons that reabsorbs electrolytes, nutrients, and 65% of water

J. Portion of the nephron located in the cortical region of the kidney where urine is isoosmotic to body fluids

True/False

If the statement is true, write a "T" in the blank. If the statement is false, make it correct by changing the underlined word(s) and writing the correct word(s) in the answer blank.

86. ______________ If blood pressure falls, the juxtaglomerular apparatus decreases production and release of renin.

87. ______________ Increased blood levels of renin produce an increase in angiotensin, which produces a(n) decrease in blood pressure.

88. ______________ Angiotensin stimulates the adrenal cortex to produce and release aldosterone, which causes reabsorption of Na+ in the kidneys.

89. ______________ Elevated osmolarity in blood fluids is detected by osmoreceptors in the posterior pituitary.

90. ______________ ADH increases water reabsorption in the kidneys and increases thirst.

91. ______________ Elevated blood pressure results in increased levels of ANF, which has the opposite effect as aldosterone on osmolarity.

Short Answer

92. With respect to the RAAS, predict the effect of decreased blood pressure detected by the juxtaglomerular apparatus. ____

____.

93. With respect to ADH, predict the effect of decreased osmolarity in blood fluids on permeability of the distal convoluted tubule and collecting ducts. ____

____.

46.5 Kidney Function in Nonmammalian Vertebrates [pp. 1054–1056]

Section Review (Fill-in-the-Blanks)

In nonmammalian vertebrates, the environmental demands placed on the animal determines if (94) _____ or (95) ____ are retained. These vertebrates have a variety of mechanisms to maintain an internal environment that is compatible with life in their external environments. Marine teleosts live in a(n) (96) _____ environment, therefore, water is (97) _____ and salts are (98) _____. Sharks and rays retain (99) _____ and other (100) _____ wastes that maintain an internal body fluid that is (101) _____ to seawater. Freshwater fishes and aquatic amphibians maintain body fluids that are (102) _____ to their environment, but (103) _____ water and (104) _____ salts. Terrestrial amphibians must (105) _____ both water and salts. Reptiles and birds conserve water by excreting (106) _____-_____ uric acid crystals. In addition to these mechanisms, these animals have a variety of the external or body (107) ______ plays a major role in either conservation or excretion of water and salts.

Matching

Match the adaptation for water/salt conservation/excretion mechanism that is used by the various vertebrates.

108. _____Rectal salt gland
109. _____Chloride cells
110. _____Isoosmotic urine
111. _____Retain urea
112. _____Hypoosmotic urine
113. _____Head salt glands
114. _____Excrete uric acid crystals
115. _____Excrete ammonia
116. _____Active uptake of salts by gills

A. Reptiles and birds
B. Marine teleosts
C. Freshwater fish and amphibians
D. Sharks and rays
E. Terrestrial amphibians

46.6 Introduction to Thermoregulation [pp. 1056–1058]

Section Review (Fill-in-the-Blanks)

Temperature regulation is based on (117) _____ feedback mechanisms and (118) _____, which detect changes from the internal (119) _____ _____. To accomplish these regulating tasks, animals have both heat (120) _____ and heat (121) _____ adaptations. There are two major strategies to maintain heat gain and loss. (122) _____ are animals that produce heat internally from physiological sources, while (123) _____ obtain heat from the external environment.

Choice

Choose the term that describes or defines the situation.

A. Conduction   B. Convection   C. Radiation   D. Evaporation

124. An animal lying in the sun can either gain or lose heat by _____.
125. When you sweat, you are losing heat through _____.
126. A snake lying on a warm rock during the early morning is gaining heat through _____.
127. Standing in front of a fan allows heat to be lost by _____.
128. If you are naked standing on ice in the winter, you will loose heat through your feet by _____ and through your breath by _____.

46.7 Ectothermy [pp. 1058–1060]

Section Review (Fill-in-the-Blanks)

Ectotherms maintain an internal temperature which is close to the (129) _____ temperature. Most animals beside (130) _____ and (131) _____ are ectotherms. These animals typically use (132) _____ mechanisms to maintain their internal temperature. During seasonal changes, fish remain in (133) _____ water during the summer and move to the (134) _____ levels during the winter. Amphibians and reptiles utilize the warming rays from the sun to increase body temperature by (135) _____. In addition to behavioral mechanisms, these animals often have various physiological mechanisms or (136) _____ _____, which ensures survival. One such mechanism are multiple (137) _____ that can catalyze the same reaction. Each has a different optimal (138) _____ range of activity.

46.8 Endothermy [pp. 1060–1066]

Matching

Match the mechanisms with the primary function with respect to thermoregulation.

139. _____ Sweating — A. Heat producing mechanism
140. _____ Shivering — B. Heat loss mechanism
141. _____ Increased thyroid hormone
142. _____ Decreased epinephrine
143. _____ Vasodilation
144. _____ Increased blood flow to the skin
145. _____ Increased blood flow to the core

True/False

If the statement is true, write a "T" in the blank. If the statement is false, make it correct by changing the underlined word(s) and writing the correct word(s) in the answer blank.

146. ____________ If the external environment is cool, endotherms expend a small amount of energy for heat production.
147. ____________ Periods of torpor during the winter is called estivation, while periods of torpor during the summer is hibernation.
148. ____________ Gills of fish and the ears of artic animals have countercurrent exchanges between arterial and venous blood flow to prevent heat loss to the environment.
149. ____________ Blubber prevents heat loss due to conduction.
150. ____________ The hypothalamus is the primary thermoregulatory center in birds and mammals.

SELF-TEST

1. If the osmolarity of the intracellular fluids is maintained to be isoosmotic to extracellular fluids, the animal is most likely an _______. [pp. 1040–1041]
   a. osmoregulator
   b. osmoconformer
   c. osmoequilibrator
   d. none of thesse

2. Cells involved with reabsorption or secretion are involved in ______ of materials either into or out of the filtrate. [pp. 1042–1043]
   a. active or facilitated transport
   b. diffusion
   c. osmosis
   d. filtration

3. Protonephridia and metanephridia are _____ systems, which either removes or adds materials to the hemolymph. [pp. 1043–1045]

a. active or facilitated transport

b. filtration

c. osmosis

d. either A or B

4. If an animal secretes uric acid crystals, you could conclude that the environment in which this animal lives is _______. [pp. 1043, 1054–1056]

a. freshwater

b. marine

c. terrestrial and very arid (dry)

d. terrestrial and very wet

5. Which of the following segment(s) of nephron is/are lacking aquaporins? [pp. 1048–1050]

a. proximal convoluted tubule

b. descending segment of the loop

c. ascending segment of the loop

d. Both A and B

6. If body fluids are hyperosmotic to the osmoreceptors in the hypothalamus, you would expect _________. [pp. 1052–1054]

a. ADH to increase

b. water reabsorption to increase

c. ANF to decrease

d. Both A and B

7. Removal of nitrogenous wastes in salmon at sea would most likely occur by loss of ____. [pp. 1054–1056]

a. uric acid crystals

b. urea

c. trimethylamine oxide

d. ammonia

8. If blood pressure were low, predict the effect on urine output. [pp. 1052–1054]

a. It would decrease due to decreased ADH

b. It would decrease due to increased aldosterone

c. It would decrease due to increased ANF

d. None of these

9. Boa constrictors will wrap around their eggs and shiver. This is an example of thermoregulation by ______. [pp. 1057–1060]

a. evaporation

b. conduction

c. convection

d. radiation

10. If the set point were increased, you would expect which of the following to occur? [pp. 1062–1064]

a. decreased epinephrine

b. shivering

c. vasodilation

d. Both A and C

INTEGRATING AND APPLYING KEY CONCEPTS

1. Joe was so happy that the biology final was over. He decided to party at the local pub. He was also playing pool with his friends and was going to the bathroom for a "leak" about every 15 minutes. His buddy's wanted to know why Joe was always going to the bathroom. You were able to explain why—you got this part totally correct on the final!

2. Compare and contrast the structure and function of the various excretory systems of invertebrates with respect to habitat—marine, freshwater, and terrestrial.

3. Discuss the advantages and disadvantages of being either an ectotherm or an endotherm.

47

ANIMAL REPRODUCTION

CHAPTER HIGHLIGHTS

- Asexual and sexual reproductive modes are described.
- Diversity potential of reproductive modes in animals are addressed.
- Cellular mechanisms involved in reproduction are explained.
- Focus is given to sexual reproduction in humans.
- Contraception and various methods for pregnancy prevention are presented.

STUDY STRATEGIES

- First, focus on the two major reproductive modes, sexual and asexual with respect to diversity potential.
- Examine the cellular mechanisms associated with sexual reproduction.
- Evaluate the importance of sexual reproduction as a key component of genetic diversity in animals.
- Next, focus on sexual reproduction in humans, making comparisons with other animals with this reproductive mode.
- Understand and evaluate the various types of contraception available and used for prevention of pregnancy.

INTERACTIVE EXERCISES

Why It Matters [pp. 1069–1070]

Section Review (Fill-in-the-Blanks)

The (1) _____ clock in palolo worms is a(n) (2) _____, which allows the (3) _____ and (4) _____ to be present at the same time for fertilization to occur. This type of (5) _____ allows animals the ability to (6) _____ in a large variety of environments.

47.1 Animal Reproductive Modes: Asexual and Sexual Reproduction [pp. 1070–1071]

Section Review (Fill-in-the-Blanks)

The primary difference between (7) _____ and (8) _____ reproduction is (9) _____ diversity. Asexual reproduction can also be viewed as (10) _____ reproduction, since the offspring is genetically (11) _____ to the parent. The three major mechanisms of this type of reproduction are (12) _____, (13) _____, and (14) _____. A special type of asexual reproduction occurs when a new individual develops from a(n) (15) _____ without (16) _____. Due to chromosome movement during (17) _____, offspring produced by (18) _____ are genetically (19) _____ and either (20) _____ or sterile. In comparison, sexual reproduction ensures (21) _____ _____ by two mechanisms during gamete formation, (22) _____ _____ and (23) _____ _____.

Matching

Match each of the following types of asexual reproduction with its correct definition.

24. ____	Parthenogenesis	A. Parent divides into two offspring
25. ____	Budding	B. Separate pieces develop into offspring
26. ____	Fission	C. Egg develops without fertilization
27. ____	Fragmentation	D. Offspring develops directly off parent

47.2 Cellular Mechanisms of Sexual Reproduction [pp. 1071–1078]

Section Review (Fill-in-the-Blanks)

The production of male and female (28) _____, or (29) _____ involves the cellular division, known as (30) _____. During embryonic development a cell line, known as the (31) _____ cell line, remains distinct from other body cells. These germ cells become located in the (32) _____, either (33) _____ or (34) _____. The process of (35) ____ production, or (36) _____, occurs in the male testes, while the process of (37) _____ production, or (38) _____ occurs in the female ovary.

Matching

Match each of the following components of an egg or sperm with its correct definition.

39. ____	Zona pellucida	A. Nonfunctional cells with the correct # of chromosomes but less cytoplasm
40. ____	Oviduct	B. Embryo develops within mother's body
41. ____	Polar bodies	C. Contains enzymes necessary for penetration of the surface coating of an egg
42. ____	Head of sperm	D. Gel-like matrix covering the egg in mammals
43. ____	Midpiece of sperm	E. Egg-laying animals
44. ____	Acrosome	F. Contains organelles necessary for movement of flagellum
45. ____	Viviparous	G. Contains the chromosomes
46. ____	Oviparous	H. Transport tube for egg

Short Answer

47. Explain the difference between internal and external fertilization. ________________________________

____.

48. Explain the importance of the acrosome reaction and the fast/slow block of fertilization. ________________

____.

47.3 Sexual Reproduction in Humans [pp. 1078–1087]

Section Review (Fill-in-the-Blanks)

In humans, like all vertebrates, the gonads have a(n) (49) _____ function, gamete and (50) _____ production. (51) _____ are the primary hormone in females, produced by the developing (52) ____ cells. Two other hormones, (53) _____ and (54) _____, are produced from the follicular cells after (55) _____ has occurred. In males, the (56) _____ cells in the (57) _____ tubules produce the (58) _____, primarily testosterone. Several hormones from the hypothalamus and pituitary gland play a role in both males and females. Two hormones from the pituitary, (59) _____-_____ hormone and (60) _____ hormone are under the control of (61) _____-_____ hormone from the hypothalamus. In females, FSH stimulates (62) _____ to undergo stimulation, and one will develop into a mature follicle. The follicular cells secrete (63)_____, which stimulate further development of the follicle and also prepares the (64) _____ lining for possible implantation. The 2nd hormone from the pituitary, (65) _____, increases just prior to and is thought to be responsible for (66) _____. After ovulation, the follicular cells continue to secrete estrogens but significantly increase the secretion of (67) _____, which further prepares the uterus for implantation and inhibits uterine (68) _____. In males, FSH stimulates (69) _____ cells to produce materials that are required for (70) _____, and LH stimulates (71) _____ cells to secrete (72) _____, which also plays a major role in spermatogenesis. Like in females, FSH and LH are under the control of (73) _____ in males. Levels of the reproductive hormones, estrogens, progesterone, testosterones, inhibin, (74) _____, (75) _____, and (76) _____ are regulated by (77) _____ between the reproductive organs and the hypothalamus and pituitary.

Choice

For each of the following statements, choose the most appropriate definition from the list below.

A. FSH B. LH C. Androgens D. Estrogens E. Progesterone F. GnRH

78. ____ Leydig cells produce this
79. ____ Corpus luteum produces large amounts of this
80. ____ Without this hormone, FSH and LH will not be released
81. ____ Responsible for development male secondary sex/reproductive structures
82. _____. Stimulates the Leydig cells
83. _____ Stimulates primary oocytes to undergo development
84. ____ Supportive cells that surround developing spermatocytes are stimulated by this
85. ____ Hormone that is predominate during the Luteal phase
86. ____ Progesterone will inhibit this hormone from being released
87. ____ Burst of this hormone results in ovulation
88. ____ Has a positive effect on the hormone from the pituitary

47.4 Methods of Preventing Pregnancy: Contraception [pp. 1087–1090]

Section Review (Fill-in-the-Blanks)

Pregnancy prevention or (89) _____ can be accomplished by one of the following methods: preventing (90) _____; preventing the (91) _____ from reaching the site of (92) _____; or if pregnancy occurs, preventing (93) _____. A more permanent means includes (94) _____ for the male and (95) _____ _____ for the female.

Matching

Match each of the following types of contraception with its correct definition.

96. _____ Rhythm method
97. _____ Condom
98. _____ Diaphragm
99. _____ Pill or implants
100. _____ IUD

A. Sheath or pouch of latex that captures sperm or prevents entry into the uterus
B. Device inserted into the uterus that prevents implantation
C. Hormonal means of preventing ovulation or implantation
D. Avoiding intercourse during time of ovulation
E. Device which covers the opening to the uterus

True/False

If the statement is true, write a "T" in the blank. If the statement is false, make it correct by changing the underlined word(s) and writing the correct word(s) in the answer blank.

101. _______________ Total abstinence from sex is the best way to prevent pregnancy, however, the human sex drive makes this an unlikely means to prevent pregnancy.

102. _______________ Measurement of LH would be a good means of contraception because low levels is an indication that ovulation is about to occur.

103. _______________ The IUD prevents implantation, while condoms or diaphragms prevent fertilization from occurring.

104. _______________ Most hormonal contraceptives allow ovulation to occur but prevent implantation.

SELF-TEST

1. A sea star was cut into two complete pieces; each developed into a complete sea star. This is an example of _____. [p. 1070]
   a. external fertilization
   b. fission
   c. budding
   d. fragmentation

2. Which of the following is an advantage of asexual reproduction? [pp. 1070–1071]
   a. independent assortment
   b. synchronization of gamete release
   c. genetic diversity
   d. finding a mate is not necessary

3. If the cortical reaction that occurs during the fast block reaction after fertilization were blocked, what might occur? [pp. 1075–1077]
   a. external fertilization
   b. polyspermy
   c. a greater than diploid number of chromosomes
   d. both b and c are correct

4. The duck-billed platypus is ___. [pp. 1077–1078]
   a. placental
   b. oviparous
   c. viviparous
   d. ovoviviparous

5. In order for progesterone to have its full effect during the luteal phase, which of the following must occur? [pp. 1080–1081]
   a. low levels of FSH
   b. ovulation
   c. the secretory phase
   d. stimulation of a primary oocyte

6. If the levels of FSH and LH were not high enough to inhibit the release of GnRH, what could happen? [pp. 1080–1082]
   a. nothing
   b. no fertilization
   c. possibility of twins
   d. no menstrual phase

7. If the prostate gland secretion were inhibited, other than a decreased volume, what is the effect on the semen? [pp. 1082–1083]
   a. more basic semen
   b. more acidic semen
   c. more alkaline semen
   d. both a and c are correct

8. With respect to fertility, if someone was born without a pituitary gland, what would be the result? [pp. 1082–1084]
   a. sterility
   b. they would be female
   c. multiple births with every pregnancy
   d. GnRH would be extremely low

9. Home pregnancy tests evaluate which of the following hormones ? [pp. 1084–1087]
   a. FSH
   b. LH
   c. relaxin
   d. HCG

10. With respect to contraception, which of the following allows fertilization but prevents implantation? [pp. 1087–1090]
   a. IUD
   b. diaphragm
   c. combination pill
   d. condoms

INTEGRATING AND APPLYING KEY CONCEPTS

1. Support or not support the following statement: "An ameba never dies."

2. Explain why a successful pregnancy must have fertilization occurring in the 1st one-third of the oviduct.

3. Assume that pregnancy has occurred. Predict the effect if progesterone levels started to drop instead of increasing.

48

ANIMAL DEVELOPMENT

CHAPTER HIGHLIGHTS

- Major developmental patterns of animals from the fertilized egg through organogenesis is presented.
- Embryonic development of mammals, including humans, is described.
- Cellular mechanisms of development are explained, with a focus on genetic and molecular controls.

STUDY STRATEGIES

- First, focus on major stages of embryological development from fertilization, cleavage patterns, and gastrulation.
- Examine developmental patterns from gastrulation to organogenesis, including overall structure of the adult form.
- Next, make comparisons between the embryonic development of humans and other mammals.
- Evaluate the mechanisms of development from a cellular perspective, focusing on genetic and molecular controls.

INTERACTIVE EXERCISES

Why It Matters [pp. 1093–1094]

Section Review (Fill-in-the-Blanks)

The developmental process from the (1) _____ _____ to the (2) _____ and further growth and development to the (3) _____ is a phenomenal process. From changes in (4) _____, shape and form, to cell (5) _____, our hereditary blueprint is written on our (6) _____. Understanding (7) _____ processes can increase our understanding of the complexity of living organisms and their relationships to each other and their environment.

48.1 Mechanisms of Embryonic Development [pp. 1094–1097]

48.2 Major Patterns of Cleavage and Gastrulation [pp. 1097–1101]

Section Review (Fill-in-the-Blanks)

All instructions for development are located in the nucleus of the (8) _____ or fertilized egg. Since the contribution of the sperm is primarily (9) _____ material, the cytoplasm, all organelles and mRNA of the (10) _____ or (11) _____ determinants play key roles in the initial activities of the zygote. The cytoplasm of the egg contains all necessary nutrients, which are often stored in granules of the (12) _____. The amount of yolk will often deterrmine the rate and location of cellular divisions, or determine the (13) _____ and (14) _____ poles of the embryo. At this early stage in development, egg (15) _____ plays a role in determining body (16) _____, which will dictate overall form of the animal. During development cellular divisions are very rapid with very little production of new cytoplasm; the early divisions are called (17) _____ divisions. During early development there are three major processes, (18) _____, (19) _____, and (20) _____. Cleavage progresses through three developmental stages, a solid ball of cells, the (21) _____; hollowing of the ball of cells with fluid in the cavity or (22) _____; and finally a(n) (23) _____. Next, the embryo enters the 2nd phase, (24) _____, where indentations and rearrangement of dividing cells result in the basic primary tissues, (25)_____, (26) _____, and (27) _____. A new cavity forms in the embryo, the (28) _____, or gut with an opening at one end, the (29) _____. This eventual use of the opening, either (30) _____ or (31) _____, is a key characteristics used in determining animal classification. Animals are either classified as (32) _____, where the blastopore becomes the (33) _____, or (34) _____, where the blastopore becomes the (35) _____. Several major mechanisms play key roles in these critical early processes: cleavage divisions and cellular movements; (36) _____ _____ _____; (37) _____; (38) _____; and (39) _____.

Choice

For each of the following statements, choose the most appropriate gastrulation pattern from the list below.

A. Gastrulation with even distribution of yolk

B. Gastrulation with uneven distribution of yolk

C. Gastrulation at one side of yolk

40. _____ Blastodisc is composed of the epiblast and hypoblast
41. _____ Pattern found in amphibians
42. _____ Invagination begins as the vegetal pole of the blastula
43. _____ Blastopore becomes the anus
44. _____ Pattern typically found in amniotes
45. _____ Symmetrical body form is common
46. _____ Dorsal lip cells control blastopore formation
47. _____ Primitive streak defines the right and left sides of the embryo
48. _____ Extraembryonic membranes are common
49. _____ Hypoblast cells become germ cell line
50. _____ Involution results in the pigmented cells of the animal half enclosing the vegetal half

Matching

Match each of the following with its correct definition.

51. _____	Chorion	A. Stores nitrogenous wastes produced by the embryo
52. _____	Primitive groove	B. Cells of early developmental cleavage
53. _____	Determination	C. Means for cells to move into the blastocoel
54. _____	Blastomeres	D. Layers that gives rise to primary tissues
55. _____	Amnion	E. Membrane that surrounds the embryo and yolk sac
56. _____	Allantois	F. Membrane that encloses the embryo
57. _____	Epiblast	G. Developmental fate of the cell is set

Choice

For each of the following organs, choose the primary tissue type from the list below.

A. Endoderm B. Mesoderm C. Ectoderm

58. ____ Muscles
59. ____ Bone
60. ____ Lining of respiratory tract
61. _____ Spinal cord
62. _____ Heart and kidneys
63. ____ Lining of mouth
64. ____ Lining of digestive tract
65. ____ Cornea of eye

48.3 From Gastrulation to Adult Body Structures: Organogenesis [pp. 1101–1104]

Section Review (Fill-in-the-Blanks)

The process of the three primary tissues, (66) _____, (67) _____, and (68) _____ develop into (69) _____. This is the third major developmental stage, (70) _____. In addition to cellular mechanisms used in cleavage and gastrulation, organogenesis has programmed cell death or (71) _____. Organogenesis involves several major changes that are occurring in rapid succession. A rod of tissue, the (72) _____, providing organization along the entire length of the embryo, is derived from (73) _____. Very rapidly, ectoderm will give rise to the (74) _____ system in a process called (75) _____. Above the notochord, cells undergo (76) _____ and develop into the (77) _____ _____, which gives rise to the central nervous system. At the same time, blocks of (78) _____ develop into (79) _____, which will develop into organs and tissues associated with the embryo.

Sequence

80. Put the following steps of eye develop in sequence, starting with # 1

A. ___ Ball of cells forms the lens vesicle

B. ___ Optic vesicles grow outward

C. ___ Out-pocket forms optic cup

D. _1_ Neural tube forms

E. ___ Lens placode forms from thickened ectoderm

F. ___ Ectoderm closes over lens and forms the cornea

G. ___ Lens synthesizes crystallin

48.4 Embryonic Development of Humans and Other Mammals [pp. 1104–1109]

Section Review (Fill-in-the-Blanks)

Human gestation is divided into (81) _____ _____. The 1st trimester includes (82) _____, (83) _____, and (84) _____. After the 1st trimester, the developing embryo is called a(n) (85)_____ until birth. A successful pregnancy requires that (86) _____ must occur in the (87) _____ _____ of the oviduct. By the time that (88) _____ in the uterine wall occurs, the embryo is at the (89) _____ stage. A fluid-filled cavity, the (90) _____, has formed, and the majority of cells are pushed to one side. The (91) _____ _____ _____, which includes the developing (92) _____, and the outer layer of cells becomes the (93) _____. The cells of the (94) _____ play critical roles for (95) _____ of the blastocyst into the (96) _____. The inner cell mass develops into the (97) _____ _____, which is composed of an inner layer, the (98) _____, and the outer layer or (99) _____. The (100) _____ develops from the (101) _____, while part of the extraembryonic membranes develop from the (102) _____. (103) _____ and (104) _____ of the human embryo is similar to the pattern in (105) _____-_____.

Matching

Match each of the following with its correct definition.

106. _____	Chorionic villi	A. Connecting stalk between the embryo and placenta
107. _____	Umbilical cord	B. Birth
108. _____	Parturition	C. Develops into female reproductive system
109. _____	Wolffian duct	D. Extensions that increase surface area
110. _____	Mullerian duct	E. Develops into male reproductive system

Short Answer

111. Explain the primary determining factor of whether the embryo develops into a male or female. _____________

____.

112. Explain how the embryo produces the extraembryonic membranes. _____________

____.

48.5 The Cellular Basis of Development [pp. 1109–1115]

48.6 The Genetic and Molecular Control of Development [pp. 1115–1122]

Section Review (Fill-in-the-Blanks)

The final shape and size, as well as the location of organs in the embryo, is determined by the (113) _____ and (114) _____ of cellular divisions during development. The orientation is in reference to the angle of the new daughter cell (115) _____ with respect to daughter cells from previous cell divisions. Division rate is determined by the time of the (116) _____ period of interphase. The actual mechanisms for these events are not totally understood; however, scientists do know that (117) _____ and (118) _____, cell (119)_____, and (120) _____ signals play important roles throughout the complex processes of development.

True/False

If the statement is true, write a "T" in the blank. If the statement is false, make it correct by changing the underlined word(s) and writing the correct word(s) in the answer blank.

121. ______________ Broad regions along the anterior-posterior axis of the embryo are controlled by <u>segment polarity</u> genes.

122. ______________ The genes that control the polarity of the zygote and subsequence cleavage events of the embryo are from a <u>maternal</u> source.

123. ______________ <u>Fate mapping</u> of embryos allows a clear understanding of cell lineage.

124. ______________ Products of <u>segmentation</u> genes divide the embryo into units of two segments each.

125. ______________ Homeotic (Hox) genes have been identified in animals <u>but not</u> plants.

SELF-TEST

1. Cytoplasmic determinants have the greatest effect during which stage of development? [pp. 1094–1095]
   a. gastrulation
   b. organogenesis
   c. cleavage
   d. 2nd and 3rd trimester

2. The two major groups of animals are based on the final use of the ____. [pp. 1095–1097]
   a. archenteron
   b. blastopore
   c. grey crescent
   d. blastocoel

3. If the endoderm tissue in a developing animal were labeled with a colored marker, where would you expect to find the marked tissue in the adult? [pp. 1095–1097]
   a. nervous system
   b. muscle and bone
   c. coverings of the animal or structures
   d. linings of major organ systems

4. If you were working with a blastodisc of a bird embryo, which portion will develop into the germ cell line in the adult? [pp. 1099–1100]
   a. epiblast
   b. hypoblast
   c. primitive streak
   d. primitive groove

5. If the primitive streak were blocked or removed, predict the effect on the embryo. [pp. 1099–1100]
   a. the adult couldn't reproduce
   b. organization, including axes of the embryo, would be lost
   c. extraembryonic membranes would be lacking
   d. the blastocoel would collapse

6. The cranial nerves in an adult vertebrate are derived from _____. [pp. 1101–1104]
   a. neural crest cells
   b. ectoderm
   c. mesoderm
   d. somites

7. A test that evaluates the fluid which surrounds a developing fetus is _____. [p. 1106]
   a. chorionic villus sampling
   b. umbilical cord sampling
   c. amniocentesis
   d. blastocoel sampling

8. If the embryo is lacking the SRY protein, the _____ develop into _____. [pp. 1108–1109]
   a. Wolffian ducts, male reproductive structures
   b. Wolffian ducts, female reproductive structures
   c. Mullerian ducts, male reproductive structures
   d. Mullerian ducts, female reproductive structures

9. If you wanted to change the orientation or axes of early cleavage, which of the following would you manipulate? [pp. 1110–1113]
   a. microtubules and microfilaments
   b. Wolffian or Mullerian ducts
   c. Blastopore location
   d. the chorion

10. Somites, which are derived from mesoderm, are ultimately formed from the products of which genes? [pp. 1116–1120]
   a. maternal-effect genes
   b. segment polarity genes
   c. segmentation genes
   d. homeotic genes

INTEGRATING AND APPLYING KEY CONCEPTS

1. If each of the primary tissues were assigned a different color, predict how those colors would be distributed in any given organ or organ system.
2. Address the role and importance of apoptosis in development.
3. Explain how mitochondria can be use to trace lineage.

49

POPULATION ECOLOGY

CHAPTER HIGHLIGHTS

- Ecologists may study biology from the perspective of the individual, the population, the community, the ecosystem, or the biosphere. This chapter focuses on interactions within populations.
- Populations exhibit unique characteristics including density, dispersion, age structure, sex ratio, survivorship, and many others.
- The life history patterns of organisms involve tradeoffs in benefits and costs. No life-history pattern is optimum in all respects.
- Most populations exhibit either exponential or logistic growth. Density-dependent and independent effects play a large role in determining population size.
- Human population growth has so far escaped many natural regulatory constraints.

STUDY STRATEGIES

- Review the hierarchy of biological investigation. This chapter will concentrate on populations, but in upcoming chapters you will need to know how populations relate to individuals, communities, ecosystems, and the biosphere. Also know that each level of the hierarchy exhibits unique characteristics (emergent properties).
- Have a solid understanding of the population concept. Interactions among members of a population are intraspecific.
- Keep in mind that many of the concepts presented in this chapter are presented in a somewhat simplified way to make it easy to understand. In reality, many factors interact with one another, sometimes producing unexpected results.
- Understand why humans have been able to circumvent some of the mechanisms that control population size in other organisms.

INTERACTIVE EXERCISES

Why It Matters [pp. 1125 -1126]

Section Review (Fill-in-the-Blanks)

The study of interactions between organisms and their environments is ecology. The use of a virus in an attempt to control the population of introduced rabbits in Australia is an example of using a(n) (1) _____ component of rabbit ecology to this end. Building a "rabbit-proof fence" is an attempt to control rabbits using a(n) (2) _____ factor.

49.1 The Science of Ecology [pp. 1126 – 1127]

Section Review (Fill-in-the-Blanks)

Ecology can be studied on a number of hierarchical levels. (3) _____ is the study of the various adaptations of individual organisms to the environment. The other levels include (4) _____, the study of groups of organisms of the same species living and interacting together; (5) _____, concentrating on the interaction of populations; (6) _____, which takes into account the physical, nonliving components of the environment. At the top of the hierarchy is the study of the (7) _____, which is comprised of all the earth's interacting ecosystems

Choice

Choose the scientist most likely to be studying each of the following questions.

a. organismal ecologist b. population ecologist c. community ecologist d. ecosystems ecologist
e. biosphere scientist

8. _____ How does destruction of Brazilian rainforests affect global warming?

9. _____ What is the effect of predation by one species on species diversity in the intertidal zone?

10. _____ How does the number of redwing blackbirds in an area affect the size of their territories?

11. _____ How are the organisms in a lake affected by agricultural fertilizers that run off fields into the lake after a rain?

12. _____ What is the range of temperatures that ribbon snakes can tolerate?

49.2 Population Characteristics [pp. 1127-1129]

Section Review (Fill-in-the-Blanks)

Populations have characteristics that the individuals that make up the population do not have. One of those is a population's (13) _____, the spatial boundaries within which it lives. Another is the specific biotic and abiotic features of its environment, the population's (14) _____. The number of individuals in a population at any given moment defines its (15) _____. (16) _____, the number of individuals per unit area or volume, often has greater impact on populations than a simple count of the number of individuals in a population. The distribution in space of individuals in a population is its (17) _____, of which three types are commonly observed. The first is (18) _____, where individuals are found in clusters. The second is (19) _____, where individuals in the population are spaced evenly; all individuals are roughly the same distance from their nearest neighbor. The third is (20) _____, in which individuals are distributed without regard to the presence or absence of others.

Other population characteristics include: (21) _____, a numerical description of the number of individuals in a population of different ages; (22) _____, the span of time between an organism's birth and the birth of its offspring; and (23) _____, the proportion of males and females in the population.

Complete the Table Fill in the blanks in the following table with the proper term or description.

Description of population	Probable dispersion pattern
A territorial bird species, e.g., redwing blackbirds, in which areas of roughly equal size are defended against entry by others	24. _____
A weed, e.g., dandelion, in which individuals develop from windblown seeds	25. _____
A carnivore, e.g., coyotes, in which hunting is done over a large area in small groups of 3–7 related individuals	26. _____

49.3 Demography [pp. 1129-1132]

Section Review (Fill-in-the-Blanks)

The movement of individuals into a preexisting population is called (27) _____, and the movement of individuals out of a population is referred to as (28) _____. Both types of movement can change a population's size and density. The study of these and other changes in population characteristics is (29) _____. These changes may be summarized in a(n) (30) _____, which is constructed from data on the life span of individuals of similar age, a group known as a(n) (31) _____. Data on this group is expressed in several ways, (32) _____, the proportion of individuals alive at the beginning of an age interval that survive to its end, (33) _____, the proportion dying during that period, and (34) _____, the average number of offspring produced by surviving females in each age interval. These data may also be depicted graphically as (35) _____, which show survivorship patterns over the average lifespan of organisms in a population.

Matching

Match each of the following populations with the survivorship curve below that best describes it.

36. ____ Humans. Infant survivorship is relatively high, and individuals surviving past their first 6 months typically survive for many years until mortality increases sharply in old individuals.

37. ____ *Hydra.* A small, stationary freshwater organism often eaten by larger organisms. The chances of being preyed upon are fairly constant from day to day.

38. ____ White throated sparrows. A small, migratory bird. Mortality is generally low except twice a year during the long, stressful migration period.

39. ____ Starfishes. A species with few predators. The numerous larvae produced when the eggs hatch are vulnerable in several ways and mortality is high. Those that survive and metamorphose into adults typically live for a long time.

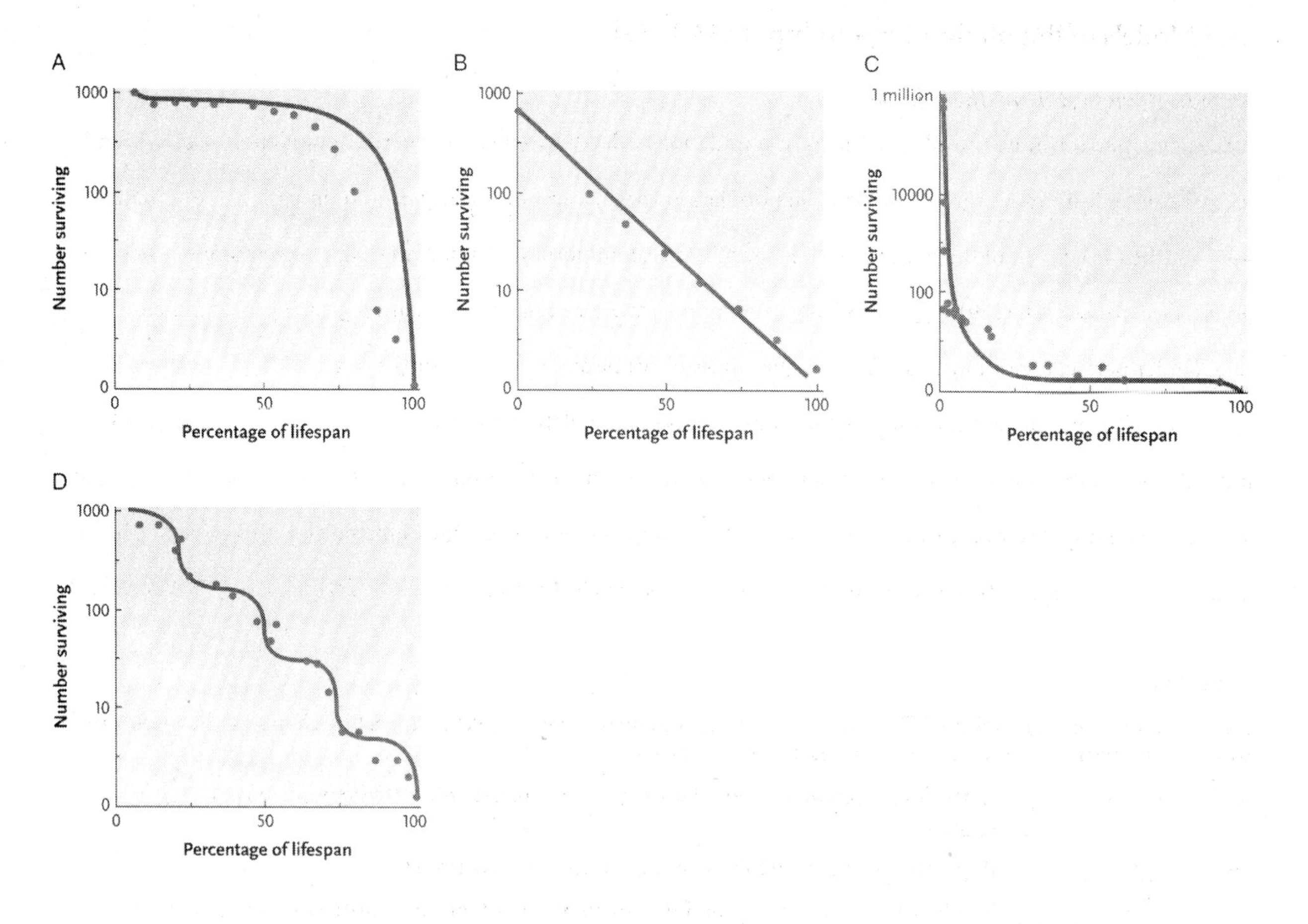

49.4 The Evolution of Life Histories [pp. 1132-1133]

Section Review (Fill-in-the-Blanks)

The lifetime patterns of growth maturation and reproduction is a species' (40) _____. These patterns must be greatly influenced by an organism's (41) _____, the total amount of energy that an organism acquires in its lifetime to drive the endergonic processes of life. Because of the constraints this finite amount of energy places on its maintenance, growth, and reproduction, tradeoffs must be made in at least some of these functions; e.g., if much energy is spent on a high reproductive rate, the organism may not have enough energy left to grow to a large size. Or if many offspring are produced, they may be provided with less energy than a small number of young. The energy invested in young before birth or hatching is referred to as (42) _____, while energy spent after birth or hatching to raise offspring is referred to as (43) _____.

Choice

For each of the following population characteristics, choose the more likely time of reproduction.

a. early reproduction   b. delayed reproduction

44. _____ High survivorship among adults

45. _____ Long life span

46. _____ Relatively low fecundity of larger individuals

47. _____ Size of individuals increases with age

49.5 Models of Population Growth [pp. 1133-1139]

Section Review (Fill-in-the-Blanks)

If the per capita birthrate (number of births in population within a specified time period/population size) = *b*, and the per capita death rate (number of births in population within a specified time period/population size) = *d*, then ($b - d$) = *r*, the (48) _____ of the population. This result, *r*, is an important component of the two common models of population growth. The (49) _____ model of population growth results in a population growing at an increasing rate. Under perfect conditions, *r* attains is highest possible value and is referred to as r_{max}, or (50) _____. In the (51) _____ model of population growth, *r* becomes smaller as the population approaches its (52) _____, the number of individuals that the environment can sustain indefinitely. This model takes into account the effect of (53) _____, the striving for a limited resource by two or more members of the same species. One of the flaws of this model is that it assumes that this fecundity and survivorship respond immediately. In reality, there is a delay or (54) _____ involved.

True/False

If the statement is true, write a "T" in the blank. If the statement is false, make it correct by changing the underlined word(s) and writing the correct word(s) in the answer blank.

55. _______________ In the logistic model of population growth, when dN/dt = 0, <u>the population has gone extinct</u>.

56. _______________ Populations <u>cannot</u> exhibit exponential growth indefinitely.

57. _______________ The logistic model of population growth predicts that *r* <u>increases</u> with population size.

58. _______________ The logistic model of population growth accounts for increasing <u>interspecific competition</u> as population size increases.

49.6 Population Regulation [pp. 1139-1145]

Section Review (Fill-in-the-Blanks)

As population density increases, it becomes easier for parasites to spread from host to host. This is an example of a(n) (59) _____ effect on population size. Low temperatures, on the other hand, may kill a fixed percentage of a population regardless of density. This is an example of a(n) (60) _____ effect.

The type of population growth a species exhibits is often correlated with other life history strategies. Species that show exponential growth when conditions are favorable are said to be (61) _____. Species exhibiting logistic growth and that utilize density dependent mechanisms to maintain population size near carrying capacity are referred to as (62) _____.

Choice

For each of the following population characteristics, choose the more likely time of reproduction.

a. *r*-selected b. K-selected

63. _____ Small size

64. _____ Long lifespan

65. _____ Type III survivorship curve

66. _____ Multiple reproductive events in an individual's lifetime

67. _____ Substantial parental care of offspring

49.7 Human Population Growth [pp. 1145-1149]

Section Review (Fill-in-the-Blanks)

The relationship between a country's population growth pattern and economic development can be depicted graphically in a(n) (68) _____. Many countries with rapid population growth are attempting to lower birthrates by developing (69) _____.

Labeling

Indicate whether each age structure pyramid predicts increasing, decreasing, or stable population size.

70. ______________

71. ______________

72. ______________

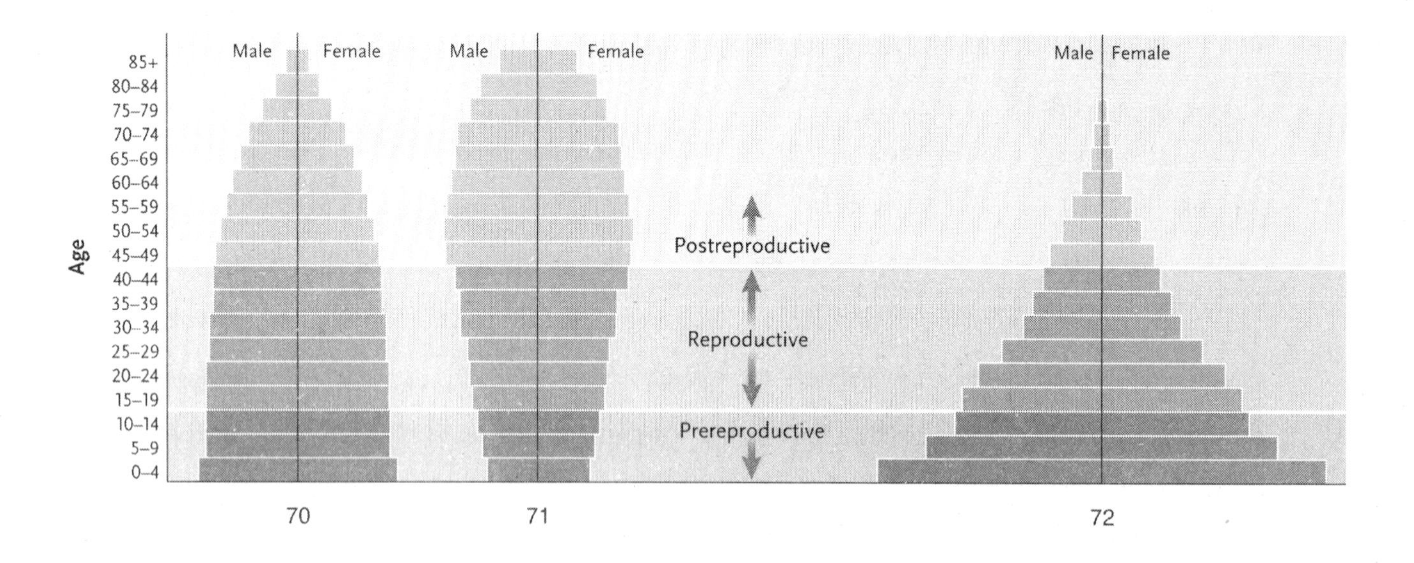

SELF-TEST

1. Which of the following is NOT a branch the science of ecology? [p. 1126]

a. abiotic ecology

b. community ecology

c. organismal ecology

d. population ecology

2. Which dispersal pattern best describes the distribution of individuals across the U.S.? [p. 1128]

a. clumped

b. even

c. random

d. uniform

3. The statistical study of the factors that affect population size and density is called ________. [p. 1130]

a. biogeography

b. demography

c. community ecology

d. logistics

4. Which of the following pairs represents opposing life history strategies? [p. 1133]

a. fecundity vs. parental care

b. geographic range vs. habitat

c. intraspecific competition vs. population density

d. b and c

5. The exponential model of population growth assumes ________. [p. 1136]

a. birthrate is higher than deathrate

b. density dependent factors are involved

c. ideal conditions with no limits to growth

d. that species are K-selected

6. Which of the following is a key component of the logistic model of population growth? [p. 1139]

a. carrying capacity

b. density dependent factors

c. density independent factors

d. a and b

7. Which of the following is NOT a density dependent factor affecting population size? [p. 1140]

a. crowding

b. intraspecific competition

c. nighttime air temperature

d. spread of disease

8. ________ cycles in population size may be due to predation or food availability. [p. 1145]

a. Exponential

b. Extrinsic

c. Intrinsic

d. Logistic

9. It took ________ years for the human population of the world to reach 1 billion, but only ________ years for it to grow from 5 to 6 billion. [p. 1147]

a. 3.5 billion; 1.5 million

b. 1.0 billion; 5 million

c. 250 million; 500

d. 2.5 million; 12

10. Which population characteristic would be most important in determining whether a community should spend money on schools vs. nursing homes? [p. 1147]

a. age structure

b. carrying capacity

c. population density

d. sex ratio

INTEGRATING AND APPLYING KEY CONCEPTS

1. *Unanswered Questions:* Compare the impact of commercial fishing on fish populations to the effect of pike cichlids on guppy populations in Trinidad. Explain any similarities or differences you would expect. Besides ignoring the effects of selection in population studies, what are some other problems involved in studying populations in a natural setting vs. in a laboratory environment?

2. *Insights from the Molecular Revolution:* Using the techniques described to track the migration of nine-banded armadillos (as well as others that you can think of), design an experiment to determine what factors influence long-range migration in armadillos.

3. *Focus on Research:* Recall the discussion of modes of natural selection from Chapter 20. Speculate on what might happen if a population of guppies was placed in a stream that contained both killifish and pike cichlids. Explain your reasoning.

50

POPULATION INTERACTIONS AND COMMUNITY ECOLOGY

CHAPTER HIGHLIGHTS

- Populations adapt not only to the physical characteristics of their environment, but also to the other organisms they encounter, a phenomenon known as coevolution.
- Communities may differ in the number of species present and the relative abundance of each species. Trophic structure may also be different in different communities.
- Symbioses, such as interspecific competition and predation, often shape community characteristics.
- Species richness in a community may actually be increased by predation and/or moderate levels of disturbance. The theory of island biogeography is a useful model to explain variations in species richness.
- Communities undergo a characteristic sequence of changes after disturbance, a process known as ecological succession.

STUDY STRATEGIES

- The emphasis of this chapter is on the interactions of populations of different species. These interactions often have a major impact on the species involved. If so, the species are said to be symbiotic. Symbiotic relationships may affect populations in positive or negative ways. The effect of the abiotic part of the environment will be addressed in the next chapter.
- Again, keep in mind that many of the concepts presented in this chapter are presented in a somewhat simplified way to make them easier to understand. In reality, many factors interact with one another, sometimes producing unexpected results.

INTERACTIVE EXERCISES

Why It Matters [pp. 1151 -1152]

Section Review (Fill-in-the-Blanks)

Cowbirds are (1) _____, an animal that lays its eggs in the nest of an individual of another species. The host species is "tricked" into raising the cowbird young as its own. On the surface, this would seem to affect the host species negatively, since cowbird chicks get fed food that would otherwise go to the hosts' young. But in Central America, giant cowbird chicks not only take food from the young of their oropendola hosts, but in some cases they also protect the young from parasitic botflies. The net effect is beneficial to the oropendolas, an example of one of the many counterintuitive results of community interactions. If the oropendolas nest near colonies of bees or wasps that repel botflies, oropendolas will eject cowbird eggs from their nests since there is no need for protection from botflies. The relationships between these different organisms is characteristic of the (2) _____, an assemblage of species living and interacting in an area.

50.1 Population Interactions [pp. 1152 – 1160]

Section Review (Fill-in-the-Blanks)

The evolutionary adaptation of two species to each other is (3) _____. Often these adaptations are obvious, e.g., predators evolving features that help them capture prey and prey evolving features that help them avoid predation. Other adaptations are less obvious.

Animals are heterotrophic and obtain energy for living from the environment. The consumption of one living animal by another is called (4) _____, while the consumption of a living plant is called (5) _____. These consumer animals have evolved many adaptations to increase the efficiency of food acquisition. (6) _____ is a model that predicts the behavior of animals based on the costs and benefits associated with different foods. Prey species have evolved many means to foil predators. One of those is (7) _____, a type of camouflage to make detection by predators difficult. Predators may also use this strategy to avoid detection and flight by prey. Prey may also protect themselves against predation by evolving various defense mechanisms, such as spines or a hard shell. Some species accumulate high concentration compounds in their tissues that make them toxic or distasteful to predators. These species often advertise their unpalatability with conspicuous, bright (8) _____ coloration. Palatable species may add another layer of complexity to the mix by evolving a resemblance to the unpalatable species, a phenomenon known as (9) _____. The unpalatable species is referred to as the (10) _____, while the palatable one is the (11) _____. If two species are both unpalatable and have evolved a similar appearance, (12) _____ has occurred.

Two or more individuals of different species striving to obtain the same limited resource is (13) _____. This phenomenon takes two forms; (14) _____, in which individuals interact directly and attempt to harm members of the other species; and (15) _____, where simple use of the resource by one species reduces its availability to the other. The experiments of G. F. Gause indicated that when two species are striving for the same limited resource in the same way, the more efficient population eliminates the other from the system. Gause used this observation to define his (16) _____. Resource utilization and the environmental conditions an organism requires are major components of an organism's (17) _____. The range of conditions and resources that an organism can potentially use defines its (18) _____, while the range of conditions and resources it actually uses in nature is its (19) _____. The fact that two or more species are using the same resource does not necessarily imply that competition is occurring. (20) _____, the use of different resources or using them in different ways can minimize interspecific competition. Often, closely related species living in the same area will specialize on different resources and evolve structures to help them exploit the specific resource, a phenomenon known as (21) _____.

Two or more species may have an unusually intimate ecological association with each other, a condition known as a(n) (22) _____. There are three kinds of this type of association: (23) _____, where both species benefit; (24) _____, where one species benefits and the other is unaffected; and (25) _____, where one species benefits and the other is harmed by the association. In the case of parasitic relationships, the harmed organism is the (26) ______.

If the parasite lives on the surface of the host, e.g., a leech, it is referred to as a(n) (27) ______. If it lives within the body of the host, e.g., a tapeworm, it is a(n) (28) ______. True parasites usually do not kill their hosts. (29) ______ are animals such as insects that lay eggs within the larvae of host insects. When the eggs hatch, they consume the host from within, killing it in the process.

Choice

Choose the correct pair of effects on two species involved in the relationships listed.

a. Species A benefits; Species B is harmed b. Species A benefits; Species B is unaffected
c. both species benefit d. both species are harmed

30. ____ commensalism
31. ____ competition
32. ____ herbivory
33. ____ mutualism
34. ____ parasitism
35. ____ predation

Short Answer

Imagine 4 species: A and B are Mullerian mimics, C is a noxious model, and D is its Batesian mimic. Do you think each of the following scenarios would be beneficial or detrimental to each species? Explain your answers.

36. A greatly outnumbers B ______________________________

37. C greatly outnumbers D ______________________________

38. D greatly outnumbers C ______________________________

50.2 The Nature of Ecological Communities [pp. 1160 – 1163]

Section Review (Fill-in-the-Blanks)

The American ecologist Frederic Clements hypothesized that in a mature community the exact mix of species, or its (39) ______, was at equilibrium. If the community was disturbed, it would return to its equilibrium state. The transition zone between adjacent communities is referred to as a(n) (40) ______.

Matching

Match each of the following persons with the correct concept.

41. ____ Frederic Clements

42. ____ Harry Gleason

43. ____ Robert Whittaker

A. Proposed the individualistic hypothesis of community structure. Species are frequently found together because they are bound by complex interactions.

B. Proposed the interactive hypothesis of community structure. Species are frequently found together simply because they are adapted to similar environmental conditions.

C. Tested the interactive and individualistic models in nature.

50.3 Community Characteristics [pp. 1163 – 1166]

50.4 Effects of Population Interactions on Community Characteristics [pp. 1166-1167]

50.5 Effects of Disturbance on Community Characteristics [pp. 1167-1170]

Section Review (Fill-in-the-Blanks)

The number of species found in a community comprises that community's (44) _____. The relative commonness or rarity of a species in a community is its (45) _____. Combining these two components of a community produce a statistic called (46) _____. Communities also have feeding relationships among their species that can be arranged in a hierarchy of (47) _____. At the base of this hierarchy are photosynthetic organisms or (48) _____ also known as (49) _____. Animals and other nonphotosynthesizers are (50) _____. Herbivores occupy the second level and are called (51) _____. Carnivores that feed on the herbivores occupy the third level and are called (52) _____. Carnivores on the fourth level are called (53) _____ and so on. Animals that feed on organisms from several levels are called (54) _____. Relatively large animals that feed on dead animals or organic matter are called (55) _____, while smaller bacteria and fungi are (56) _____. Because they are nonphotosynthetic, all consumers in the community are called (57) _____; they must eat other organisms to obtain energy and nutrients.

The trophic structure of a community is often depicted in the form of a(n) (58) _____, a simple, linear diagram that shows who eats who. But a more realistic depiction is the (59) _____, since it allows for organisms to feed on different foods and from different trophic levels. Robert MacArthur drew a parallel, since confirmed, between species diversity in a community and trophic complexity. He argued that the more links between trophic levels, the greater the (60) _____ of the community; loss of one or a few species did not have a great impact on richness or relative abundance of species.

Interspecific competition and predation are important factors affecting community structure. In the case of predation, a predatory species may enhance species richness by keeping the best competitors at bay and allowing weaker competitors to get established. This type of predator is called a(n) (61) _____.

In the late 1970s, Joseph Connell hypothesized that communities that are subject to disturbances of moderate frequency and intensity show greater diversity than those that are disturbed severely and/or with great frequency or those that are rarely disturbed. The former are dominated by *r*-selected species during recovery from disturbance. The latter are dominated by K-selected species. This idea became known as the (62) _____.

Matching

Match each of the organisms with an appropriate term.

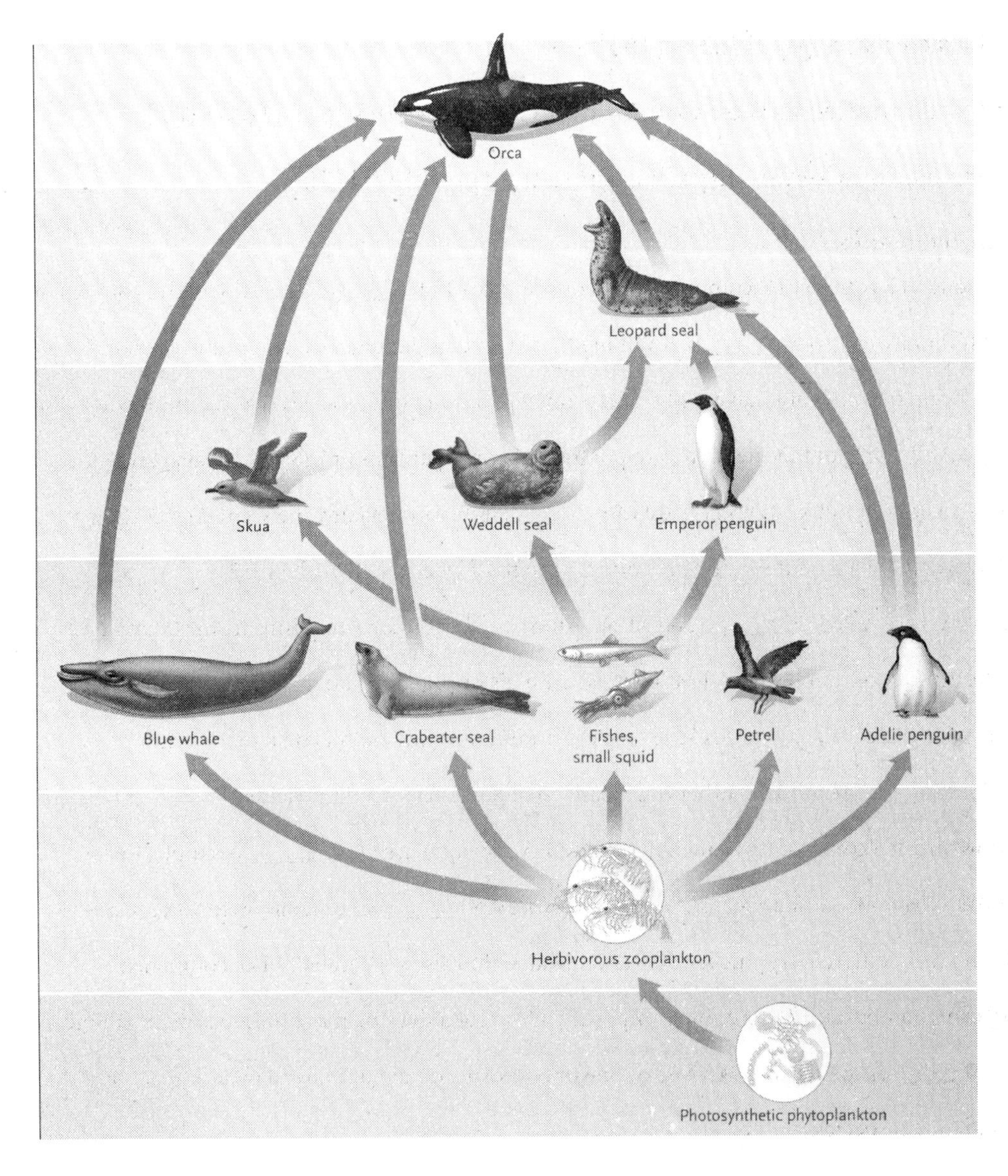

63.	____ autotroph		A.	crabeater seal
64.	____ carnivore		B.	emperor penguin
65.	____ herbivore		C.	orca
66.	____ primary consumer		D.	phytoplankton
67.	____ primary producer		E.	zooplankton

50.6 Ecological Succession: Responses to Disturbance [pp. 1170 – 1174]

50.7 Variations in Species Richness among Communities [pp. 1174 – 1176]

Section Review (Fill-in-the-Blanks)

When a community is disturbed, it does not immediately return to its equilibrium state, but rather goes through a sequence of changes known as (68) _____. If an area has never supported a community before, e.g., a volcanic island, the sequence of changes is referred to as (69) _____. The final, relatively stable community in the sequence is the (70) _____. When an existing community is disturbed, the recovery process is called (71) _____. The process of ecosystem development is not restricted to terrestrial communities. If the process occurs in, say, a lake ecosystem, we refer to it as (72) _____.

Several hypotheses have been proposed to explain the causes of ecological succession. The oldest of these is the (73) _____ hypothesis, which states that earlier communities alter conditions in ways that make it easier for later communities to get established. The (74) _____ hypothesis is based on the notion that earlier communities resist invasion by new species, but disturbances and life span of species ultimately lead to changes in community structure. The (75) _____ hypothesis is based on competitive superiority of some species over other. The three models are not mutually exclusive and all three may play a role in succession. A(n) (76) _____ may occur when biotic or abiotic factors prevent succession from proceeding to true climax.

MacArthur and Wilson's (77) _____ predicts species richness on islands based on island size, distance from the mainland, and immigration and extinction rates.

Matching

Match each island with the appropriate characteristics.

78.	_____	large, distant island	A.	high immigration rate, high extinction rate
79.	_____	large island, close to mainland	B.	high immigration rate, low extinction rate
80.	_____	small, distant island	C.	low immigration rate, high extinction rate
81.	_____	small island, close to mainland	D.	low immigration rate, low extinction rate

SELF-TEST

1. Which of the following is most likely a result of coevolution? [p. 1152]
   a. ability to tolerate extreme low temperatures
   b. Batesian mimicry
   c. specialized digestive organs
   d. territoriality

2. Which of the following is NOT a symbiosis? [p. 1159]
   a. commensalism
   b. intraspecific competition
   c. mutualism
   d. predation

3. The transition zone between adjacent communities that often has a high species diversity is called a(n) ________. [p. 1163]
   a. climax community
   b. ecotone
   c. ecotype
   d. trophic zone

4. Which of the following are components of species diversity in a community? [p. 1165]
   a. number of species
   b. relative abundance of the different species
   c. uniqueness of density
   d. a and b.

5. Organisms can be classified as ________ based on how they obtain the energy necessary for life. [p. 1165]
   a. consumers
   b. decomposers
   c. detritivores
   d. all of the above

6. A ________ is an organism that has a greater impact on community structure than its numbers might suggest. [p. 1169]
   a. keystone species
   b. Mullerian mimic
   c. mutualist
   d. primary producer

7. Which of the following community characteristics would be predicted by the intermediate disturbance hypothesis? [p. 1171]
   a. a preponderance of K-selected species
   b. a preponderance of *r*-selected species
   c. extreme sensitivity to slight disturbances
   d. high species diversity

8. In which of the following areas would you expect to observe primary succession? [p. 1172]
   a. a clear-cut forest with all plant material removed
   b. a prairie after a fire
   c. an abandoned farm
   d. sand dunes created by a receding shore

9. Which of the following is NOT a hypothesis that attempts to explain the processes that drive succession? [p. 1174]

a. equilibrium hypothesis

b. facilitation hypothesis

c. inhibition hypothesis

d. tolerance hypothesis

10. Which assumption is necessary in MacArthur's and Wilson's equilibrium theory of island biogeography? [p. 1176]

a. Islands close to the mainland have relatively high immigration rates.

b. Large islands have relatively low extinction rates.

c. Large islands have relatively low immigration rates.

d. Small islands have relatively low extinction rates.

INTEGRATING AND APPLYING KEY CONCEPTS

1. *Unanswered Questions:* Now that you have a basic understanding of coevolution, speculate on the relationship between an organism's evolutionary history and how that might relate to answering questions about the relative importance of positive and negative interactions on community structure.

2. *Insights from the Molecular Revolution:* The plasmid in the *Rhizobium* bacterium that promotes its mutualistic association with various legumes is similar to the plasmid in another bacterium that causes crown gall disease in deciduous trees. Speculate on the evolutionary relationship between the two bacteria.

3. *Focus on Research:* Review Wilson and Simberloff's test of the equilibrium theory of island biogeography in terms of island size and distance from the mainland. Which predictions of the theory were supported by their results? Besides testing the theory, how else was this research important to field ecologists?

51

ECOSYSTEMS

CHAPTER HIGHLIGHTS

- The interaction of the biologic community with the physical, nonliving part of the environment is an ecosystem.
- Ultimately the energy to maintain ecosystem structure comes from the sun. Ecosystems are usually open systems with respect to energy. In other words, energy enters an ecosystem as solar radiation and leaves as entropy. Energy flows through ecosystems.
- Unlike energy, the materials that are used for ecosystem structure are in finite supply and must be recycled between the living and nonliving parts of the system.
- Ecosystems are extremely complex systems with many components. Ecologists often attempt to understand ecosystem function by cutting through the noise in the system and creating models that incorporate only those variables that have the greatest impact on ecosystem function.

STUDY STRATEGIES

- It is important to understand that energy FLOWS THROUGH ecosystems. When the source of that energy, the sun, dies, so will ecosystems and, indeed, all life on earth. Food webs and ecological pyramids are simply ways to track the flow of energy through an ecosystem.
- It is equally important to realize that the supply of elements that make up the organisms in an ecosystem are in finite supply. When an organism dies, the substances that make up its cells and other structures must be broken down and used by other organisms. Thus, material CYCLES BETWEEN the living and non-living parts of the system. This is the basic premise behind the study of nutrient cycling.
- Modeling is a method of reducing a complex system to what are believed to be its most important components. It is a compromise. The goal is to create a concept that helps understand the general operation of the system, while keeping in mind the fact that in specific cases, factors not incorporated into the model can have an impact.

INTERACTIVE EXERCISES

Why It Matters [pp. 1181 -1182]

Section Review (Fill-in-the-Blanks)

The biologic community and its interactions with the physical part of its environment is a(n) (1) _____. The Lake Erie ecosystem suffered severe damage due to human activity and, while it has recovered to some extent, it will never return to its former condition.

51.1 Energy Flow and Ecosystem Energetics [pp. 1182 – 1190]

Section Review (Fill-in-the-Blanks)

The conversion of solar energy into chemical energy by autotrophs is known as (2) _____. Primary producers are, however, complex organisms, and the energy required to maintain themselves is referred to as (3) _____. Any energy left after deducting the amount spent on maintenance is (4) _____, and this energy can be used for growth and/or reproduction. There are several ways to estimate these different forms of energy. One is by measuring (5) _____, the dry weight of organic matter per unit area or volume of a habitat. (6) _____ is the total amount of organic matter in an area, not the rate of synthesis of that matter. Like all organisms, plants need a variety of materials to live and function. The element in shortest supply is called a(n) (7) _____, because its scarcity determines productivity.

Heterotrophic organisms acquire some of the energy accumulated by autotrophs when they consume the autotrophs. Any energy left over after consumer maintenance is stored as (8) _____. That energy is available to organisms at the next trophic level. The ratio of net productivity at one trophic level to that of the level below it is known as (9) _____. The flow of energy through a food web is often represented diagrammatically in the form of (10) _____, of which there are three types. The (11) _____ depicts the amount of energy temporarily stored in a trophic level as dry weight. This pyramid is often inverted in aquatic systems if the producers reproduce rapidly, i.e., they have a high (12) _____. The amount of energy at a trophic level often affects the population size of the organisms at the next higher level, thus the (13) _____ uses the number of individuals at a particular trophic level to represent energy. This pyramid can also be inverted if the producer organisms are large and not too numerous relative to the next trophic level, e.g., trees whose leaves support insect herbivores. The only pyramid that cannot be inverted is the (14) _____ since it is the only one that actually measures energy flow, and the second law of thermodynamics dictates that it must be upright. Some interactions between two trophic levels have indirect impact on other levels, a phenomenon known as (15) _____.

Matching

Match each of the following ecological efficiencies with the correct definition.

16.	____	assimilation efficiency	A. Ratio of energy absorbed from food consumed to total energy content of food
17.	____	harvesting efficiency	B. Ratio of energy content of food consumed to energy content of available food
18.	____	production efficiency	C. Ratio of energy content of new tissue produced to energy assimilated from food

True/False

If the statement is true, write a "T" in the blank. If the statement is false, make it correct by changing the underlined word(s) and writing the correct word(s) in the answer blank.

19. ____________ A pyramid of biomass is sometimes inverted because it sometimes <u>overestimates</u> the importance of large organisms.

20. ____________ A pyramid of energy can never be inverted because the <u>transfer of energy from one trophic level to the next is never 100% efficient.</u>

21. ____________ A pyramid of numbers is sometimes inverted because it underestimates the importance of <u>small</u> organisms.

51.2 Nutrient Cycling in Ecosystems [pp. 1191 – 1199]

50.3 Ecosystem Modeling [pp. 1199 - 1200]

Section Review (Fill-in-the-Blanks)

The circulation of nutrients from the living to the nonliving part of an ecosystem and back again is referred to as a(n) (22) _____. The (23) _____ model describes nutrient cycling based on whether molecules and ions are available or unavailable for assimilation and whether they are in organic or inorganic form.

Although not technically a nutrient, water cycles through ecosystems much as minerals do. This cycling is referred to as the (24) _____ cycle. It is based on evaporation and precipitation of water.

The (25) _____ cycle is based on the activity of prokaryotic organisms, since they are the only organisms capable of converting atmospheric nitrogen (N_2) into organic form that can be utilized by eukaryotes, a process known as (26) _____. The breakdown of organic nitrogen in detritus into ammonia by various bacteria and fungi is (27) _____. This ammonia can be assimilated by plants. The ammonia may also be converted to nitrites and nitrates by other bacteria in a process called (28) _____. Still other bacteria convert unused nitrites and nitrates into N_2 through the process of (29) _____. Other important nutrient cycles include the (30) _____ and (31) _____ cycles.

Ecosystem modeling helps in the understanding of ecosystem dynamics. Conceptual models tend to be more general in scope than (32) _____ models which, if well constructed, can give precise answers to specific questions about ecosystems without disturbing them.

Choice

Choose the biogeochemical cycle associated with each item in the following list.

a. carbon cycle b. hydrologic cycle c. nitrogen cycle d. phosphorus cycle

33. ____ ammonification
34. ____ fixation into organic form mainly by photosynthesis
35. ____ fixation into organic form mainly by prokaryotes
36. ____ largest available reservoir is bicarbonate ions in the oceans
37. ____ largest available reservoir is the atmosphere
38. ____ largest available reservoir is the earth's crust
39. ____ largest available reservoir is the earth's oceans

SELF-TEST

1. Energy in ecosystems usually flow through which two types of food webs? [p. 1182]
   a. detrital and decomposer food webs
   b. grazing and detrital food webs
   c. predator and prey food webs
   d. producer and consumer food webs

2. The rate of conversion of solar energy into chemical energy by autotrophs is called ________. [p. 1183]
   a. gross primary productivity
   b. net primary productivity
   c. respiration
   d. standing crop biomass

3. Which of the following is used to estimate the amount of energy being processed each trophic level in a food web? [p. 1188]
   a. biomass at each trophic level
   b. number of individual organisms at each trophic level
   c. total solar input
   d. a and b

4. Energy stored by consumers as it is transferred to them from producers is ________. [p. 1185]
   a. gross primary productivity
   b. net primary productivity
   c. respiration
   d. secondary productivity

5. The two types of nutrient cycles are ________. [p. 1191]
   a. atmospheric and sedimentary
   b. biotic and abiotic
   c. organic and inorganic
   d. terrestrial and aquatic

6. Carbon enters the biotic (living) component of the ecosystem via ________. [p. 1192]
   a. ammonification
   b. photosynthesis
   c. precipitation
   d. respiration

7. Which of the following would NOT cause levels of atmospheric carbon dioxide to rise? [p. 1194]
   a. burning fossil fuels
   b. destruction of terrestrial forests
   c. increased rates of organismal respiration
   d. increased rates of photosynthesis

8. Which of the following processes returns N_2 to the atmosphere? [p. 1195]
   a. ammonification
   b. denitrification
   c. nitrification
   d. nitrogen fixation

9. Which of the following is true regarding the phosphorus cycle? [p. 1197]

a. Excess phosphorus from fertilizers is a pollutant of lakes and ponds.

b. Phosphorus cycles slowly through terrestrial ecosystems.

c. Phosphorus is made available to plants through fixation by prokaryotes.

d. The atmosphere is the largest reservoir of phosphorus.

10. Which of the following is NOT true of ecosystem modeling? [p. 1200]

a. It attempts to identify the most important factors that are involved in ecosystem function.

b. It describes ecosystem processes but has no predictive value.

c. It simplifies events that occur in nature.

d. It uses mathematical equations to define relationships between populations and the environment.

INTEGRATING AND APPLYING KEY CONCEPTS

1. *Unanswered Questions:* If human activity is increasing CO_2 levels in the atmosphere and causing a rise in temperature, what might be the effect on ecosystem processes such as carbon fixation, decomposition, productivity and nutrient cycling?

2. *Insights from the Molecular Revolution:* Based on what you have learned about the use of mtDNA to determine migration routes of loggerhead turtles, can you think of any other interesting or important questions regarding the movement of other animals (including humans) that might be solved using this methodology?

3. *Focus on Research:* Summarize the environmental impact of deforestation based on what you have learned about the carbon cycle in this chapter and the Bormann and Likens experiments in the Hubbard Brook Forest watershed.

52

THE BIOSPHERE

CHAPTER HIGHLIGHTS

- The earth's ecosystems interact with each other to comprise the biosphere.
- Organisms use a wide variety of homeostatic mechanisms to cope with variations in environmental conditions, which may occur on daily or annual cycles or randomly.
- Biomes are characterized on the basis of the types of plants they contain and the heterotrophs associated with them. Climatic factors often determine the flora and fauna of a specific biome.
- The three broad categories of biomes are terrestrial, freshwater, and marine.

STUDY STRATEGIES

- The sun is the ultimate source of energy to sustain all biomes. Temporal and geographic differences in solar input are responsible for variations in productivity as well as determining climatic and meterological conditions. Understand how and why these differences exist and what their impact on specific biomes is.
- Learning the different kinds of biomes and the organisms that comprise them is, to some degree, a memory exercise. It will help if you think of the differences outlined above and how they might influence productivity patterns, species diversity, and the types of organisms found in a biome.

INTERACTIVE EXERCISES

Why It Matters [pp. 1203-1204]

Section Review (Fill-in-the-Blanks)

Periodic changes in Pacific Ocean currents called El Niño and La Niña illustrate how events occurring in one ecosystem can affect other, distant ecosystems. The sum of the earth's interacting ecosystems is called the (1) _____, which includes anyplace where living things are found. Its components are (2) _____, which includes all of the earth's bodies of water and polar ice caps, the (3) _____, which includes the rocks, sediments and soils of the earth's crust, and the (4) _____, which is made up of the gases and airborne particulates that surround the earth.

52.1 Environmental Diversity of the Biosphere [pp. 1205-1209]

Section Review (Fill-in-the-Blanks)

A region's (5) _____ is the sum of numerous abiotic factors, such as sunlight, precipitation, and temperature, that occur over an extended period and are somewhat predictable. Because of the tilt of the earth on its axis, the intensity of solar radiation differs at different points on the earth's surface. For example, the only places on earth where the sun is ever directly overhead are the (6) _____, the portion of the earth's surface between 23.5° N and S latitude. As the surface heats, the air above it is heated and rises. As it rises it cools, a phenomenon known as (7) _____ cooling, an important factor in creating atmospheric currents such as prevailing winds. Oceans and ocean currents also affect local climatic conditions so that coastal cities often have a(n) (8) _____ climate which is milder than that of an inland city that will have a(n) (9) _____ climate, even at a similar latitude. Seasonal reversals in wind direction can cause (10) _____. Topography can also affect local climate. Moist air that is forced up a mountain slope loses its water vapor to precipitation on the upslope. After passing the crest of the range, the air descends, compresses, and warms, leaving the leeward side warm in a(n) (11) _____. For many organisms, the abiotic conditions in their immediate proximity, known as the (12) _____, play an important role in their distributions.

Matching

Match each of the following phenomena with the correct definition.

13.	____	adiabatic cooling	A. A decrease in air temperature without loss of heat energy
14.	____	Coriolis effect	B. A force produced by the rotation of the earth that alter the direction of movement of air masses
15.	____	upwelling	C. The replacement of surface waters pushed out to sea by winds with deeper, nutrient-rich waters

Choice

For each of the following latitudes, choose the biome most likely to be found there.

a. desert b. temperate forest c. tropical rain forest

16. ____ equator
17. ____ 30° N and S latitude
18. ____ 60° N and S latitude

52.2 Organismal Responses to Environmental Variation [pp. 1209-1211]

52.3 Terrestrial Biomes [pp. 1211-1219]

Section Review (Fill-in-the-Blanks)

The main determinant of biome distribution is climate. Two aspects of climate, temperature and precipitation, may be displayed in the form of a(n) (19) _____ to give a visual representation of environmental variation.

(20) _____ is a biome characterized by frequent rains (> 250 cm/year) and high temperatures (annual mean temperature > 25° C.) and high humidity. Because water and temperature are not limiting, productivity is high. Habitats near 20° N and S latitude with rainy summers and dry winters often have (21) _____ as a common biome. Higher elevations in tropical regions often have (22) _____, or "cloud forests" as a common biome. Because the high humidity and lack of sunlight limit production, trees often do not grow very tall. (23) _____, a tropical grassland with few trees, is characterized by distinct rainy and dry seasons, for which the grasses are well adapted. At the margins of these ecosystems are (24) _____, where plants store much energy in extensive root systems. At approximately 30° N and S latitude, where descending air masses, depleted of moisture and heated by compression, reach the surface, most of the world's (25) _____ are found. A biome found in coastal regions with cool, wet winters

and hot, dry summers in which scrubby trees and shrubs are typical is (26) _____. (27) _____, such as the American prairie, are disclimax communities where periodic fires are common and important. (28) _____ is a biome commonly found in regions with warm summers and cold winters that force broadleaf plants to shed their leaves. As these leaves decompose, they return nutrients to the soil. (29) _____ or (30) _____ is a biome found in polar regions of the northern hemisphere characterized by various evergreen trees such as white spruce and balsam fir. The northernmost North American biome is the (31) _____, so cold and windy that it is treeless. Even in summer only the topmost layers of soil thaw and deeper layers are frozen year round and known as (32) _____. A similar biome, the result not of high latitude but of high altitude, is (33) _____.

Matching

Match the species groupings with the terrestrial biome in which they are likely to be common.

34.	____	desert	A. Cacti, mesquite, scorpions, lizards
35.	____	temperate grassland	B. Grasses, sunflowers, jackrabbits, wolves
36.	____	savanna	C. Grasses, shrubby tress, zebras, giraffes
37.	____	taiga (boreal forest)	D. Oak and maple tress, squirrels, deer
38.	____	temperate deciduous forest	E. Shade tolerant shrubs, lianas, epiphytes, orchids
39.	____	tropical rain forest	F. Mosses, lichens, lemmings, lynx
40.	____	tundra	G. White spruce, balsam fir, moose, bears

52.4 Freshwater Biomes [pp. 1219-1221]

Section Review (Fill-in-the-Blanks)

Characteristics of freshwater ecosystems are strongly affected by the depth to which light penetrates. The depth to which light can reach is the (41) _____, while depths below which light can penetrate constitute the (42) _____. Marshes and swamps that border freshwater ecosystems, or (43) _____, are highly productive. The (44) _____ is the region of a pond or lake where light can reach the bottom and rooted plants grow. The (45) _____ is the region of deeper water to which light reaches. The (46) _____ zone is the deeper water where sunlight can never penetrate.

Two times a year in lakes, seasonal changes in water temperature and winds cause the (47) _____ and (48) _____, which brings nutrient-rich water to the surface and oxygen-rich surface water to the deepest part of the lake. During the summer, lake waters are thermally stratified into a warm upper layer called the (49) _____ and a deeper, colder (50) _____. The sharp dividing line between these layers is the (51) _____, where water temperature changes rapidly with depth. Lakes can also be classified according to their productivity. Nutrient-poor, highly oxygenated lakes are often crystal clear due to lack of primary productivity and are called (52) _____. Nutrient-rich lakes with much decomposing organic matter and low levels of oxygen are said to be (53) _____. They are highly productive but often unattractive for recreation due to algal blooms.

Matching

Match each of the indicated regions of the lake in the diagram with its correct term.

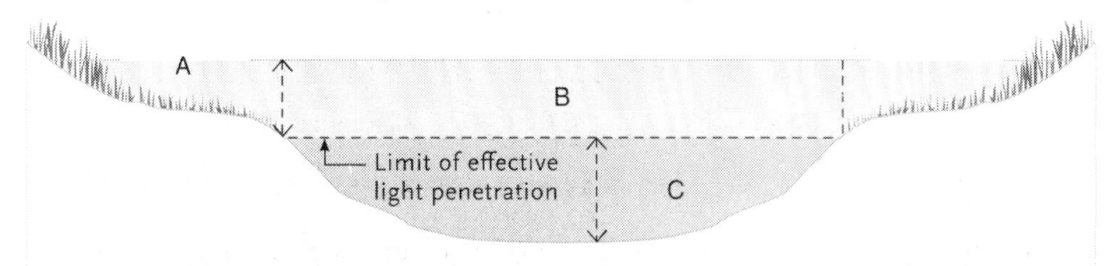

54. ____ Aphotic zone
55. ____ Limnetic zone
56. ____ Littoral zone
57. ____ Photic zone
58. ____ Profundal zone

52.5 Marine Biomes [pp. 1221-1225]

Section Review (Fill-in-the-Blanks)

Marine biomes share some features with freshwater biomes, but obvious differences are the salinity and size of marine biomes. Like freshwater systems, the oceans can be divided into zones or "provinces." Among those are the ocean floor, or the (59) _____ province. This region may be divided into the (60) _____ zone, a region alternately submerged by the tides and exposed to the air and the (61) _____ zone, the sediments that lie below deeper waters. The free water above the ocean floor, or the (62) _____ province, is subdivided into the (63) _____ zone, which includes open waters to the depth of light penetration, and the deep waters of the (64) _____ zone beyond the shelf. Other unique marine biomes include (65) _____ where freshwater from rivers mixes with salty ocean water. Salinity changes over a matter of hours as tides rise and fall; the organisms in this habitat must posses mechanisms to cope with these changes. These biomes are often bordered by (66) _____, tidal wetlands dominated by grasses and reeds. Warm, nutrient-poor tropical waters often support (67) _____, which are often very structurally complex, highly productive, and possess a high species diversity. In the waters of the open ocean are numerous organisms capable of swimming against the currents. These are called (68) _____, as opposed to plankton, which simply drift with the currents. Other species which live on or in the sediments of the ocean floor are called (69) _____.

Matching

Match the species groupings with the marine biome in which they are likely to be common.

70. ____ coral reefs
71. ____ deep benthos
72. ____ estuaries
73. ____ intertidal zone
74. ____ neritic zone

A. Algae, barnacles , mussels, sea stars
B. Endosymbiotic dinoflagellates, corraline algae, diverse invertebrates
C. Kelp forests, lobsters, large, commercially important fishes
D. Salt-tolerant grasses, shorebirds, marine arthropods
E. Worms, mollusks, diverse detritovore species

SELF-TEST

1. A changing the tilt of the earth's axis from 23.5° to 0° would most affect which of the following? [p. 1206]
   a. direction of prevailing winds
   b. length of a day
   c. length of a year
   d. the seasons

2. At 30° latitude many of the earth's ________ are found as ________. [p. 1207]
   a. deserts; dry air is heated as it descends are compresses
   b. grasslands; water evaporates from the surface and warms as it rises
   c. rain forests; water precipitates from rising, moist air
   d. taigas; cold, dry air descends to the surface

3. Which of the following is a homeostatic mechanism used by organisms to cope with environmental variation? [p. 1212]
   a. behavior
   b. torpor
   c. both a and b are correct
   d. neither a nor b is correct

4. Which of the following biomes would most likely contain the greatest number of different species? [p. 1214]
   a. desert
   b. savanna
   c. temperate deciduous forest
   d. tropical rain forest

5. Which biome would have the shortest growing season? [p. 1120]
   a. arctic tundra
   b. prairie
   c. savanna
   d. tropical rain forest

6. An example of a lotic ecosystem is a(n) ________. [p. 1221]
   a. cloud forest
   b. lake
   c. river
   d. savanna

7. During the spring and fall turnovers, ________. [p. 1223]

a. nutrient rich water sinks to the bottom

b. oxygen rich water rises to the top

c. a thermocline develops

d. none of the above

8. Which of the following describes a clear lake with few nutrients and low productivity? [p. 1223]

a. eutrophic

b. heterotrohic

c. oligotrophic

d. phototrophic

9. Where would you expect to find the most phytoplankton? [p. 1227]

a. abyssal zone

b. aphotic zone

c. benthic province

d. neritic zone

10. Organisms that are large and/or strong enough to swim against ocean currents are known as ________? [p. 1227]

a. benthos

b. flotsam

c. nekton

d. plankton

INTEGRATING AND APPLYING KEY CONCEPTS

1. *Unanswered Questions:* Populations of animals and plants have always evolved in response to the environmental challenges that they face. With that in mind, why is global warming considered such a serious threat? Isn't it reasonable to assume that organisms will simply evolve in response to the new conditions created by global warming?

2. *Insights from the Molecular Revolution:* Review Watson and Crick's use of Rosalind Franklin's x-ray diffraction data from Chapter 14 and discuss how that technique helped them to discern the structure of DNA and also led Sicheri and Yang to their discoveries about fish antifreeze proteins.

3. *Focus on Research:* Often unknown species and unusual species interactions are discovered only when we get access to hard-to-reach parts of an ecosystem, e.g., the rainforest canopy. Can you think of other places that have not been well studied, but which might yield valuable information about ecological interactions? What methods would you use to gain access to these sites?

53

BIODIVERSITY AND CONSERVATION BIOLOGY

CHAPTER HIGHLIGHTS

- Preservation of biodiversity is not merely an exercise in aesthetics; biodiversity directly benefits humans.
- Human activities are threatening many species, and extinction rates are rising as a result of these activities.
- Recent advances in systematics and population genetics have provided useful tools to conservation biologists.
- Preserving and restoring habitat is a major focus of conservation biologists.

STUDY STRATEGIES

- Most of the threats to biodiversity are human in origin. Based on what you have learned about ecological interactions in previous chapters, think about how human activity affects those interactions and how attempts at preservation or remediation can strengthen or reestablish those interactions.

INTERACTIVE EXERCISES

Why It Matters [pp. 1229-1230]

Section Review (Fill-in-the-Blanks)

In the broadest sense, (1) _____ refers to the richness of living systems. In a more narrow sense, it may refer to genetic diversity, species richness, or ecosystem interactions, all of which have implications for humans and humanity.

53.1 The Benefits of Biodiversity [pp. 1230-1232]

53.2 The Biodiversity Crisis [pp. 1232-1239]

Section Review (Fill-in-the-Blanks)

Humans benefit directly from various ecosystem processes such as decomposition of wastes, nutrient recycling, and purification of air and water. Collectively, these processes can be referred to as (2) _____.

Disruption of these processes can wreak havoc on ecosystems. For example, human activity can lead to destruction of large chunks of habitat, leaving only small "islands" undisturbed. This is the phenomenon of (3) _____. Because these islands are surrounded by human activity, they may have to cope with (4) _____, the negative consequences of those activities that impinge on the habitat islands. Among the ecosystems most affected by human intervention are forests. When subtropical forests are cleared and overused, they often undergo (5) _____ as the water table sinks to deeper levels and winds and water cause erosion of topsoil. The release of substances or energy that are harmful to ecosystems, (6) _____, is another manmade threat to ecosystems. For example, in the northeast U.S., emission of sulfur dioxide from coal-fired power plants is converted to sulfuric acid in the atmosphere and returns to earth in the form of (7) _____, which has severely damaged both aquatic and terrestrial ecosystems. Yet other ecosystems have been harmed by (8) _____, the excessive harvesting of the plants or animals in an ecosystem.

Matching

Match each of the following organisms with the ecological threat that harmed it.

9.	____	Atlantic cod	A. Introduction of exotic species
10.	____	various woodpecker species	B. Overexploitation
11.	____	vultures in south Asia	C. Pollution

53.3 Biodiversity Hotspots [pp. 1239-1241]

53.4 Conservation Biology: Principles and Theory [pp. 1241-1247]

53.5 Conservation Biology: Practical Strategies and Economic Tools [pp. 1247-1251]

Section Review (Fill-in-the-Blanks)

In 2000, a group of British and American scientists identified 25 (12) _____ based on the distribution of local species and the degree to which they are threatened by human activity. These local species, which are found nowhere else, are called (13) _____, and the area in which they are found must have lost much of its natural vegetation to human activity. In 1973 the U.S. Congress adopted the (14) _____ Act, defining and protecting species "in danger of extinction throughout all or a significant portion of its range."

(15) _____ is the branch of biology concerned with the maintenance and preservation of biodiversity. It incorporates concepts from systematics, population biology, behavioral biology, and ecology. As the size of a population dwindles, the danger of extinction increases. Conservation biologists often perform (16) _____ that evaluate factors such as habitat suitability and probability of catastrophe to determine how large a population must be to ensure its long-term survival. These analyses may determine (17) _____, the smallest number of individuals in a population that meets predetermined specifications of a conservation plan. Often, conservation biologists are concerned with the dynamics of (18) _____, a collection of neighboring populations of a species that commonly exchange individuals. These subpopulations may be referred to as (19) _____ if they are growing and potentially provide individuals to other subpopulations, and (20) _____ if they are shrinking in size and represent a potential drain on individuals from other subpopulations. The study of how the plants, topography, and human activity in a region influence local populations is known as (21) _____. One of the most important debates in this field goes by the acronym SLOSS, which stands for (22) _____. It asks whether one large preserve or a number of smaller ones of the same total area is the better strategy for conservation biology. There is probably no single solution that applies to all species.

One approach toward conservation and preservation of species is (23) _____, where visitors from wealthier countries pay a fee to visit nature preserves. There is some disagreement about whether this is more harmful than helpful. Another approach is (24) _____, in which ecosystem services provided by intact ecosystems are given an economic value and private companies or conservation organizations pay a fee to the local government to maintain the ecosystem.

Choice

Choose the correct branch of biology that would most likely be concerned with the problems listed below.

a. Landscape ecology b. Population genetics c. Systematics

25. ____ estimating the number of living species
26. ____ genetic bottleneck in endangered species
27. ____ landscape corridors and habitat islands
28. ____ metapopulation dynamics
29. ____ the size of nature preserves
30. ____ species inventory of a particular habitat

Matching

Match the approach to habitat conservation with the terrestrial biome in which they are likely to be common.

31. ____ purchasing tracts of habitat and strictly enforcing rules of usage
32. ____ protection of some habitat parcels while allowing limited development of others
33. ____ undoing (remediating) human disturbances to habitat

A. conservation preservation
B. conservation through restoration
C. mixed-use conservation

SELF-TEST

1. Which individuals would be most interested in maintaining biodiversity? [p. 1232]
   a. ethicists
   b. food scientists
   c. medical researchers
   d. all of the above

2. Which of the following is NOT a consequence of habitat fragmentation in songbird populations in Illinois? [p. 1233]
   a. increased brood parasitism
   b. increased nest predation
   c. increased rate of spread of disease
   d. lack of specific habitat types

3. Which of the following is NOT a result of desertification? [p. 1234]
   a. erosion of topsoil by wind and water
   b. loss of biodiversity
   c. loss of salts from the soil
   d. water table receding to deeper levels

4. Which compound, released from coal burning power plants, is most responsible for acid precipitation? [p. 1235]
   a. carbon dioxide
   b. methane
   c. nitrous oxide
   d. sulfur dioxide

5. Which of the following is NOT an exotic species? [p. 1239]
   a. European starling
   b. hemlock wooly adelgid
   c. kudzu
   d. ruffed grouse

6. The reduction in the Atlantic cod fishery is the result of ________. [p. 1239]
   a. competition from exotic species
   b. habitat destruction
   c. overexploitation
   d. pollution

7. Which of the following is true of endemic species? [p. 1241]
   a. they are generalists in terms of habitat
   b. they have good dispersal abilities
   c. they have restricted geographic distribution
   d. all of the above

8. A group of neighboring populations that exchange individuals (and genes) is called a(n) ________. [p. 1244]
   a. dispersal group
   b. fragmentation community
   c. habitat chain
   d. metapopulation

9. Which type of biologist would be most concerned with the SLOSS (Single Large Or Several Small) debate? [p. 1247]

   a. economic ecologist
   b. landscape ecologist
   c. pollution biologist
   d. systematist

10. Conservation biologists seek to ________ habitats. [p. 1248]

   a. conserve
   b. preserve
   c. restore
   d. all of the above

INTEGRATING AND APPLYING KEY CONCEPTS

1. *Unanswered Questions:* To come

2. *Insights from the Molecular Revolution:* On the surface, DNA barcoding of organisms seems to be an objective way of determining species identity. But it does not solve the question of exactly how to define species. Closely related species are likely to have more similarities in their "barcodes." How great must the differences be for two individuals to be considered different species? Do you think that the degree of difference will be the same for all species within a taxonomic grouping? How would you propose resolving these questions?

3. *Focus on Research:* Conservation efforts on yellow gliders in Australia may require transplantation of individuals by humans in order to facilitate gene flow and reduce the area needed to maintain a viable population. Based on what you have learned in this chapter about the intervention of conservation biologists to assist other species, can you think of any other ways to ensure survival of the yellow glider population?

54

THE PHYSIOLGY AND GENETICS OF ANIMAL BEHAVIOR

CHAPTER HIGHLIGHTS

- Animals must cope with their environment, and behavior is a way to accomplish this. The behaviors may be instinctive or learned. Most behaviors have both learned and instinctive components.
- An animal's neural circuitry and genome are both involved in the control of behavior.
- The physiological condition of an animal is important to the behaviors it displays. Hormonal state is one of the most important of these.
- The nervous system of many animals is specifically adapted to process sensory information likely to be important to it and produce adaptive motor responses.

STUDY STRATEGIES

- Remember that an animal's behavior is a combination of instinctive and learned components. The relative contribution of each to total behavior varies between species. Instinctive behaviors tend to be less variable than learned behaviors and usually play a greater role in the behavior of animals with simple nervous systems. Learned behaviors are usually more common in animals with more complex nervous systems.
- Instinctive behavior is generally thought to be more "hardwired" and often assumed to be more subject to the forces of evolution, but remember the ability to learn is almost certainly the result of evolution.
- Keep in mind that animal behavior, even instinctive behavior, is often highly variable. The number of
- different stimuli impinging on an individual and the intensity of those stimuli may combine to produce variability in behavior.

INTERACTIVE EXERCISES

Why It Matters [pp. 1253 -1254]

Section Review (Fill-in-the-Blanks)

The sum of actions that a white-crowned sparrow (or any other animal) displays in response to environmental stimuli is its (1) _____. The study of how and why animals respond to those stimuli is (2) _____, which may include researchers from many disciplines. The biological study of animal behavior in natural environments, founded by Lorenz, Tingbergen, and von Frisch, is called (3) _____. The mechanisms of behavioral responses and the development of those mechanisms is the focus of (4) _____, a branch of biology that has greatly aided our understanding of the proximate causes of animal behavior in recent years.

54.1 Genetic and Environmental Contributions to Behavior [pp. 1254-1255]

54.2 Instinctive Behaviors [pp. 1255-1257]

54.3 Learned Behaviors [pp. 1257-1259]

Section Review (Fill-in-the-Blanks)

Behavioral scientists have long debated whether an animal's behavior is (5) _____, that is, innate and performed correctly the first time it's used, or whether behavior is (6) _____, that is, modified by experience. Most attempts to answer this question involved "isolation experiments" wherein animals were not allowed exposure to experiences that could enable them to copy or learn a behavioral response. The consensus today is that instinct and learning are both required for most behaviors to be performed correctly.

Instinctive behaviors are often performed in rather invariant fashion and are called (7) _____. Simple stimuli that elicit these behaviors are called (8) _____.

On the other hand, behaviors that are more variable and are modified by experience are a result of (9) _____. Many types of learning have been identified. One is (10) _____, in which an animal learns parents' and/or species' identity. For this type of learning to occur, an appropriate stimulus must be presented during a specific time early in life. This time span is known as the (11) _____. Another form of learning in animals occurs when a previously neutral stimulus becomes associated with another stimulus that causes a behavioral response. Ivan Pavlov taught dogs to associate the sound of a bell with food which elicited the behavior of salivation. This type of learning is called (12) _____. Another type of associative learning is (13) _____, where behavior can be modified by linking a stimulus with a positive result called a reinforcement. Yet another kind of learning exhibited by some animals is (14) _____, wherein the behavioral solution to a problem appears without a period of trial and error. Finally, (15) _____ is a kind of learning where repeated, unreinforced stimuli are no longer responded to.

Choice

Choose the type of causation appropriate to the research described.

a. Proximate causation b. Ultimate causation

16. ____ comparing the evolution of similar behaviors in closely related species
17. ____ determining whether polygamous male redwing blackbirds produce more offspring than monogamists
18. ____ finding the specific cues that trigger a fixed action pattern
19. ____ measuring the reproductive success of territory-holding individuals vs. nonterritorials
20. ____ tracing neural pathways involved in specific behaviors

Matching

Match each of the following types of behavior with the appropriate example.

21. ____ classical conditioning
22. ____ habituation
23. ____ imprinting
24. ____ insight learning
25. ____ instinct
26. ____ operant conditioning

A. A cat comes running when it hears the sound of an electric can opener.

B. A chick follows a toy car after being exposed to it 24 hours after hatching.

C. A chimp stacks two boxes in order to reach some bananas hanging beyond its reach.

D. A fish exhibits typical courtship behavior despite being raised from a zygote isolated from other fishes.

E. A hungry mouse learns to press a bar to get a food pellet.

F. A hydra stops contracting when the aquarium it lives in is continually tapped gently.

True/False

If the statement is true, write a "T" in the blank. If the statement is false, make it correct by changing the underlined word(s) and writing the correct word(s) in the answer blank.

27. ______________ In Pavlov's experiments, the food was the conditioned stimulus.
28. ______________ The act of pressing a bar by a mouse to obtain food is the reinforcement.
29. ______________ The loss of responding by a snail to repeated light touches is imprinting.
30. ______________ A fixed action pattern is a stereotypical, instinctive behavior

54.4 The Neurophysiological Control of Behavior [pp. 1259 – 1260]

54.5 Hormones and Behavior [pp. 1260-1263]

54.6 Nervous System Anatomy and Behavior [pp. 1263-1267]

Section Review (Fill-in-the-Blanks)

Many songbirds, such as zebra finches, use a distinctive song to help in defense of their (31) _____, an area of land defended by males or couples within which they have exclusive access to resources. Proper development of the song and recognition of the species' typical song in other birds is controlled by specific nuclei in the brain. Proper development of these nuclei is under hormonal influence.

Matching

Match the species with correct behavioral adaptation.

32. ____ crickets
33. ____ fiddler crabs
34. ____ honeybees
35. ____ star-nosed mole
36. ____ zebra finches

A. Direction of visual stimuli (from above or below) results in different behaviors.

B. Hormonal differences between young males and females results in differences in neural development.

C. Increasing levels of a hormone causes changes in specific social tasks.

D. Most of the cerebral cortex devoted to processing information from the nose and forelimbs.

E. Ultrasonic vocalizations of predatory bats causes a neural response resulting in moving away from the source of the sound.

SELF-TEST

1. Peter Marler demonstrated that the white crowned sparrow song is a product of ________. [p. 1254]
   a. habituation
   b. instinct
   c. learning
   d. the interaction of learned and instinctive components

2. Which of the following would an ethnologist most likely be interested in? [p. 1254]
   a. classical conditioning
   b. instinctive behavior
   c. insight learning
   d. operant conditioning

3. Birds that lay their eggs in the nests of other species and whose young are raised by the "host" species are called ________. [p. 1256]
   a. brood parasites
   b. ectoparasites
   c. endoparasites
   d. parental hijackers

4. The cues that trigger fixed action patterns are called ________. [p. 1256]
   a. conditioned stimuli
   b. instinctive stimuli
   c. sign stimuli
   d. unconditioned stimulus

5. Stevan Arnold's experiments on newborn garter snakes and banana slugs demonstrated that ________. [p. 1257]
   a. banana slugs exhibit stereotyped avoidance behavior when they detect a garter snake in their vicinity
   b. food preferences in garter snakes have a genetic component
   c. food preferences in garter snakes are learned
   d. the aposematic coloration of banana slugs deters garter snakes from striking

6. A rat presses a bar in its cage and receives a food pellet. Which of the following is true? [p. 1260]
   a. Insight learning has occurred.
   b. Pressing the bar in exchange for the food pellet is an example of classical conditioning.
   c. Pressing the bar is the operant; the food pellet is the reinforcement
   d. Pressing the bar is the reinforcement; the food pellet is the operant.

7. Cell clusters in the brains of white crowned sparrows, zebra finches, and other bird species that are involved in song learning and recognition are called ________. [p. 1261]
   a. ganglia
   b. nuclei
   c. tracts
   d. ventricles

8. Changes in the types of behaviors honeybees perform as they get older are caused by changes in ________. [p. 1262]

a. diet
b. hormone levels
c. photoperiod
d. savanna

9. The evasive behavior of crickets when they hear the ultrasonic sounds of a bat is an example of ________. [p. 1264]

a. habituation to bat vocalizations
b. hardwiring between the sensory and motor systems
c. hormonal control of behavior
d. imprinting on bat vocalizations during a critical period

10. Star-nosed moles live in dark tunnels and vision is not a useful sensory modality. Which sense is most important to star-nosed moles? [p. 1266]

a. hearing
b. smell
c. taste
d. touch

INTEGRATING AND APPLYING KEY CONCEPTS

1. *Unanswered Questions:* The popular press and the media often report that researchers have isolated a gene "for" a particular type of behaviors, e.g., a gene "for" alcoholism, a gene "for" schizophrenia, etc. Since a gene only codes for the production of a protein, how can one make sense of these statements? That is, how can a gene code for a complex pattern of behaviors? Assuming that the research has been properly done, can you think of a better way of summarizing the results?

2. *Insights from the Molecular Revolution:* Some people have questioned the importance of investing time and money studying the genetics of fruit flies and other nonhuman species. Given the apparent similarity of the *disheveled* gene in *Drosophila* and the gene associated with O/C disorder, Huntington's disease, and schizophrenia (bipolar disorder) in humans, is this argument justified?

3. *Focus on Research:* Herring gull chicks will direct their food-begging behavior at a cardboard cutout of an adult's head as long as it bears a red spot on the bill. This visual stimulus triggers food-begging behavior. Can you think of any ways to manipulate this artificial situation that might enhance responding?

55

THE ECOLOLGY AND EVOLUTION OF ANIMAL BEHAVIOR

CHAPTER HIGHLIGHTS

- Migration is a way for animals to increase their fitness and avoid stressful environmental conditions, but it is usually stressful and involves risks.
- Another way to increase fitness is to select a favorable habitat. This may be instinctive or involve learning. In some instances, defense of a territory may be a good strategy.
- Animals use virtually every sensory modality to communicate. The fitness consequences of clear information transfer are great.
- The reproductive strategies that lead to maximum fitness may be different for males and females. The basis of these differences is usually rooted in differences in investment in reproduction between the sexes. These differences are often correlated with anatomical and physiological differences between the sexes.
- Social groupings provide both costs and benefits to their members. Altruistic behavior, which seems to contradict Darwinian selfishness, is observed in many species. Haplodiploidy in some insect species may be a key to understanding social behavior.
- Understanding the ecological and evolutionary roots of animal behavior may lead to a better understanding of human behavior.

STUDY STRATEGIES

- Always be aware that behavior may have a genetic component and is thus subject to the forces of natural selection.
- Like morphology or anatomy, behavior can have fitness consequences. Just as structures can evolve in response to selection pressures, so can behaviors. Keep in mind that humans are animals, and human behavior is subject to natural selection just as it is in other species.
- As you might imagine, communication between individuals can have a very significant impact on fitness. The communication methods used by animals are incredibly diverse.
- The "battle of the sexes" is real. Behaviors that may increase fitness in one sex may not be the same behavior that would maximize fitness in the other. Thus, the sexes must reach some "compromise" that may not be ideal for both sexes but is better than any other strategy.
- This chapter is devoted largely to the ultimate causation of behavior, the "why" questions about animal behavior. Chapter 54 concentrated on proximate causation, the "how" questions.

INTERACTIVE EXERCISES

Why It Matters [pp. 1269–1270]

Section Review (Fill-in-the-Blanks)

The environmental cues that trigger migration in white-crowned sparrows (principally day length causing hormonal changes) represent the (1) _____ causes of migration. The fitness consequences to individuals that migrate vs. those that do not make up the (2) _____ causation of migration.

55.1 Migration and Wayfinding [pp. 1270–1274]

Section Review (Fill-in-the-Blanks)

The large scale, predictable movement of animals that occurs on a seasonal basis is (3) _____. There are a number of mechanisms that enable animals to find their way during these travels. One is (4) _____, in which animals use landmarks to guide them. Another method is (5) _____, in which animals use some environmental cue, such as the sun or stars to enable them to travel in a specific direction and often for a specific period of time. The most complex type of wayfinding is (6) _____, which requires both a compass and a "mental map."

Matching

Match each of the animals with the cues it uses for migration. You may choose more than one cue per animal.

7. ____ digger wasps
8. ____ homing pigeons
9. ____ indigo buntings
10. ____ many migratory birds

A. landmarks
B. position of the stars in the sky
C. position of the sun in the sky
D. olfaction (smell)

55.2 Habitat Selection and Territoriality [pp. 1274–1276]

Section Review (Fill-in-the-Blanks)

One of the simplest behavioral responses to environmental cues that animals use to locate suitable habitat is (11)_____, which involves a change in the rate of movement or frequency of turning. Another simple response, but one that involves direct orientation toward or away from a stimulus, is called (12) _____. Often, after finding a suitable habitat, animals will defend an area from other individuals of the same species, giving it exclusive use of resources in that area, a behavior known as (13) _____.

Choice

Choose the type behavior appropriate to the behavior described.

a. Kinesis   b. Taxis   c. Territoriality

14. ____ male yellow-headed blackbirds defending an area against intrusion by other yellow-headed blackbirds
15. ____ moths flying directly toward a light
16. ____ wood lice increasing their rate of locomotion in dry areas

55.3 The Evolution of Communication [pp. 1276–1278]

Section Review (Fill-in-the-Blanks)

Animals use many sensory "channels" or modalities to communicate. Specific sounds or vocalizations are examples of (17) _____, while movements that convey information are examples of (18) _____. The use of scents to communicate with others is an example of (19) _____ signaling, and the compounds used are referred to as (20) _____. (21) _____ signaling involves physical contact between individuals. Finally, some fishes use (22) _____ because they live in murky waters that make other forms of signaling ineffective.

Choice

Choose the communication channel most appropriate to the situation described. More than one channel may be appropriate to a given situation.

a. Acoustical b. Chemical c. Electrical d. Tactile e. Visual

23. ____ animals living in a cluttered environment
24. ____ animals living in dimly lit areas
25. ____ animals living in open areas
26. ____ animals that need long lasting, yet energetically inexpensive signals

55.4 The Evolution of Reproductive Systems and Mating Behavior [pp. 1279–1281]

Section Review (Fill-in-the-Blanks)

Because of the anatomical and physiological differences between males and females, behaviors that affect their respective fitness, their (27) _____, may differ markedly between the sexes. The time, energy, and resources spent on reproduction, termed (28) _____, is usually greater for females. As a result, females of most species tend to be more selective in choosing a mate. This female choosiness often leads to a form of natural selection known as (29) _____, which often leads to sexual dimorphism between the sexes, with elaborate ornamental structures often evolving in the male. These ornaments are frequently used in (30) _____, behaviors designed to attract the attention of females. In some cases, these male structures are used for male-male competition rather than for attracting females. Sometimes males of a species gather for competition in a common area known as a(n) (31) _____. Females assess the quality of males and select one for mating. Successful males may thus wind up mating with numerous females. This is only one example of the ways that males and females of a species may pair up, a part of their life history known as a(n) (32) _____. One male mating with multiple females is known as (33) _____, while the rarer condition of a single female having multiple male mates is known as (34) _____. These are both types of (35) _____. Other patterns of male-female pairing include (36) _____, where a single male pairs with a single female, and (37) _____, where both sexes have multiple mates without any lasting pair bond forming between them.

Matching

Match the mating system with the correct life history pattern.

38. ____ monogamy
39. ____ polyandry
40. ____ polygyny
41. ____ promiscuity

A. Solitary species with little or no real social structure. Individuals encounter each other randomly.

B. Species in which females are physically drained after egg laying, but resources are temporarily abundant allowing them to recover quickly and mate with a second male while the first male incubates the eggs.

C. Species in which resource availability is so poor that both parents are required to successfully rear offspring.

D. Territorial species in which quality of territory is highly variable. High quality territories provide abundant resources such that bi-parental care is unnecessary.

55.5 The Evolution of Social Behavior [pp. 1281–1285]

55.6 An Evolutionary View of Human Social Behavior [pp. 1285–1286]

Section Review (Fill-in-the-Blanks)

The interactions of animals with members of their own species represent (42) _____. Living in groups has costs as well as benefits associated with it. Often groups form a pecking order known as a(n) (43) _____. Subordinate individuals may stay in the group despite their low status because survival is difficult for solitary individuals. Many social species exhibit (44) _____, behaviors that help others while putting the performer at somewhat increased risk. This seems to contradict the concept of selfishness inherent in Darwin's view of natural selection, but William Hamilton demonstrated how, by helping relatives, these behaviors could evolve. His theory of (45) _____ shows how helping a relative can get copies of the performer's genes into the next generation. (46)_____, the unique genetics of sex determination in ants, bees, and wasps, lends credence to Hamilton's hypothesis. Under certain circumstances, altruistic behavior can evolve among nonrelated individuals if there is a good chance that the roles of helper and helped may be reversed in the future. This form of cooperative behavior is known as (47) _____.

Short Answer

Calculate the degree of relatedness, r, for the following pairs of individuals.

(48) Siblings_______________

(49) Parent and offspring_______________

(50) Uncle and nephew_______________

(51) First cousins_______________

SELF-TEST

1. Which of the following is NOT an environmental cue used by migrating animals to reach their destination? [p. 1271]
   a. odors
   b. position of the stars in the sky
   c. position of the sun in the sky
   d. sounds

2. You place several planaria in a pan of water with one side covered so that it is dark while the other side is open and well lit. You notice that their rate of movement slows on the dark side. What phenomenon have you observed? [p. 1275]
   a. compass orientation
   b. kinesis
   c. operant conditioning
   d. taxis

3. Food preferences of many vertebrates ________. [p. 1275]
   a. are innate
   b. are learned
   c. have both innate and learned components
   d. Animals do not show food preferences; diet is determined by the abundance of food only.

4. Which resource would likely justify territorial defense? [p. 1276]
   a. an abundant, localized food supply
   b. food distributed evenly over a large area
   c. oxygen
   d. prey that move over great distances, e.g., a school of fish

5. What channel would an animal likely use to communicate, day and night, in a highly cluttered environment? [p. 1276]
   a. acoustical
   b. chemical
   c. tactile
   d. visual

6. The most favorable reproductive strategies are often different for males and females of a species due to differences in ________. [p. 1279]
   a. body temperature
   b. emotional makeup
   c. parental investment
   d. sex ratio

7. Special ornaments or structures of males that increase their likelihood of attracting females are probably the result of [p. 1279]
   a. artificial selection.
   b. disruptive selection.
   c. sexual selection.
   d. stabilizing selection.

8. The mating system in which one female has multiple male mates is ________. [p. 1280]
   a. monogamy
   b. polyandry
   c. poylgyny
   d. promiscuity

9. The behavior of which animals supports the idea of reciprocal altruism? [p. 1284]
   a. honeybees
   b. musk oxen
   c. naked mole rats
   d. vampire bats

10. Wilson and Daly found that criminal aggression by an adult toward a child was most common between ________. [p. 1285]
   a. adult and juvenile siblings
   b. fathers and daughters
   c. fathers and sons
   d. stepparents and stepchildren

INTEGRATING AND APPLYING KEY CONCEPTS

1. *Unanswered Questions:* Is there any reason to believe that human behavior is not affected by natural selection? Besides the obvious ethical concerns that make this difficult to test, what other aspects of human behavior make it difficult to get at the roots of the evolution of human behavior? Richard Dawkins has written about a culturally inherited unit of behavior called the "meme," which is passed from one generation to the next by nongenetic means. Speculate on the plausibility of this type of evolution.

2. *Insights from the Molecular Revolution:* Despite the fact that members of naked mole rat colonies are more closely related, on average, than siblings due to inbreeding, wouldn't individuals enjoy a greater fitness benefit by mating themselves with a closely related individual rather than helping a relative? What plausible explanations can you come up with to explain the presence of a sterile "caste" among naked mole rats?

3. *Focus on Research:* Further support for the ability of birds to navigate by the stars might be provided if one could manipulate the position of the stars in the sky. One might be able to "fool" the birds into taking an inappropriate but predictable direction. Can you think of any way Emlen's experiments on stellar navigation could be modified to provide the desired data?

ANSWERS

Chapter 1 Introduction to Biological Concepts and Research

Why It Matters [pp. 1-2]

1. life

1.1 What is Life? [pp. 2-7]

2. hierarchy; 3. cell; 4. unicellular organism; 5. multicellular organism; 6. population; 7. community; 8. ecosystem; 9. biosphere; 10. emergent properties; 11. DNA; 12. RNA; 13. protein; 14. metabolism; 15. primary producers; 16. consumers; 17. decomposers; 18. external environment; 19. homeostasis; 20. reproduction; 21. inheritance; 22. development; 23. life cycle; 24. biological evolution; 25. biology; 26. ecology; 27. homeostasis; 28. ecosystem; 29. biosphere; 30. protozoa; 31. organized chemical system surrounded by a membrane; 32. multicellular organism; 33. population; 34. the collection of all of the populations of different organisms living in the same place; 35. ecosystem; 36. Living systems a) are organized in a hierarchy, b) contain chemical instructions that govern their structure and function, c) engage in metabolic activities, d) have energy flows and cycle matter through them, e) compensate for changes in the external environment, f) reproduce and undergo development, g) change from one generation to the next; 37. b; 38. a; 39. a; 40. c; 41. b; 42. b; 43. d; 44. a; 45. c; 46. g; 47. f; 48. e

1.2 Biological Evolution [pp. 7-9], **1.3 Biodiversity** [pp. 9-13].

49. artificial selection; 50. natural selection; 51. genes; 52. mutations; 53. adaptations; 54. species; 55. scientific name; 56. genus; 57. family; 58. order; 59. class; 60. phylum; 61. kingdom; 62. domain; 63. Archaea; 64. Bacteria; 65. Eukarya; 66. prokaryotes; 67. eukaryotes; 68. nucleus; 69. organelles; 70. Protoctista; 71. Plantae; 72. Fungi; 73. Animalia; 74. c; 75. e; 76. d; 77. a; 78. b; 79. b; 80. a; 81. c; 82. a; 83. b; 84. d; 85. f; 86. a; 87. e; 88. c; 89. b; 90. h. 91 g

1.4 Biological Research [pp. 13-19]

92. biological research; 93. scientific method; 94. basic research; 95. applied research; 96. observational data; 97. experimental data; 98. hypothesis; 99. predictions; 100. alternative hypothesis; 101. control; 102. experimental variable; 103. replicates; 104. null hypothesis; 105. model organisms; 106. biotechnology; 107. scientific theory; 108. b; 109. a; 110. c; 111. a; 112. b; 113. d; 114. a; 115. b; 116. g; 117. f; 118. c; 119. e; 120. Model organisms have rapid development, short life cycle, small adult size, and other characteristics that make them amenable to laboratory research. The fruit fly is an example of a model organism; 121. A scientific theory has been exhaustively tested and is not likely to be contradicted by future research.

Self-Test

1. b [The initial observations indicated that small size is accompanied by low IGF levels and that large size is accompanied by high IGF levels. The experiment is aimed at demonstrating a causal relationship between size and IGF level; therefore, b is correct]

2. a [a is correct because the biologist is manipulating the system; such manipulation results in experimental data]

3. a [The experiment tests the effects of saline (control) and IGF injection; therefore, the experimental variable is IGF level (none vs. 100ng/g body weight) and a is correct; all other parameters were held constant]

4. a

5. a

6. a

7. c

8. c

9. c [c is correct because species names are two-part names with a genus and a specific epithet]

10. d

11. b

Chapter 2 Life, Chemistry, and Water

Why It Matters [pp. 21-22],

1. plants; 2. animals; 3. atoms; 4. chemical bonds; 5. biology; 6. chemical substances

2.1 The Organization of Matter: Elements and Atoms [pp. 22-23]

7. elements; 8. atoms; 9. pure; 10. cannot; 11. carbon; 12. hydrogen; 13. oxygen; 14. nitrogen; 15. atom; 16. symbol; 17. atomic number; 18. atomic mass; 19. molecules; 20. compounds

2.2 Atomic Structure [pp. 23-28]

21. B; 22. C; 23. A; 24. A; 25. A; 26. B; 27. B; 28. A; 29. F; 30. C; 31. E; 32. D; 33. 2; 34. 2; 35. 2s; 36. 2p; 37. 2; 38. 3; 39. 2; 40. 2; 41. 8; 42. 8; 43. 8; 44. 16; 45. O or oxygen; 46. 11; 47. 11; 48. 22 or 23 (most common form has 12 neutrons); 49. Na or sodium; 50. 17; 51. 17; 52. 34; 53. Cl or chlorine

2.3 Chemical Bonds [pp. 28-32]

54. O or oxygen; 55. 6; 56. Na or sodium; 57. 1; 58. Cl or chlorine; 59. 7; 60. Na; 61. Cl; 62. ionic; 63. 1; 64. Na; 65. positive; 66. Cl; 67. negative; 68. charge; 69. ionic bond; 70. electrochemical; 71. Na^+; 72. Cl^-; 73. covalent; 74. hydrogen; 75. van der Walls forces; 76. covalent; 77. electrons; 78. equal; 79. unequal; 80. electronegativity; 81. nonpolar covalent; 82. polar covalent; 83. polar; 84. attracted; 85. repealed; 86. polar; 87. hydrophilic; 88. nonpolar; 89. hydrophobic

2.4 Hydrogen Bonds and the Properties of Water [pp. 32-36]

90. Polar water molecules have a negative end (oxygen) and positive ends (hydrogen). Hydrogen bonds form between the oxygen of one water molecule and the hydrogen of another water molecule. Thus a lattice type structure will form between adjacent water molecules; 91.Polar ends of a molecule will associate with water (hydrophilic), while the nonpolar ends will be repealed (hydrophobic). The nonpolar ends will be attracted to each other and thus a bilayer structure will form. The nonpolar ends of two lipids will be attracted to each other and the polar ends will associate with the water lattice.

92A. phosphate; 92B. fatty acid; 92C. polar; 92D. hydrophobic; 92E. polar; 92F. membrane;

93. E; 94. C; 95. B; 96. A; 97. D; 98. True; 99. False, heat of vaporization; 100. False, cohesion; 101. True; 102. True

2.5 Water Ionization and Acids, Bases, and Buffers [pp. 36-39]

103. hydrogen (H^+); 104. hydroxide (OH^-); 105. acid; 106. base; 107. buffer; 108. accepting; 109. releasing; 110. hydrogen (H^+); 111. A; 112. B; 113. B; 114. A; 115. C

Self-Test

1. d [an isotope has variable numbers of neutrons; electrons can be gained or lost; proton number is unique to each type of atom for a given element]
2. a [isotopes differ in the number of neutrons]
3. b [the positive charge of the nucleus attracts the electrons. In outer orbitals, the negative charge of the electrons contributes to the reactivity of the atom]
4. a [polar covalent bonds in a molecule cause the molecule to have a positive and a negative end, thus hydrogen bonds can form between two adjacent polar molecules]
5. d [molecules with ionic or polar covalent bonds produce ions or atoms with a positive or negative end which are more likely to dissolve in water]
6. b [hydrogen-bond lattice becomes rigid and the spaces are farther apart, the volume increases, but the density decreases]
7. a [molarity is the concentration of solute per unit solvent, the solute concentration is directly related to the molarity]
8. c [high specific heat allows water to freeze from the top down. A large of amount of heat loss is required for water to change from a liquid to a solid. Cold air temperature next to the surface of the water will cause heat to be lost at the top first]
9. c [pH 9.3 is the only basic pH. If enzyme activity occurs in an acidic pH, then a basic pH would be expected to decrease activity.]
10. a [buffers either release or bind to hydrogen ions, thus the pH would not be expected to greatly change].

Chapter 3 Biological Molecules: The Carbon Compounds of Life

Why It Matters [pp. 41-42],

1. carbon; 2. carbon dioxide, CO_2; 3. carbon; 4. organisms; 5. photosynthesis;

3.1 Carbon Bonding [pp. 42-43],

3.2 Functional Groups in Biological Molecules [pp. 43-45],

6. organic; 7. carbon; 8. inorganic; 9. four electrons; 10. covalent polar; 11. carbon; 12. hydrogen; 13. hydrocarbons; 14. carbon; 15. hydrogen; 16. oxygen; 17. nitrogen; 18. carbohydrates; 19. lipids; 20. proteins; 21. nucleic acids; 22. hydroxyl; 23. amino; 24. amino acids or proteins; 25. organic acids; 26. phosphate; 27. ketones; 28. dehydration synthesis; 29. hydroxyl; 30. hydrogen; 31. oxygen; 32. synthesis; 33. water; 34. dehydration; 35. hydrolysis; 36. water; 37. hydroxyl; 38. hydrogen; 39. hydrolysis

3.3 Carbohydrates [pp. 45-50],

40. poly; 41. mono; 42. carbon; 43. hydrogen; 44. oxygen; 45. 1 (carbon): 2 (hydrogen): 1 (oxygen); 46. Polymerization; 47. dehydration synthesis; 48. polysaccharides or polymers; 49. Molecules with the same chemical formula (same ratio of atoms) but different molecular structures. Optical isomers are mirror images, while structural isomers have different arrangement of the atoms. 50. Both glycogen and cellulose are polymers of glucose. Glycogen is branched with alpha-linkages, while cellulose is unbranched with beta-linkages. Most animals lack the enzyme necessary to break beta-linkages, thus digestion of cellulose is not possible.

3.4 Lipids [pp. 50-55],

51. neutral; 52. phospholipids; 53. steroids; 54. glycerol; 55. three; 56. fatty acid; 57. triglycerides; 58. carboxyl; 59. dehydration synthesis; 60. carboxyl; 61. hydroxyl; 62. saturated; 63. hydrogen; 64. unsaturated; 65. double; 66. monounsaturated; 67. polyunsaturated. 68. Double bonds cause fatty acids to have bends, thus the chain has less organization and is more likely to be fluid and melt at lower temperatures. Lack of double bonds cause fatty acid chains to be solid (due to more uniformity and organization) with higher melting temperatures; 69. One of the fatty acid chains is replaced with a polar phosphate group. 70A. phosphate; 70B. fatty acids, monounsaturated, polyunsaturated; 70C & E. hydrophilic or polar; D. hydrophobic or nonpolar; F. membrane; 71. Phospholipids will orient so that the hydrophobic portions (fatty acid chains) away from the polar or hydrophilic solution. The phosphate group will associate with the polar or hydrophilic solution. 72. The membrane would like it were inside-out—the phosphate ends (polar) would orient to be away from the nonpolar solution. The fatty acid chains would be faced or associate with the nonpolar solution; 73. Cholesterol is the structural unit of steroids. Steroids play important roles in membrane structure as well as hormones.

3.5 Proteins [pp. 55-64],

74. proteins; 75. amino acids; 76. amino; 77. carboxyl; 78. R; 79. N-terminal; 80. amino; 81. C-terminal; 82. carboxyl; 83. peptide; 84. dehydration synthesis; 85. polypeptide; 86. B; 87. D; 88. C; 89. A; 90. Hydrogen bonds: 91. Hydrogen bonds are broken and the protein looses its three-dimensional structure and often function. Extreme temperatures as well as acid or basic conditions; 92. F; 93. C; 94. B; 95. A; 96. D; 97. E

3.6 Nucleotides and Nucleic Acids [pp. 64-68]

98. nucleic acids; 99. deoxyribonucleic acid; 100. DNA; 101. ribonucleic acid; 102. RNA; 103. nucleotide; 104. nucleotide; 105. nitrogenous base; 106. 5 carbon; 107. phosphate; 108. two; 109. pyrimidines; 110. purines; 111. sugar; 112. phosphate; 113. phosphodiester bond; 114. two; 115. single; 116. complementary; 117. B; 118. A; 119. A; 120. B; 121. C; 122. E; 123. B,E; 124. A,C,D; 125. C; 126. A; 127. F; 128. G; 129A. 2; 129B. deoxyribose; 129C. adenine; 129D. thymine; 129E. guanine; 129F. cytosine; 129G. 1; 129H. ribose; 129I. adenine; 129J. uracil; 129K. guanine; 129L. cytosine;

Summary

130. carbohydrate; 131. lipid—triglyceride; 132. lipid—phospholipid; 133. lipid—steroid; 134. protein; 135. nucleic acid;

Self-Test

1. a [dehydration synthesis reactions produce water]
2. a [Cell membranes have a polar portion and nonpolar portion; phospholipids with a phosphate group are the primary lipid type in membranes.]
3. d [Microbes in the GI tract contain enzymes to breakdown cellulose. In addition, the GI tract of these animals is considerably longer than humans.]
4. d [phospholipids and steroids have polar side groups and are hydrophilic; triglycerides are nonpolar and hydrophobic]
5. c [unsaturated means the carbons are not saturated with hydrogen, thus double bonds are present; monounsaturated refers to one double bond, two or more double bonds refers to polyunsaturated]
6. c [chlorophyll is a lipid]
7. b [peptide bonds form between the amino and carboxyl groups of adjacent amino acids]
8. d [the primary level of protein structure is the amino acid sequence. If the sequence is wrong, both the secondary and tertiary level could be greatly altered]

9. b [a zipper is a specific type of motif, chaperones are separate proteins that are thought to assist in protein folding, domains are large structural divisions within a protein]
10. a [since there are 20 amino acids, the building blocks of proteins, the possible combination of the primary sequence is extremely large; carbohydrates are essentially carbon, hydrogen, and oxygen; lipids are similar to carbohydrates and nucleic acids; while the expression of the code is astronomical, there are 5 different nucleotides]
11. a [the two strands of DNA are held together by hydrogen bonds; each strand is complementary to the other]
12. d [hydrogen bonds form between the complementary bases].

Chapter 4 Energy, Enzymes, and Biological Reactions

Why It Matters [pp. 75–76]

1. metabolism; 2. enzymes; 3. energy

4.1 Energy, Life, and the Laws of Thermodynamics [pp. 76–79]

4. Energy; 5. energy; 6. radiation energy; 7. Kinetic; 8. Potential; 9. potential; 10. kinetic; 11. catabolic; 12. exergonic; 13. anabolic; 14. endergonic; 15. Thermodynamics; 16. system; 17. surroundings; 18. system; 19. open system; 20. created; 21. destroyed; 22. constant; 23. sun; 24. photosynthesis; 25. potential; 26. potential; 27. kinetic; 28. heat; 29. entropy; 30. spontaneous; 31. Free; 32. positive; 33. negative; 34. exergonic; 35. negative; 36. positive;37. C; 38. J; 39. K; 40. D; 41. A; 42. E; 43. G; 44. F; 45. H; 46. B; 47. I

48. Catabolism: Breakdown reactions Fat → glycerol + fatty acids	49. Anabolism: Building reactions Amino acids join → protein
50. Kinetic energy: Energy released by breaking down glucose to support cellular functions	51. Potential energy: Energy stored in glucose molecules
52. Endergonic reactions: Photosynthesis reaction CO_2 + H_2O + Light energy → glucose + O_2	53. Exergonic reaction: Cellular respiration glucose + O_2 → CO_2 + H_2O + Light energy
54. Positive ΔG: Hydrolysis of sucrose to form two monosaccharides	55. Negative ΔG Combining of two monosaccharides to form sucrose

4.2 How Living Organisms Couple Reactions to Make Synthesis Spontaneous [pp. 79–81]

4.3 Thermodynamics and Reversible Reactions [pp. 81–82]

56. Anabolic; 57. endergonic; 58. positive; 59. exergonic; 60. negative; 61. coupled; 62. coupling; 63. ribose; 64. adenine; 65. phosphate; 66. free; 67. phosphate; 68. potential; 69. phosphorylation; 70. endergonic; 71. exergonic; 72. potential; 73. ATP; 74. phosphate; 75. equilibrium; 76. reversible; 77. equilibrium; 78. products 79. C; 80. A; 81. B; 82. E; 83. D; 84. Coupled reactions are reactions that are connected; one reaction may require energy (endergonic) while the other reaction releases energy (exergonic). 85. ATP is made of ribose sugar to which adenine and 3 phosphates are attached. It helps in transfer of energy from exergonic to endergonic reactions.

4.4 The Role of Enzymes in Biological Reactions [pp. 82–86]

4.5 Conditions and Factors Affecting Enzyme Activity [pp. 86–89]

4.6 RNA-based Biological Catalysts: Ribozymes [pp. 89-90]

86. increase; 87. activation; 88. catalysts; 89. proteins; 90. active; 91. specificity; 92. -ase; 93. cofactors; 94. coenzymes; 95. product; 96. transition; 97. active; 98. reactants; 99. active; 100. reactants; 101. transition; 102. reactants; 103. transition; 104. pH; 105. temperature; 106. collision; 107. 3-dimensional; 108. denature; 109. 3-dimensional; 110. collision; 111. saturation; 112. inhibitors; 113. active; 114. competitive inhibitors; 115. active; 116. 3-dimensional; 117. active; 118. noncompetitive inhibitors; 119. inhibitors; 120. active;

121. allosteric; 122. feedback inhibition; 123. phosphorylation; 124. dephosphorylation; 125. protein kinases; 126. protein phosphotases; 127. RNA; 128. proteins; 129. proteins; 130. B; 131. A; 132. N; 133. D; 134. I.; 135. E; 136. J; 137. L; 138. A; 139. M; 140. B; 141. G; 142. H; 143. C; 144. K; 145. F

A. Enzymes use their active site to increase the chances of bringing the reactants closer together.
B. Enzymes use their active site to orient the reactants correctly to increase the chance of forming the transition state.
C. Enzymes provide the reactants the appropriate ionic environment for forming the transition state.

Inhibitor	*Explanation*
A. Competitive Inhibitor	Where the inhibitor compete with the reactants for the active site
B. Noncompetitive Inhibitor	Where the inhibitor binds to sites other than the active site of the enzyme, causing a change in 3D structure of the enzyme
C. Allosteric Regulation	Where an activator or an inhibitor binds to an enzyme, changing its 3D structure, and thereby activating or inhibiting its activity
D. Feedback Inhibition	Where the ends product of a chain reaction acts as the inhibitor of the first enzyme

148. False—Each enzyme has its optimal range. 149. False—Competitive inhibitor binds to the active site while noncompetitive inhibitor works by binding to site other than the active site.150. True

Self-Test

1. b [Enzymes are proteins that act as chemical catalysts]
2. c [Heat is a form of kinetic energy]
3. a [Glucose has potential energy]
4. d [Entropy is defined as the state of disorder in the system]
5. c [Endergonic reactions have a negative ΔG]
6. c [Active site is where the reactant/s fit]
7. d [Enzymes decrease the activation energy—energy required to start a reaction]
8. a [Competitive inhibitors inhibit by fitting into the active site—preventing the reactants from entering the active site]
9. c [Cofactors are inorganic chemicals—such as minerals—that help enzymes]
10. a [coenzymes are organic—often vitamins—that help enzymes]
11. Enzymes are protein molecules whereas ribozymes are RNA molecules]
12. c [Enzymes speed up chemical reactions]
13. d [Enzymes are affected by temperature, pH, and substrate concentration]
14. d [ATP is made of ribose sugar to which are attached the adenine and 3 phosphates]
15. b [Coupled reactions have an overall negative ΔG]

Chapter 5 The Cell: An Overview

Why It Matters [pp. 91-92]

1. cell theory

5.1 Basic Features of Cell Structure and Function [pp. 92-96], **5.2 Prokaryotic Cells** [pp. 96-97], **5.3 Eukaryotic Cells** [pp. 97-110]

2. cytoplasm; 3. cytosol; 4. organelles; 5. plasma membrane; 6. prokaryotes; 7; nucleoid; 8. eukaryotes; 9. nucleus; 10. cell wall; 11. capsule; 12. bacterial chromosome; 13. ribosome; 14. prokaryotic flagellum; 15. cell wall; 16. nuclear envelope; 17. nuclear pores; 18. nucleoplasm; 19. chromatin; 20. eukaryotic chromosome; 21. nucleolus; 22. endomembrane system; 23. endoplasmic reticulum; 24. Golgi complex; 25. lysosomes; 26. vesicles; 27. cisternae; 28. ER lumen; 29. rough ER; 30. smooth ER; 31. secretory vesicles; 32. exocytosis; 33. endocytosis; 34. phagocytosis; 35. mitochondria; 36. outer mitochondria membrane; 37. inner mitochondria membrane; 38. cristae; 39. mitochondrial matrix; 40. microbodies; 41. peroxisomes; 42. cytoskeleton; 43. microtubules; 44. intermediate filaments; 45. microfilaments; 46. centromer; 47. centrioles; 48. flagella; 49. cilia; 50. basal body; 51. cytoplasm; 52. cytoskeleton; 53. glyoxisome; 54. peroxisome; 55. lysosome; 56. microfilament; 57. microtubule; 58. cytosol; 59. prokaryotes; 60. eukaryotes; 61. a; 62. b; 63. c; 64. a; 65. c; 66. b; 67. b; 68. a; 69. a; 70. l; 71. h; 72. i; 73. d; 74. c; 75. b; 76. e; 77. j; 78. k; 79. f; 80. g; 81. secretory vesicles fuse with the plasma membrane and release their contents to the outside of the cell, whereas endocytotic vesicles form when material from the outside is surrounded by a small section of plasma membrane which pinches off and enters to cytoplasm of the cell; 82. The nuclear envelope is a double membrane that surrounds the nucleus, whereas nuclear pores are perforation in the envelope and the nucleoplasm is the contents of the nuleus; 83. bacterial flagellum; 84. cell wall; 85. capsule; 86. rough ER; 87. smooth ER; 88. cisternae; 89. ER lumen; 90. Golgi complex; 91. mitochondria; 92. outer membrane; 93. inner membrane; 94. cristae; 95. matrix; 96. centriole; 97. nuclear envelope

5.4 Specialized Structures of Plant Cells [pp. 110-113], **5.5 The Animal Cell Surface** [pp. 113-115].

98. chloroplasts; 99. outer boundary membrane; 100. inner boundary membrane; 101. stroma; 102. thylakoids; 103. grana; 104. plastids; 105. amyloplats; 106. chromoplasts; 107. central vacuole; 108. tonoplast; 109. primary cell wall; 110. secondary cell wall; 111. middle lamella; 112. plasmodesmata; 113. extracellular matrix; 114. cell adhesion molecules; 115. cell junctions; 116. anchoring junctions; 117. desmosomes; 118. adherens junctions; 119. tight junctions; 120. gap junctions; 121. chlorophyll; 122. chromoplast; 123. amyloplast; 124. a; 125. b. 126. c; 127. c; 128. b; 129. f; 130. a; 131. e; 132. d; 133. both are channels that connect the cytoplasms of adjacent cells; plasmodesmata are membrane-lined channels that perforate cell wall material between plant cells, whereas gap junctions are protein-lined channels through the plasma membrane that align between two animal cells; 134. chloroplast; 135. outer membrane; 136. inner membrane; 137. stroma; 138. thylakoid; 139. grana; 140. central vacuole; 141. tonoplast

Self-Test

1. a, d

2. c, d [d is correct because surface area increases with the square of a dimension, whereas volume increases with the cube of a dimension; therefore, as the dimension increases, volume increases more rapidly than surface area; c is correct because the ability of a cell to take up nutrients and to eliminates wastes depends on diffusion, which is limited by surface area]

3. a, b

4. a, b, c, d

5. a, b, c, d

6. a, c, d

7. a, b, c, e

8. a, b, c [a is correct because the cytoskeleton is constantly be built up and broken down in various regions of the cell; b is correct because protein are what make up microfilaments, intermediate filaments, and microtubules; c is correct because microtubles are one of the three types of cytoskeletal elements]

9. a, c, d [a, c, and e are correct because plastids are found in plants (which are eukayotes), fungi (which is a eukaryote), and protoctistins, including algae (which also are eukaryotes), but not in animals (which also are eukaryotes), which means that some but not all eukaryotes contain plastids; e is incorrect because no prokaryotes have membrane-bound organelles, including plastids]

10. a, d

11. a, b, c

12. c

Chapter 6 Membranes and Transport

Why It Matters [pp.123-124], **6.1 Membrane Structure** [pp.124-128]

1. environment; 2. sterols; 3. fatty acids; 4. phosphate group; 5. bilayer; 6. end-to-end; 7. fluid-mosaic model; 8. unsaturated; 9. asymmetric; 10. functions; 11. integral protein; 12. peripheral proteins; 13. intermixed; 14. b; 15. a; 16. d; 17. c; 18. b; 19. a; 20. phospholipid; 21. integral protein; 22. peripheral protein; 23. cholesterol; 24. hydrophobic region; 25. hydrophilic region; 26. transport, recognition, receptor, cell adhesion; 27. The double bonds produce kinks in the fatty acid chains that prevents close packing;

6.2 The Functions of Membranes in Transport: Passive Transport [pp.128-131], **6.3 Passive Water Transport and Osmosis** [pp.132-135], **6.4 Active Transport** [pp.135-137], **6.5 Exocytosis and Endocytosis** pp.137-139]

28. passive transport; 29. active transport; 30. diffusion; 31. selectively permeable; 32. simple diffusion; 33. facilitated diffusion; 34. channel proteins; 35. carrier proteins; 36. osmosis; 37. active transport; 38. membrane potential; 39. symport; 40. antiport; 41. exocytosis; 42. endocytosis; 43. endocytosis; 44. exocytosis; 45. hypertonic; 46. isotonic; 47. phagocytosis; 48. pinocytosis; 49. b; 50; c; 51. a; 52. c; 53. d; 54. a; 55. b; 56; net movement of a substance from a region of higher concentration to a region of lower concentration; 57. net movement of water across a selectively permeable membrane by passive diffusion; 58. active transport; 59. movement of polar and charged molecules across membrane down their concentration gradient aided by transport protein; 60. symport; 61. antiport; 62. substance binds to specific cell surface receptor before being taken into cell.

Self-Test

1. b

2. a

3. d

4. c

5. d [The passage of charged/polar molecules is generally impeded because of the membrane's hydrophobic core.]

6. b

7. d [Osmotic removal of water can lead to extensive cell shrinkage and retraction from cell walls.]

8. c

9. c

10. c [Because total solute concentration is greater in compartment B, it is considered hypertonic compared to A.]

11. c [Because solutes moved passively down their concentration gradient, there is net movement of solutes from compartment B to compartment A until equilibrium is reached.]

12. b[Because water moves passively from a solution of lesser solute concentration to a solution of greater solute concentration, there is net movement of water from compartment A to compartment B until equilibrium is reached.]

Chapter 7 Cell Communication

Why It Matters [pp. 143–144]

1. activities

7.1 Cell Communication: An Overview [pp. 144–146], **7.2 Characteristics of Cell Communication Systems with Surface Receptors** [pp. 147–149], **7.3 Surface Receptors with Built-In Protein Kinase Activity: Receptor Tyrosine Kinases** [pp. 149–150], **7.4 G-Protein Coupled Receptors** [pp. 150–155], **7.5 Pathways Triggered by Internal Receptors: Steroid Hormone Receptors** [pp. 155–156], **7.6 Integration of Cell Communication Pathways** [pp. 157–158]

2. reception; 3. transduction; 4. response; 5. peptide hormones; 6. neurotransmitters; 7. protein kinases; 8. protein kinase cascade; 9. protein phosphatase; 10. amplification; 11. receptor tyrosine kinases; 12. insulin; 13. epidermal growth factor (EGF); 14. platelet-derived growth factor (PDGF); 15. nerve growth factor (NGF); 16. G-protein coupled receptor; 17. first messenger; 18. effector; 19. second messenger; 20. adenylyl cyclase; 21. cAMP; 22. phospholipase C; 23. inositol triphosphate (IP_3); 24. diacylglycerol (DAG); 25. steroid hormone receptors; 26. cross talk; 27. reception; 28. transduction; 29. response; 30. a; 31. b; 32. a; 33. b; 34. b; 35. c; 36. a; 37. c; 38. b; 39. a; 40. b; 41. b; 42. b; 43. a; 44. c; 45. c; 46. g; 47. f; 48. a; 49. d; 50. b; 51. e; 52. activates protein kinases; 53. IP_3; 54. DAG; 55. steroid hormone receptor; 56. undergo autophosphorylation, then phosphorylates other proteins; 57. G-protein coupled receptor

Self-Test

1. a, b, c

2. a, b, c

3. c, d

4. b, c

5. c

6. a, b, d [a, b, and d are correct because activated G-protein complex can link with several different effectors that stimulated cytoplasmic pathways; c is not correct because there is no evidence that G-proteins associate with steroid hormone receptors]

7. b

8. b, c, d, e [b, c, d and e are correct because RTKs do bind peptide hormones, and upon binding, the receptor undergoes autophosphorylation before phosphorylating other cellular protein, including RAS; a is incorrect because there is no evidence that RTKs directly bind to DNA]

9. a, b

10. a, b, c [a, b, and c are correct because as the production of these second messenger is stimulated, their concentration in the cytoplasm of the target cell increases, creating a diffusion gradient to move from the original target cell through the gap junction to the adjacent cell, in which biological response will be evoked]

11. d

12. c

Chapter 8 Harvesting Chemical Energy: Cellular Respiration

Why It Matters [pp. 157–158]

1. carbon dioxide; 2. water; 3. ATP; 4. heat; 5. metabolic; 6. mitochondria; 7. Luft syndrome; 8. age-related; 9. plants; 10. protists; 11. prokaryotes; 12. light; 13. water; 14. carbon dioxide; 15. oxygen; 16. carbohydrates; 17. oxygen; 18. energy

8.1 Overview of Cellular Energy Metabolism [pp. 158–162]

19. oxidation; 20. oxidized; 21. protons; 22. energy; 23. reduction; 24. reduced; 25. protons; 26. redox; 27. electrons; 28. water; 29. light; 30. sugar; 31. sugars; 32. oxidative; 33. ATP; 34. electrons; 35. photosynthesis; 36. cellular respiration; 37. G; 38. F; 39. D; 40. J; 41. K; 42. I; 43. E; 44. B; 45. A; 46. C; 47. H;

48.

Events	*Photosynthesis*	*Cellular Respiration*
A. Sugars (makes/breaks)	makes	breaks
B. O_2 (uses/releases)	releases	uses
C. CO_2 (uses/releases)	uses	releases
D. Net energy (stores/releases)	stores	releases

8.2 Glycolysis [pp. 162–165], **8.3 Pyruvate Oxidation and the Citric Acid Cycle** [pp. 165–168], **8.4 The Electron Transfer System and Oxidative Phosphorylation** [pp. 168–172]

49. oxidative; 50. cytoplasm; 51. glucose; 52. pyruvates; 53. two; 54. four; 55. two; 56. NAD^+; 57. NADH; 58. ATP; 59. NADH; 60. phosphofructokinase; 61. third; 62. matrix; 63. CO_2; 64. electrons; 65. acetyl; 66. coenzyme A; 67. NADH; 68. eight; 69. matrix; 70. acetyl-CoA; 71. two; 72. three; 73. one; 74. NADH; 75. $FADH_2$; 76. substrate-level; 77. two; 78. glucose; 79. ATP; 80. citric synthase; 81. monosaccharides; 82. amino acids; 83. amino; 84. glycerol; 85. fatty acids; 86. six; 87. electrons; 88. electron transfer; 89. mitochondria; 90. Electrons; 91. oxygen; 92. electron; 93. water; 94. electrons; 95. electron transfer; 96. protons; 97. mitochondria; 98. proton; 99. mitochondria; 100. protons; 101. ATP synthase; 102. oxidative; 103. 32; 104. 38; 105. four; 106. two; 107. 32; 108. 32; 109. ATP; 110. heat; 111. C; 112. A; 113. E; 114. D; 115. B; 116. Oxidative phosphorylation is where phosphate is transferred using energy released during electron transfer to an electron acceptor; substrate-level phosphorylation is where phosphate is transferred from one substrate to another; 117. NADH and $FADH_2$ are nucleotide-based electron carrier molecules. Upon oxidation, they release two electrons and protons in the mitochondrial matrix; 118. Glycolysis is regulated by feedback inhibition of phosphofructokinase, an enzyme that catalyses the third reaction, by ATP levels. The citric acid cycle is regulated in a similar manner by feedback inhibition of the enzyme, citrate synthase, by ATP levels; 119. The electron transfer proteins, located in the inner mitochondrial membrane, use energy from electrons to pump protons into the intermembrane space. This creates a gradient that will eventually help in the generation of a large number of ATP; 120. ATP synthase is an enzyme that is located in the inner mitochondrial membrane. It is proton-motive force that moves protons through this enzyme, back into the matrix, and powers the synthesis of ATP; 121. From each molecule of glucose, there are four, zero, two, and thirty-two ATP molecules produced during glycolysis, pyruvate oxidation, the citric acid cycle, and the electron transfer system, respectively; 122. C, D, A, B;

123. *Events of glycolysis* *Steps number(s) in which the event takes place*

Events of glycolysis	Steps number(s) in which the event takes place
A. ATP is used	**Steps 1, 3**
B. ATP is produced	**Steps 7, 10**
C. NADH and H^+ are produced	**Step 6**
D. Key step for regulation	**Step 3**

124. *Events of citric acid cycle* *Steps number(s) in which the event takes place*

Events of citric acid cycle	Steps number(s) in which the event takes place
A. ATP is produced	**Step 5**
B. NADH and H^+ are produced	**Steps 3, 4**
C. $FADH_2$ is produced	**Step 6**
D. Key step for regulation	**Step 1**

125.

Steps of cellular respiration	*Location in the cell*	*# of ATP produced*	*# of NADH and $FADH_2$ produced*	*O_2 used*	*# of CO_2 produced*
A. Glycolysis	**Cytoplasm**	4	2, 0	None	None
B. Pyruvate oxidation	**Matrix**	0	2, 0	None	2
C. Citric acid cycle	**Matrix**	2	6, 2	None	4
D. Electron transfer system	**Inner membrane**	32	0, 0	Yes	None

126. B; 127. C; 128. D; 129. A; 130. E; 131. C; 132. E; 133. A; 134. B; 135. D; 136. H; 137. F; 138. I; 139. G; 140. A. Outer mitochondrial membrane; B. Intermembrane space; C. Inner mitochondrial membrane; D. Matrix ; 141. C; 142. E; 143. E; 144. D; 145. D; 146. False—Glycolysis takes place in the cytoplasm, and it produces a total of four ATP molecules; 147. False—Part of the energy is stored in ATP, and the remaining energy is released as heat; 148. False—Fats and proteins can also be broken down to produce energy; 149. True—This step generates as many as 32 molecules of ATP per glucose molecule; 150. False—Cellular respiration takes place in prokaryotes and eukaryotes.

8.5 Fermentation [pp. 172–175]

151. oxidative; 152. cytoplasm; 153. pyruvates; 154. ATP; 155. NADH; 156. oxygen; 157. mitochondria; 158. oxygen; 159. cytoplasm; 160. fermentation; 161. glycolysis ; 162. lactate fermentation; 163. carbon dioxide; 164. alcohol fermentation; 165. strict anaerobes; 166. oxygen; 167. facultative anaerobes; 168. oxygen; 169. strict aerobes ; 170. C; 171. E; 172. D; 173. B; 174. F; 175. A

176.

Steps of anaerobic respiration	*Location in the cell*	*# of ATP produced*	*# of NADH and $FADH_2$ produced*	*O_2 used*	*CO_2 produced*
A. Glycolysis	**Cytoplasm**	4	2, 0	No	No
B. Fermentation	**Cytoplasm**	0	0	No	Yes/No

177. False—During anaerobic respiration, the glycolysis step produces four ATP molecules. The fermentation step does not produce any ATP; 178. False—In anaerobic respiration, both the glycolysis and fermentation steps take place in the cytoplasm; 179. False—In anaerobic respiration, neither step requires oxygen; 180. False—Bacteria can be aerobic or anaerobic.

Self-Test

1. a [During glycolysis, two ATP molecules are used to produce a total of four ATP molecules.]
2. b [During glycolysis, one molecule of glucose is broken down to form two molecules of pyruvate.]
3. a [Glycolysis takes place in the cytoplasm of the cell.]
4. b [Fats are broken down to form glycerol and three fatty acids, which can then be further processed to produce energy.]
5. b [The pyruvate oxidation step produces carbon dioxide and NADH, but not ATP.]
6. d [The electron transfer system uses high energy electrons from NADH and $FADH_2$ to generate 32 ATP molecules.]
7. a [As one molecule of acetyl-CoA enters the citric acid cycle, one molecule of ATP is produced.]
8. b [When glucose is broken down, some of the energy is released as heat.]
9. b [During exercise, most ATP are produced by aerobic respiration, but additional ATP are produced by anaerobic respiration.]
10. a [The entire process of anaerobic respiration, glycolysis and fermentation, takes place in the cytoplasm.]
11. d [Bacteria involved in making of yogurt break down glucose by anaerobic respiration to produce lactic acid and ATP.]
12. b [Yeast breaks down glucose by anaerobic respiration to produce alcohol, carbon dioxide, and ATP.]
13. a [During anaerobic respiration, from each molecule of glucose, a total of four and a net of two ATP molecules are produced.]
14. a [From each molecule of glucose, 36–38 ATP molecules are produced during aerobic respiration, whereas 2–4 ATP molecules are produced by anaerobic respiration.]
15. b [The presence or absence of oxygen determines aerobic or anaerobic breakdown of glucose.]

Chapter 9 Photosynthesis

Why It Matters [pp. 177–178]

1. light; 2. inorganic; 3. oxygen; 4. Theodor Engelmann; 5. oxygen; 6. violet, blue, and red; 7. action spectrum

9.1 Photosynthesis: An Overview [pp. 178–180]

8. autotrophs; 9. producers; 10. heterotrophs; 11. producers; 12. consumers; 13. decomposers; 14. light; 15. inorganic; 16. Consumers; 17. decomposers; 18. cellular activities; 19. heat; 20. light dependent; 21. light independent; 22. Calvin; 23. light; 24. chemical; 25. ATP; 26. NADPH; 27. ATP and NADPH; 28. carbohydrates; 29. fixation; 30. electrons; 31. water; 32. water; 33. oxygen; 34. oxygen; 35. chloroplast; 36. inner; 37. outer; 38. thylakoids; 39. stroma; 40. granum; 41. thylakoids; 42. thylakoids; 43. stroma; 44. oxygen; 45. stomata; 46. roots; 47. Carbohydrates48. C; 49. H; 50. G; 51. L; 52. D; 53. M; 54. A; 55. K; 56. B; 57. F; 58. J; 59. I; 60. E

61.

Events	*Photosynthesis*	*Cellular Respiration*
A. Sugars (Makes/breaks)	**Makes**	**Breaks**
B. O_2 (Uses/releases)	**Releases**	**Uses**
C. CO_2 (Uses/releases)	**Uses**	**Releases**
D. Net energy (Stores/releases)	**Stores**	**Releases**

9.2 The Light-Dependent Reaction Of Photosynthesis [pp. 180–189]

62. light; 63. chemical; 64. ATP; 65. NADPH; 66. radio; 67. gamma; 68. Visible; 69. 700; 70. 400; 71. wave; 72. photon; 73. wavelength; 74. wavelength; 75. lesser; 76. wavelength; 77. greater; 78. thylakoid; 79. carotenoids; 80. transmitted; 81. chlorophylls; 82. green; 83. absorbed; 84. electrons; 85. ground; 86. excited; 87. electrons; 88. fluorescence; 89. electrons ; 90. acceptor; 91. bacteriochlorophylls; 92. magnesium; 93. hydrophobic; 94. a and b; 95. a; 96. Carotenoids; 97. chlorophyll a; 98. absorption; 99. action; 100. oxygen; 101. photosystems; 102. thylakoid; 103. photosystem; 104. photosystem; 105. antenna; 106. reaction; 107. I; 108. II; 109. water; 110. II; 111. light; 112. reaction; 113. acceptor; 114. mitochondria; 115. H^+; 116. ATP; 117. photophosphorylation; 118. I; 119. light; 120. reaction; 121. acceptor; 122. NADPH; 123. reductase; 124. electrons; 125. noncyclic; 126. cyclic; 127. ATP; 128. NADPH; 129. light; 130. P700; 131. reaction; 132. acceptor; 133. reductase; 134. NADPH; 135. H^+; 136. ATP; 137. ATP; 138. ATP; 139. light-independent, 140. C; 141. F; 142. G; 143. A; 144. P; 145. B; 146. M; 147. I; 148. O; 149. R; 150. Q; 151. N; 152. L; 153. E; 154. D; 155. H; 156. J; 157. K; 158. C; 159. E; 160. B; 161. A; 162. D,163. C B E A D,164. Noncyclic is a one way flow of electrons that use photosystem I and II and results in production of ATP and NADPH. Cyclic is a circular flow of electrons through photosystem I only and produces additional ATP, 165. Absorption spectrum is to compare the absorption of different colors of light by the photosynthetic pigments. Action spectrum is to compare the rates of photosynthesis when exposed to different color light., 166. Water splits to replace electrons in photosystems I and II. During this process, oxygen is released., 167. Light dependent reaction uses light energy to produce ATP and NADPH that are eventually used by light independent reaction to make glucose., 168. Antenna complex refers to the cluster of chlorophyll and carotenoids pigments. Chlorophyll a is the reaction center molecule to which all excited electrons are directed., 169. Each type of pigment absorbs certain wavelength. Therefore, by having different types of pigments, plants maximize the absorption of light energy.,

170.

Events	*Product/s generated*
A. Splitting of water	O_2 e^- H^+
B. Electron transfer proteins and ATP synthase	ATP
C. Photosystem I and NADH+ reductase	NADPH

9.3 The Light-Independent Reactions Of Photosynthesis [pp. 189–194]

171. light; 172. chemical; 173. ATP; 174. NADPH; 175. CO_2; 176. glucose; 177. Calvin; 178. thylakoids; 179. stroma; 180. ribulose biphosphate; 181. carbon-fixation; 182. rubisco; 183. 3-phosphoglycerate; 184. ATP; 185. NADPH; 186. light-dependent; 187. 3-phosphoglyceraldehyde; 188. three; 189. six; 190. five; 191. ribulose biphosphate; 192. one; 193. glucose; 194. six; 195. two; 196. sucrose; 197. ATP; 198. sucrose; 199. starch, 200. A; 201. B

202.

Chemicals	It's role in light-independent reaction
A. CO_2	CO_2 fixation—combines with RuBP—carbon source for making glucose
B. 3PGAL	3 carbon chemical formed after CO_2 fixation
C. Glucose	Formed during the reaction as an end product
D. Sucrose	Glucose may be converted to sucrose as a storage form
E. Starch	Glucose may be converted to starch as a storage form

203.

Steps of Photosynthesis	Location in chloroplast	ATP used/produced	NADPH used/produced	O_2 produced	CO_2 used	Glucose produced
A. Light-dependent reaction	Thylakoid membranes	Produced	Produced	Produced	NA	NA
B. Light-independent reaction/ Calvin cycle	Stroma	Used	Used	NA	Used	Produced

204. Rubisco is an enzyme complex that combines CO_2 with ribulose biphosphate to form a six-carbon chemical., 205. During each Calvin cycle, one molecule of CO_2 is fixed with one molecule of ribulose biphosphate to form two molecules of G3P. If the cycle goes around 6 times, 12 molecules of G3P are formed, out of which two G3P are used to get one molecule of glucose., 206. A. Outer/Inner membranes; B. Thylakoids; C. Granum; D. Stroma, 207. B; 208. D, 209. False. The Calvin cycle produces only two molecules of G3P which go back in the cycle. It takes 6 cycles to make enough G3P molecules so that one molecule of glucose can be generated., 210. False. Some of the glucose is used by the plant to support its own functions while the excess is stored in fruits, seeds, stems, and roots., 211. False. Light reaction generates ATP and NADPH to support the Calvin cycle., 212. False. Certain bacteria, protists, and most plants photosynthesize.

9.4 Photorespiration And The C4 Cycle [pp. 194–198]

213. CO_2 fixation; 214. CO_2; 215. oxygen; 216. phosphoglycolate; 217. glycolate; 218. CO_2; 219. oxygen; 220. CO_2; 221. photorespiration; 222. close; 223. oxygen; 224. CO_2; 225. photorespiration; 226. carboxylase; 227. oxaloacetate; 228. PEP carboxylase; 229. C4; 230. CO_2; 231. mesophyll; 232. bundle sheath; 233. CAM, 234. B; 235. E; 236. A; 237. D; 238. E

SELF-TEST

1. e {Fungi are heterotrophic]
2. b [Light dependent reaction produces ATP and NADPH]
3. b [Chlorophyll a acts as the reaction center molecule]
4. d [A photosystem has chlorophylls a and b, and carotenoids]
5. a [Water provides electrons to the light dependent reaction]
6. d [Photosynthetic pigments are present in the thylakoid membranes of the chloroplast]
7. c [ATP and NADPH are produced by the non-cyclic electron flow]
8. a [Calvin cycle requires rubisco, RUBP and 3PGAL but not the pigments]
9. d [Calvin cycle must go around 6 times before 2 molecules of 3PGAL are donated towards making of glucose]; 10. b [Each Calvin cycle makes 2 molecules of 3PGAL]

11. c [Most 3PGAL go back into the cycle to form RUBP]

12. a [Photosynthetic organisms are referred to as autotrophs because they are able to make their own organic chemicals]

13. d [Stomata present on the under surface of the leaves allows gas exchange]

14. a [C3 plants undergo photorespiration that reduces photosynthesis and growth]

15. d [CO_2 is stored by C4 and CAM plants at night when stomata are open].

Chapter 10 Cell Division and Mitosis

Why It Matters [pp. 201-202],

1. cell division; 2. two; 3. oxygen; 4. reduced;

10.1 The Cycle of Cell Growth and Division: An Overview [pp. 202-203]

5. three; 6. replication; 7. DNA; 8. two; 9. controls; 10. cell; 11. growth; 12. mitosis; 13. cytokinesis; 14. mitosis; 15. identical; 16. meiosis; 17. not identical; 18. DNA; 19. chromosomes; 20. replication; 21. two; 22. sister chromatids; 23. number; 24. type; 25. chromosome; 26. segregation; 27. one; 28. two; 29. identical; 30. chroma (chromosome); 31. mito (mitosis); 32. kinesis (cytokinesis); 33. d; 34. a; 35. c; 36. b

10.2 The Mitotic Cell Cycle [pp. 203-209]

37. growth; 38. division; 39. interphase; 40. mitosis; 41. interphase; 42. growth; 43. mitosis; 44. end; 45. beginning; 46. mitosis; 47. three; 48. G1 phase; 49. S phase; 50. G2 phase; 51. G1; 52. G0; 53. continuous; 54. prophase; 55. metaphase; 56. anaphase; 57. telophase; 58. four; 59. mitosis; 60. telophase; 61. cytokinesis; 62. F; 63. D; 64. E; 65. A; 66. B; 67. H; 68. M; 69. K; 70. L; 71. G; 72. I; 73. C; 74. Chromatid is the term that is used to describe the chromosome after replication has occurred, i.e., sister chromatids. Sister chromatids are held together at the centromere. Once the centromere area is separated and the sister chromatids are moving towards opposite poles (anaphase), they can be called chromosomes; 75C/D. G2, 4N DNA; 75E/F. Prophase, 4N DNA; 75G/H. Prometaphase, 4N DNA; 75I/J. Metaphase, 4N DNA; 75K/L. Anaphase, 4N DNA; 75M/N. Telophase, 4N DNA; 75O/P. G1, 2N DNA for each cell; 76. In both cell types the cytoplasm is divided into two daughter cells. In animal cells, a cell furrow forms due to microtubules and microfilaments. The microfilaments undergo sliding, and a constriction forms which deepens until the cytoplasm is divided. In plant cells a cell plate forms from microtubules which organize vesicles produced by the ER and Golgi complex. The vesicles fuse to be incorporated into the developing cell wall which divides the cytoplasm into two daughter cells. Vesicle membranes fuse to form the new cell membrane which lines the cell wall.

10.3 Formation and Action of the Mitotic Spindle [pp. 209-212],

77. mitotic spindle; 78. two; 79. centrosome; 80. animal; 81. centrosome; 82. plants; 83. centrosome; 84. microtubule organization center; 85. pair; 86. centrioles; 87. microtubules; 88. spindle; 89. generated (produced); 90. prophase; 91. replicate; 92. old; 93. new; 94. two; 95. centrosomes; 96. microtubules; 97. centrioles; 98. without; 99. nucleus; 100. spindle; 101. microtubules; 102. kinetochore; 103. kinetochore; 104. attach; 105. nonkinetochore; 106. motor; 107. kinetochore; 108. walk

10.4 Cell Cycle Regulation [pp. 212-216],

109. checkpoints (control); 110. hormones; 111. growth factors; 112. cyclin; 113. cyclin-dependent kinase; 114. CDK; 115. cyclin; 116. phosphatase; 117. cyclin; 118. CDK; 119. two; 120. mitosis; 121. S; 122. B; 123. 1; 124. mitosis; 125. E; 126. 2; 127. G1; 128. B; 129. D; 130. E; 131. A; 132. C

10.5 Cell Division in Prokaryotes [pp. 216-218]

133. False, large (major); 134. True; 135. False, middle

Self-Test

1. b [Assuming mistakes don't occur, mitosis results in two daughter cells that are genetically identical to the parent cell.]

2. a [G1 phase is the most variable and could lead to cell arrest. All other phases of interphase and mitosis are uniform in length for a given species.]

3. d [breakdown of the nuclear envelope is primary indicator of prometaphase]

4. c [cells lacking a nucleus is unable to undergo cell division, the cell would be in the arrest phase or G0 phase]

5. c [all statements except c are correct, there is a separate kinetochore for each sister chromatid]

6. c [with respect to most plant cells, all statements are correct except c, they lack centrioles]

7. d [at the checkpoint prior to mitosis, many of the proteins are phosphorylated, a phosphatase must remove the inhibitory phosphate in order for CDK to become active. Decreasing the level of phosphatase would prevent the cell from undergoing mitosis.]

8. a [Contact inhibition occurs when cell surface receptors are in contact with adjacent cells or extracellular matrix. If blocked, cells would continue to undergo division and a tumor or mass of cells would develop.]

9. b [of the given selections, b, or lacking checkpoints is the primary difference]

10. a [only cells that undergo binary fission have an origin of replication site, these cells typically have 1 circular DNA molecule]

Chapter 11 Meiosis:The Cellular Basis of Sexual Reproduction

Why It Matters [pp. 221-222],

1. gametes; 2. meiosis; 3. 1/2; 4. haploid; 5. fertilization; 6. zygote; 7. zygote; 8. equal; 9. diploid; 10. unique; 11. recombination

11.1 The Mechanisms of Meiosis [pp. 222-227],

12. diploid; 13. chromosomes; 14. pair; 15. paternal; 16. male; 17. maternal; 18. female; 19. gametes; 20. meiosis; 21. haploid; 22. one; 23. homologous; 24. genes; 25. order; 26. same; 27. different; 28. alleles; 29. alleles; 30. meiosis; 31. homologous; 32. exchange; 33. recombination; 34. four; 35. one; 36. homologous pair; 37. random; 38. gametes; 39. haploid; 40. genetically; 41. Alleles are different versions of the same gene. Each allele has a different DNA sequence. In a population, there can be 1 or 2 alleles or up to hundreds of alleles for a given gene; 42. Interkinesis is the short time period between meiosis I and meiosis II—very important to note that DNA replication doesn't occur; 43. Synapsis—homologous chromosomes that have replicated come together and pair up with the potential to exchange alleles. Tetrad is when Synapsis occurs. For each homologous pair, there are four sister chromatids; 44. A; 45. A; 46. B; 47. A; 48. B; 49. A; 50A. two, diploid; 50B. occurred; 50C. recombination or synapsis; 50D. homologous pairs, random, No; 50E. sister chromatids, haploid, one, homologous

11.2 Mechanisms That Generate Genetic Variability [pp. 227-230],

51. three; 52. recombination; 53. random; 54. paternal; 55. maternal; 56. opposite; 57. random; 58. gamete; 59. fertilization; 60. D; 61. C; 62. B; 63. E; 64. A; 65. M; 66. N; 67. A; 68. D; 69. F; 70. E; 71. J; 72. B; 73. G; 74. L; 75. K; 76. I; 77. C; 78. H; 79. E/F; 80. D; 81. B; 82. C; 83. A or G; 84. A or G

11.3 The Time and Place of Meiosis in Organismal Life Cycles [pp. 230-233].

85. three; 86. diploid; 87. meiosis; 88. gametes; 89. haploid; 90. unique; 91. generations; 92. diploid; 93. haploid; 94. haploid; 95. dominant; 96. gametes; 97. mitosis; 98. identical

Self-Test

1. c [2N or diploid number represents the total number of chromosomes. Chromosomes are in homologous pairs with one member of each pair from the maternal source and the other member from the paternal source. In this example, 8 chromosomes are maternal and 8 chromosomes are paternal];
2. c [alleles on homologous chromosomes could be the same or different, depending on whether paternal and maternal alleles are identical or different]
3. a [DNA replication only occurs during interphase (S-phase) prior to prophase I]
4. a [homologous pairs separate (in a random fashion) during anaphase I]
5. d [nondisjunction is when a homologous pair doesn't separate; one daughter cell of meiosis I would have two chromosomes and the other would have 4. Each of these chromosomes would be composed of two sister chromatids]
6. b [females have two X chromosomes, so the only gamete type would be X; males have both a X and Y chromosomes, so gametes of each type could be produced]
7. d [selection d explains the various sources of genetic variability]
8. c [crossing over is only an exchange of DNA between homologous chromosomes, so the DNA amount would not increase or decrease]
9. d [all three life cycle pattern have meiosis as part of the life cycle—variability it present in all three patterns]
10. b [mitosis produces daughter cells that are genetically identical]

Chapter 12 Mendel, Genes, and Inheritance

Why It Matters [pp. 235-236],

1. traits; 2. inheritable; 3. offspring; 4. sickle; 5. Gregor Mendel; 6. genetics

12.1 The Beginnings of Genetics: Mendel's Garden Peas [pp. 236-245]

7. garden peas; 8. genetics; 9. alleles; 10. chromosome; 11. meiosis; 12. character; 13. trait; 14. self-pollinating; 15. cross-pollinated; 16. alleles; 17. numbers; 18. multiple (many); 19. parental; 20. P; 21. F1; 22. F2; 23. self-pollination; 24. F1; 25. F; 26. E; 27. D; 28. B; 29. G; 30. A; 31. C; 32. B; 33. B; 34. B; 35. A; 36. A; 37. 2; 38. R; 39. r; 40. TR; 41. TR; 42. Tr; 43. Tr; 44. 4; 45. TR; 46. Tr; 47. tR; 48. tr; 49. heterozygote; 50. red; 51. genotype; 52. phenotype; 53. tall; 54. red; 55. homozygous dominant; 56. heterozygous; 57. phenotype; 58. heterozygous; 59. both; 60. phenotype; 61. homozygous recessive; 62. rr; 63. phenotype; 64. homozygous recessive; 65. tt; 66. When two or more independent events occur, the probability they will occur in order is determined by the product. When the probability is that either events can occur, the individual probabilities are added. Formation of gametes and determination of which gamete is involved in fertilization are independent events. In the case of phenotypes with dominance and recessive traits, there are often multiple ways to produce a given phenotype, in this case the probabilities are added together; 67. Alleles (traits) from the maternal and paternal source separate when gametes are formed; 68. Homologous pairs separate independently when gametes are formed. The percentage of either maternal or paternal chromosomes that end up in any given gamete is randomly determined; 69a. Tt gametes are T and t, tt gametes are t; 69b. 50% Tt and 50% tt or 1:1 ratio of heterozygotes to homozygous recessive; 69c. 50% Tall to 50% dwarf or 1:1 Tall to dwarf; 70a. homozygous dominant gametes are TS, homozygous recessive gametes are ts; 70b. heterozygous for both traits, 100% TtSs; 70c. all Tall plants with Strong stems; 70d. 1 (TTSS): 1 (TTss): 1 (ttSS): 1 (ttss): 2 (TTSs): 2 (TtSS): 2 (Ttss): 2 (ttSs): 4 (TtSs); 70e. 9 (Tall Strong): 3 (Tall weak): 3 (dwarf Strong): 1 (dwarf weak); 71. The resulting probabilities of genes and traits can be associated with chromosome behavior and movement during meiosis.

12.2 Later Modifications and Additions to Mendel's Hypotheses [pp. 245-251].

72. complete; 73. incomplete dominance; 74. codominance; 75. two; 76. two; 77. paternal; 78. maternal; 79. multiple alleles; 80. interact; 81. epistasis; 82. inhibit; 83. expression; 84. polygenic inheritance; 85. False, Epistasis; 86. False, Pleiotrophy;

Self-Test

1. d [expression in the phenotype is one of the factors which determines recessive and dominance]
2. c [of the given genetic patterns, the heterozygous has the most variation, since one allele is dominant and the other is recessive]
3. a [the phenotype is the expression or in this example the blood type, while the genotype are the actual alleles that are present. In this example both the mom and dad are type A – phenotype and their genotypes must be AO to have type O children. Blood typing is an example of codominance, if the allele is present, it will be expressed]
4. c [with the given genotype of the two traits A and B, the dominant A will combine with the dominant B and the recessive a will combine with the dominant B, resulting in two different type of alleles, AB and aB]
5. a [the parental cross is between a homozygous dominant and a homozygous recessive. If the offspring has a different phenotype than either of the parents, the most likely explanation would be codominance or incomplete dominance. In either case the phenotypic expression is either both alleles are present and expressed, or there is a blending of the alleles]
6. c [The heterozygous for blue would have a genotype of Bb or 50% dominant allele of B and 50% recessive allele of b. The white plant would have 100% of the alleles of b. When these two plants are crossed, the phenotypic rate is 50% of the offspring will be blue and 50% will be white]
7. c [blue offspring will have a genotype of Bb and the white offspring will have a genotype of bb – it genotypic ratio will be 50% heterozygous and 50% homozygous recessive]
8. b [Type A blood could have a genotype of either AO or AA; Type B blood could have a genotype of either BO or BB; given these possibilities, then this couple could have children with any of the possible phenotypes]
9. a [In order to get a yellow lab puppy, each of the parents must have a recessive allele for both color and the E gene (epistasis gene), since the genotype of a yellow lab puppy is bbee]
10. b [There are multiple genes with multiple alleles that determine height. In addition, external factors such as nutritional factors also play a significant role in growth or height of an individual]

Chapter 13 Genes, Chromosomes, and Human Genetics

Why It Matters [pp. 255-256],

1. genetics (inheritance); 2. genotype; 3. expression; 4. phenotype; 5. alleles (genes); 6. interactions

13.1 Genetic Linkage and Recombination [pp. 256-261]

7. seven; 8. independently; 9. gamete; 10. segregated; 11. assorted; 12. chromosomes; 13. locus; 14. different; 15. independently; 16. same; 17. no always; 18. genes; 19. same; 20. unit (group); 21. linked; 22. linkage; 23. fruit fly; 24. *Drosophila*; 25. same; 26. close; 27. recombination; 28. crossing over; 29. linkage group; 30. far apart; 31. linked; 32. crossing over; 33. recombination (gene) frequencies; 34. recombination (gene) frequencies 35. maps; 36. A linkage group is a group of genes on the same chromosome that are very close in proximity which undergo recombination as a unit; 37. Phenotypes that always occur together or don't follow the typical or predicted probability patterns.

13.2 Sex-Linked Genes [pp. 261-266],

38. chromosomes(s); 39. sex; 40. males; 41. females; 42. Genes; 43. sex-linked; 44. males; 45. females; 46. autosomes; 47. homologous pair; 48. two; 49. X; 50. one; 51. X; 52. one; 53. Y; 54; two; 55. X; 56; one; 57. one; 58. Y; 59. dominant; 60. recessive; 61. expressed; 62. one; 63. condensation (inactivation); 64. barr body; 65. random; 66. paternal; 67. maternal; 68. paternal; 69. maternal; 70. A female can have a sex-linked recessive condition if both X chromosomes have the recessive gene. This can occur if the mother is either a carrier or has the condition and if the father has the condition; 71. The SRY gene is on the Y chromosome. The product (expression) of this gene determines that the embryo will develop into a male. If this gene is defective, the embryo will develop into a female phenotype with a male genotype; 72a. If Suzie's father has the condition, then Suzie must be heterozygous for the condition or a carrier with a normal phenotype; 72b. The probably that a child of this couple will have the genetic condition is 50%; 72c. Since male children will inherit the X chromosome from the mother, the male could receive the normal X chromosome and have a normal genotype and phenotype or receive the defective X chromosome and have the condition; 72d. Female children will receive 1 X chromosome from each parent. The X chromosome from the father is normal, since he doesn't have the condition. So if the female received the normal X chromosome from the mother, both genotype and phenotype are normal, if the female received the defective X chromosome, the genotype will be heterozygous (carrier) and the phenotype will be normal.

13.3 Chromosomal Alterations That Affect Inheritance [pp. 266-269]

73. F; 74. D; 75. H; 76. A; 77. B; 78. C; 79. E; 80. G

13.4 Human Genetics and Genetic Counseling [pp. 269-272],

81. pedigree (history); 82. educated; 83. inherit; 84. heterozygous; 85. recessive; 86. homozygous recessive; 87. genotype; 88. phenotype; 89. heterozygous; 90. two recessive; 91. homozygous dominant; 92. prenatal diagnosis; 93. genetic screening.

13.5 Nontraditional Patterns of Inheritance [pp. 272-274],

94. Inheritance pattern is associated with genes that are in organelles found in the cytoplasm, such as mitochondria or chloroplasts; 95. During gamete formation in most eukaryotes, the female gamete is significantly larger with more cytoplasm (and organelles) than the male gamete. When fertilization occurs, all cytoplasmic organelles are from a maternal source.

Self-Test

1. b [When traits always express together, the simplest explanation is the genes are in a linkage group.]
2. a [Given the ratio of offspring, this is sex-linked, since the variation in wing length is only seen in males and a recessive trait, because it was not present in either of the parental phenotypes, which are most likely heterozygotes.]
3. b [Red-green color blindness is a X-linked trait. Males are more likely to have this condition since they only have 1 X chromosome. Females can be carriers since they can be heterozygous with 2 X chromosomes.]
4. c [One of the X-chromosomes in females will become inactive (condensed) to form a barr body. This can be observed in a karyotype of a female.]
5. d [the pattern shows both an inversion (CDEF) as well as a duplication (EFG)]
6. d [if the 2N number is 50, then 3 complete sets or triploidy would be 150 chromosomes]
7. b [A karyotype is a chromosome preparation in which homologous pairs are grouped together. If an extra chromosome were present, such as in trisomy 21, it would be most obvious in a karyotype. The other means of evaluation do not use the actual chromosomes present.]
8. a [the most common cause of either an extra or missing chromosome is when homologous pairs don't separate correctly]

9. c [given the information, it can be concluded that both the homozygous dominant as well as the heterozygous genotypes have the phenotype, since the phenotype is not associated with the sex of the individual, this is an autosomal condition]

10. b [screening for PKU is performed after birth]

Chapter 14 DNA Structure, Replication, and Organization

Why It Matters [pp. 277-278]

1. acidic; 2. phosphorus; 3. nucleus; 4. nuclein; 5. deoxyribonucleic acid; 6. DNA

14.1 Establishing DNA as the Hereditary Molecule [pp. 278-281]

7. lipids; 8. carbohydrates; 9. nucleic acids; 10. proteins; 11. proteins; 12. 20; 13. amino acids; 14. transformation; 15. heat killing; 16. destroyed; 17. bacteriophage (phage); 18. proteins; 19. genetic (hereditary) material; 20. DNA; 21. DNA; 22. genetic (hereditary) material; 23. double helix; 24. nucleotide; 25. C; 26. B; 27. D; 28. E; 29. A

14.2 DNA Structure [pp. 281-284]

30. deoxyribose; 31. phosphate; 32. nitrogenous; 33. sugar-phosphate backbone; 34. bridge; 35. 3' OH; 36. 5' OH; 37. phosphodiester bond; 38. two; 39. parallel; 40. nitrogenous bases; 41. purines, pyrimidines; 42. Adenine, guanine; 43. guanine, cytosine; 44. Complementary; 45. purine; 46. pyrimidine; 47. base pairs; 48. adenine, thymine; 49. cytosine, guanine; 50. two; 51. antiparallel; 52. 3'; 53. carbon; 54. 5'; 55. carbon; 56. A. 5' end; B. 5 carbon; C. 4 carbon; D. 3 carbon; E. 1 carbon; F. 2 carbon; G. 1 carbon; H. 2 carbon; I. 5 carbon; J. 4 carbon; K. 3 carbon; L. 3' end ; 57. A. guanine; B. cytosine; C. cytosine; D. guanine; E. adenine; F. thymine

14.3 DNA Replication [pp. 284-292]

58. semiconservative; 59. antiparallel; 60. helicase; 61. replication fork; 62. RNA primer; 63. 10; 64. primase; 65. 3', 5'; 66. 5', 3'; 67. DNA polymerase; 68. RNA primer; 69. 3' OH; 70. 3'; 71. 5', 3'; 72. continuously; 73. leading strand; 74. discontinuously; 75. lagging strand; 76. lagging; 77. short; 78. Okazaki fragments; 79. DNA ligase; 80. telomeres; 81. end; 82. replicates; 83. telomeres; 84. shorter; 85. telomerase; 86. telomere; 87. repeating units; 88. aging; 89. cancer; 90. telomere; 91. telomerase; 92. b; 93. e; 94. c; 95. d; 96. d; 97. a; 98. Leading strand is 3'→ 5' and lagging strand is 5' → 3'; 99. The 3'→ 5' strand (leading strand) will have continuous replication, because after the RNA primer is produced, a 3' OH end is always available after each addition of a nucleotide. DNA polymerase adds nucleotides to the 3' OH end. The 5' → 3' strand (lagging strand) will have discontinuous replication because the 5' OH group is at the end. Multiple RNA primers will be added and short fragments (Okazaki) of DNA will be formed by DNA polymerase. These short fragments of DNA are linked together by DNA ligase.

14.4 Mechanisms That Correct Replication Errors [pp. 292-294],

14.5 DNA Organization in Eukaryotes and Prokaryotes [pp. 295-298]

100. deoxyribonuclease; 101. mismatched; 102. DNA repair mechanisms; 103. Hydrogen bonds; 104. accuracy; 105. chromatin; 106. histones, nonhistones; 107. negatively; 108. positively; 109. histone; 110. nucleosome; 111. Nonhistone; 112. gene expression; 113. euchromatin; 114. heterochromatin; 115. euchromatin; 116. Mutation rate would be increased. Mutations are caused by mismatched base pairs, which could significantly affect protein translation and production; 117. Both are DNA strands found in prokaryotic cells. The nucleoid is circular DNA that is located in the cytoplasm of the prokaryote. A plasmid is an additional strand of DNA. It is usually circular; however, plasmids can also be linear. During replication of prokaryotic cells, the nucleoid and plasmid (if present) are replicated and distributed to daughter cells; 118. A. DNA; B. Nucleosome (H2A, H2B, H3, H4); C. Linker; D. H1; E. Solenoid; F. Chromosome

Self-Test

1. d [Proteins were thought to be the hereditary material because of the large number of amino acids and greater diversity compared to other biological molecules. The other molecules have many repeating units, and it was thought that the hereditary material required a tremendous diversity.]

2. b [The *S* strain is transformed into the *R* strain due to the addition of the *R* strain DNA.]

3. b [Since the two possible candidates for the hereditary material were proteins and DNA, use of an atom that was unique to each—sulfur (proteins) and phosphate (DNA)—would help identify which was replicated.]

4. c [The nitrogenous base can be either a pyrimidine or a purine; the term nucleotide is a generic term; phosphodiester is the bond between the sugar and phosphate group of the DNA backbone.]

5. c [X-ray diffraction is the bending or reflection of X-rays on the crystalline form of DNA, resulting in a pattern that identified a helix; a is not a technique, but an observation of nitrogenous base pairing; b is a process of exchange between one organism and another; d is not a technique, rather it is the process of DNA duplication]

6. a [The daughter DNA is new and part is conserved from the parent DNA; b and c are two types of replication that were shown to be incorrect in the Meselson and Stahl experiments; d is a type of replication that does not exist.]

7. a [DNA polymerase can only add nucleotides in this fashion; b and c are incorrect because the DNA polymerase is not able to add nucleotides to the 5' end; d is incorrect because phosphodiester is a bond, not an end of either strand of DNA.]

8. d [d is the enzyme necessary for the unwinding of DNA; the other enzymes are involved in steps of DNA replication after the DNA unwinds.]

9. a [Only short segments of DNA are made in the 3' to 5' direction, which are then connected together; b is the type of replication that occurs in the leading strand; c is the fragments of DNA that are produced; d is the directional relationship of the two strands of DNA.]

10. c [Primase lays down short RNA primers at the replication fork; the other enzymes are used at different steps in DNA replication.]

11. b [Mutations occur if there is an error that occurs in the final product of DNA replication; a is when DNA is put into another organism; c doesn't exist; d is a mismatch.]

12. d [This is an area of loosely packed DNA that is available for expression and replication; a represents blocks of DNA that are inactive or highly condensed; nucleoids are masses of circular DNA found in prokaryotes; c is the structure that contains both histone proteins and DNA.]

Chapter 15 From DNA to Protein

Why It Matters [pp. 301–302]

1. proteins; 2. DNA; 3. DNA; 4. DNA; 5. RNA; 6. ribosomes; 7. proteins

15.1 The Connection Between DNA, RNA, and Protein [pp. 302–307]

8. Archibald Garrod; 9. William Bateson; 10. metabolic; 11. George Beadle; 12. Edward Tatum; 13. enzymes; 14. minimal; 15. nutrients; 16. auxotrophs; 17. mutants; 18. genes.19. enzymes; 20. metabolic; 21. gene; 22. enzyme; 23. amino acids; 24. polypeptides; 25. gene; 26. polypeptide; 27. transcription; 28. DNA; 29. gene; 30. RNA; 31. RNA polymerase; 32. translation; 33. RNA; 34. ribosome; 35. amino acids; 36. RNA; 37. central dogma; 38. prokaryotic; 39. pre; 40. mRNA; 41. cytoplasm; 42. adenine, guanine, cytosine, thymine; 43. adenine, guanine, cytosine, uracil; 44. uracil ; 45. adenine; 46. cytosine; 47. guanine; 48. amino acids; 49. polypeptide; 50. genetic; 51. Marshall Nirenberg; 52. Hargobind Khorana; 53. mRNA; 54. polypeptide; 55. 20; 56. 3; 57. 4; 58. 64; 59. 20; 60. codon; 61. degeneracy;

62. redundancy; 63. 64; 64. 61; 65. sense; 66. methionine; 67. start; 68. stop; 69. mRNA; 70. universal, 71. C; 72. E; 73. G; 74. J; 75. H; 76. K; 77. A; 78. L; 79. F; 80. B; 81. D; 82. I; 83. D; 84. A; 85. B; 86. C, 87A: Transcription; 87B: RNA Processing; 87C: Translation

88.

Nucleic Acids	*Types of Nitrogenous Bases*
A. DNA	Adenine, Thymine, Guanine, Cytosine
B. RNA	Adenine, Uracil, Guanine, Cytosine

15.2 Transcription: DNA-directed RNA Synthesis [pp. 307–309]

89. transcription; 90. replication; 91. thymine; 92. guanine; 93. uracil; 94. DNA; 95. DNA; 96. double; 97. single; 98. DNA; 99. mRNA; 100. primer; 101. primer; 102. transcription; 103. gene; 104. genes; 105. protein-coding; 106. non–protein-coding; 107. RNA polymerase; 108. promoter; 109. DNA; 110. 3'; 111. 5'; 112. nucleotides; 113. 5'; 114. 3'; 115. DNA; 116. mRNA; 117. DNA; 118. DNA; 119. mRNA; 120. mRNA; 121. all; 122. II; 123. I; 124. III; 125. tRNA; 126. rRNA; 127. promoter; 128. promoter; 129. TATA; 130. II; 131. transcription; 132. TATA; 133. II; 134. terminator; 135. protein; 136. mRNA, 137. C; 138. F; 139. B; 140. H; 141. G; 142. A; 143. J; 144. D; 145. K; 146. I; 147. L; 148. E, 149. BDEGACF

150.

Points of Comparison	*DNA Replication*	*Transcription*
Base pairing	A pairs with T; G pairs with C	A pairs with U; G pairs with C
Number of DNA strands copied	Both the strands	Only one of the 2 strands
Number of new strands formed	Two	One
Enzyme involved	DNA polymerase	RNA polymerase
Primer formed	RNA primer on lagging strand	None

151.

Points of Comparison	*Prokaryotic Cells*	*Eukaryotic Cells*
Transcription enzyme/s	RNA polymerase	RNA polymerases I, II, III
Binding of RNA polymerase to promoter region	RNA polymerase directly binds to promoter region	Transcription factors help RNA polymerase bind to promoter region
Termination of transcription	Protein binds to termination signal to trigger termination	No such termination signal

152. Prokaryotic cells have only one type of RNA polymerase. Eukaryotic cells have RNA polymerase to transcribe protein coding genes; RNA polymerases I and III transcribe other types of RNA such as rRNA and tRNA., 153. The TATA box is part of promoter region in eukaryotic cells. Transcription factors bind to the TATA box and thus helps in binding of RNA polymerase. , 154. Protein coding genes code for proteins a cell needs. Non–protein-coding genes code for RNA that do not need to be translated, such as rRNA and tRNA., 155. False—In DNA, A pairs with T and G pairs with C, whereas RNA polymerase places U against A. , 156. False—Prokaryotic cells have only one type of RNA polymerase that transcribes protein-coding and non–protein-coding genes, whereas in eukaryotic cells, RNA

polymerase II transcribes protein coding genes, and RNA polymerase I and III transcribes non–protein-coding genes., 157. Prokaryotic cells do not have compartments or organelles and therefore all process take place in the cytoplasm, whereas in eukaryotic cells, most of the DNA is located in the nucleus and therefore transcription takes place in the nucleus.

15.3 Production of mRNA in Eukaryotes [pp. 309–313]

158. protein; 159. untranslated; 160. nucleus; 161. pre-mRNA; 162. mRNA; 163. translation; 164. guanine; 165. ribosomes; 166. terminator; 167. adenines; 168. poly A; 169. 3'; 170. nucleus; 171. cytoplasm; 172. Richard Roberts; 173. Philip Sharp; 174. introns; 175. exons; 176. mRNA splicing; 177. introns; 178. exons; 179. small nuclear; 180. small ribonucleoprotein particles; 181. snRNPs; 182. spliceosome; 183. mRNA; 184. alternative; 185. exon shuffling, 186. C; 187. G; 188. A; 189. K; 190. B; 191. J; 192. D; 193. L; 194. E; 195. F; 196. H; 197. I

198.

Eukaryotic Transcription	Role in Transcription
A. Pre-mRNA	Larger form of mRNA made by eukaryotes that has 5' cap, 3' poly A tail, and introns—for protection during its travel from the nucleus to the cytoplasm and regulation
B. 5' GTP cap	5' end of pre-mRNA that serves as the binding site for the ribosomes and protects from enzymatic degradation
C. 3' poly A Tail	5' end of pre-mRNA that serves as the binding site for the ribosomes
D. Introns and Exons	Regulation of genetic expression
E. mRNA Splicing	Regulation of genetic expression

15.4 Translation: mRNA-Directed Polypeptide Synthesis [pp. 313–326]

199. mRNA; 200. amino acids; 201. polypeptide; 202. cytoplasm; 203. translation; 204. pre-mRNA; 205. nucleus; 206. mRNA; 207. ribosomes; 208. transfer; 209. amino acids; 210. enzymes; 211. polypeptide; 212. 75–90; 213. clover; 214. nucleotides; 215. anticodon; 216. mRNA; 217. amino acid; 218. codon; 219. anticodon; 220. aminoacylation; 221. aminoacyl-tRNA synthetases; 222. ATP; 223. large; 224. small; 225. ribosomal; 226. ribosomal; 227. smaller; 228. prokaryotic; 229. A; 230. P; 231. E; 232. Initiation; 233. large; 234. small; 235. AUG; 236. 5'; 237. UAC; 238. methionine; 239. P; 240. initiation; 241. Elongation; 242. tRNA; 243. A; 244. peptidyl transferase ; 245. amino acid; 246. RNA; 247. E; 248. P; 249. A; 250. 3'; 251. peptide; 252. Termination; 253. stop; 254. Termination/Release; 255. A; 256. polypeptide; 257. subunits; 258. mRNA; 259. polyribosome; 260. transcription; 261. translation; 262. amino acids; 263. chaperones; 264. pepsinogen; 265. free; 266. bound; 267. signal; 268. SRP; 269. DNA; 270. nucleotide; 271. codons; 272. amino acid; 273. silent; 274. missense; 275. nonsense; 276. nucleotide; 277. mRNA; 278. amino acid, 279. C; 280. G; 281. A; 282. K; 283. B; 284. J; 285. D; 286. L; 287. E; 288. F; 289. H; 290. I; 291. M; 292. O; 293. N; 294. C; 295. A; 296. E; 297. B; 298. D, 299. In prokaryotic cells, both the steps take place in cytoplasm and as mRNA is being synthesized, ribosomes in the vicinity can start translating from 5' end. In eukaryotic cells, this would not be possible since DNA is in nucleus and ribosomes are in the cytoplasm., 300. Since there are 20 types of amino acids and 64 possible codes, some of the amino acids have multiple codes. If the substitution mutation changes DNA sequence in such a way that the code still codes for the same amino acid, there will be no change in the protein. , 301. A site: aminoacyl site, where aminoacyl-tRNA (carrying specific amino acid) binds. P site: peptidyl site, where tRNA shifts after its amino acid has joined the growing peptide. E site: exit site, where tRNA leaves the ribosomes after its peptide chain has joined to the new tRNA., 302. AUG is a start codon and it codes for methionine. ,

303. A gene has a sequence of: TACTTCGCAAATCCCGCAGTCACGTTGATC Give the mRNA sequence: AUGAAGCGUUUAGGGCGUCAGUGCAACUAG. Copy the mRNA sequence and mark codons: AUG,AAG,CGU,UUA,GGG,CGU,CAG,UGC,AAC,UAG. Give amino acid sequence coded in the above mRNA: Met-Lys-Arg-Leu-Gly-Arg-Gln-Cys-Asn

Self-Test

1. a [RNA polymerase copies DNA sequence to make mRNA]
2. b [Ribosomes read the codons to put specific amino acids in a specific sequence]
3. c [AUG is a start codon in prokaryotic and eukaryotic mRNA]
4. b [Start codon codes for methionine which makes it the first amino acid in most proteins]
5. d [There are 4 types of nitrogenous bases and codons are a set of 3 nucleotides. There are 64 possible sequences using 4 types of nitrogenous bases]
6. a [Prokaryotic cells lack membrane bound organelles. Therefore, both the steps take place in the cytoplasm]
7. c [Eukaryotic cells make a larger transcript that has introns and exons—called pre-mRNA]
8. c [A transcript is a RNA. Methionine is an amino acid.]
9. a [snRNP are small nuclear ribonucleoproteins that splices pre-mRNA in eukaryotic cells]
10. c [tRNA has the anticodon that matches the codon on mRNA]
11. a [A site is the first site where tRNA enters with its specific amino acid]
12. b [P is the second site where a tRNA attaches it amino acid to the growing chain of amino acids]
13. b [Peptidyl transferase is an exceptional enzyme that is not a protein but a RNA]
14. c [Frame shift mutation shifts all the codons on a mRNA and therefore changes all the codes, hence amino acid sequence]
15. b [An mRNA can be read by several ribosomes at the same time].

Chapter 16 Control of Gene Expression

Why It Matters [pp. 329-330]

1. inactive; 2. sperm; 3. active; 4. differentiate; 5. genes; 6. off; 7. short; 8. short; 9. long

16.1 Regulation of Gene Expression in Prokaryotes [pp. 330–335]

10. environment; 11. sugars; 12. enzymes; 13. enzymes; 14. Francois Jacob; 15. Jacques Monod; 16. operon; 17. DNA; 18. prokaryotic; 19. lacZ, lacY, lacA; 20. B-galactosidase; 21. permease; 22. transacetylase; 23. mRNA; 24. enzymes; 25. promoter; 26. regulatory; 27. repressor; 28. operator; 29. enzymes; 30. B-galactosidase; 31. inducer; 32. regulatory; 33. operator; 34. promoter; 35. lacZ, lacY, lacA; 36. mRNA; 37. enzymes; 38. allolactose; 39. operator; 40. operator; 41. lacZ, lacY, lacA; 42. enzymes; 43. inducer; 44. inducible; 45. negative; 46. positive; 47. operon; 48. glucose; 49. lactose; 50. enzymes; 51. operon; 52. positive; 53. energy; 54. disaccharide; 55. allolactose; 56. repressor; 57. repressor; 58. CAP; 59. cAMP; 60. CAP; 61. promoter; 62. enzymes; 63. adenylate; 64. cAMP; 65. CAP; 66. promoter; 67. enzymes; 68. positive; 69. amino acid; 70. trpA-trpE; 71. promoter; 72. operator; 73. regulatory; 74. Regulatory; 75. operator; 76. promoter; 77. trpA-trpE; 78. tryptophan; 79. corepressor; 80. operator; 81. trpA-trpE; 82. tryptophan; 83. corepressor; 84. repressible; 85. operon; 86. negative, 87. C; 88. F; 89. G; 90. A; 91. J; 92. K; 93. H; 94. L; 95. B; 96. I; 97. D; 98. E; 99. N; 100. M

101.

Inducible Operon	Repressible Operon
An operon where the metabolite molecule enhances or increases the expression of the cluster of genes	An operon where the metabolite molecule represses or decreases the expression of the cluster of genes
Negative Gene Regulation	*Positive Gene Regulation*
Regulation mechanism where the active repressor turns off the gene expression	Regulation mechanism where a catabolite activator protein ensures the turning on or turning off of the genes

16.2 Regulation of Transcription in Eukaryotes [pp. 335–342]

102. operons; 103. protein; 104. regulatory; 105. short; 106. long; 107. post; 108. post; 109. histones; 110. nucleosome; 111. histones; 112. histone; 113. histones; 114. inactive; 115. promoter; 116. remodeling; 117. remodeling; 118. histones; 119. TATA; 120. transcription; 121. transcription initiation; 122. transcription; 123. DNA; 124. coactivators; 125. proximal; 126. DNA; 127. repressors; 128. transcription; 129. operons; 130. regulatory; 131. hormone; 132. methyl; 133. cytosine; 134. promoter; 135. hemoglobin; 136. silencing; 137. methyl; 138. hemoglobin; 139. X; 140. methylated.,141. C; 142. E; 143.D ; 144. A; 145. B; 146. G; 147. H; 148. F; 149. J; 150. I; 151. L; 152. K

153.

Mechanism	*Description*
Chromatin control	The genes are turned on or off depending upon how compactly DNA is wrapped around the histones.
Transcription initiation control	The genes are turned on or off depending upon the binding of activators or repressors to the transcription factors-RNA polymerase II complex at the promoter.
Coordinated control	All the genes related to a specific function are transcribed at the same time by activating the same regulatory sequences associated with all the related genes.
DNA methylation control	The genes are turned on by adding methyl group to cytosine of DNA or turned off by removing methyl group.

16.3 Posttranscriptional, Translational, and Posttranslational Regulation [pp. 342–345]

154. mRNA; 155. pre-mRNA; 156. splicing; 157. exons; 158. introns; 159. mRNA; 160. mRNA; 161. mRNA; 162. masking; 163. masking; 164. mRNA; 165. casein; 166. mi; 167. RNA; 168. translation; 169. mRNA; 170. nucleosome; 171. proteins; 172. enzymes; 173. ubiquitin, 174. C; 175. D; 176. A; 177. E; 178. B; 179. F

180.

Mechanism	*Description*
Posttranscriptional control	Splicing exons out and joining different combinations of introns from a pre-mRNA → different types of mRNA → different proteins are formed. Binding of masking proteins to mRNA blocks its translation. Binding of mi-RNA to mRNA blocks their translation.
Translational control	3'poly(A) tail can be increased or decreased to increase or decrease the translation
Posttranslational control	Protein function can be regulated by chemically modifying, processing, or its degradation.

16.4 The Loss of Regulatory Control in Cancer [pp. 345–348]

181. genes; 182. tumors; 183. dedifferentiation; 184. benign; 185. malignant; 186. cancer; 187. metastasis; 188. proto-oncogenes; 189. oncogenes; 190. tumor-suppressor; 191. p53, 192. C; 193. D; 194. B; 195. E; 196. A; 197. G; 198. F; 199. I; 200. H

201.

Benign tumors	*Malignant tumors*
Tumor that remains at its original site	The tumor whose cells separate and spread to other tissue and organs
Proto-oncogenes	*Oncogenes*
Genes that regulate cell division in normal cells	Modified proto-oncogenes that stimulate the cell to become cancerous
Differentiation	*Dedifferentiation*
When unspecialized embryonic cells become specialized	When specialized cells become embryonic in a tumor

Self-Test

1. c [Operon refers to the cluster of genes and associated sequences in prokaryotes]

2. b [RNA polymerase binds directly to the promoter sequence in prokaryotes]

3. d [Lactose acts as an inducer by binding to the active repressor, making it inactive, therefore allowing transcription of the cluster of genes]

4. b [Tryptophan acts as a corepressor and binds to the repressor, making it active and then blocking the transcription of the cluster of genes]

5. b [In eukaryotes, RNA polymerase binds only after the transcription factors have attached to the promoter sequence]

6. b [In prokaryotes, most of the regulation of gene expression takes place at the transcription level, whereas in eukaryotes, control can be at transcription, posttranscription, translation, or posttranslation level]

7. c [Histones are known to be the proteins that pack DNA into nucleosome]

8. a [Systems where multiple genes need to be controlled at the same time; it is now known that they have the same regulatory sequence to turn them on]

9. a [The process of adding methyl group to cytosine in DNA is called methylation]

10. b [TATA box in promoter region of eukaryotes is recognized by the transcription factors]

11. b [Masking proteins inactivate mRNA in an unfertilized egg]

12. b [The increase or decrease in the length of the poly(A) tail can increase or decrease translation]

13. b [Benign tumor remains in its original site whereas malignant tumor spreads]

14. b [Proto-oncogenes stimulate normal cells to divide whereas oncogenes are altered proto-oncogenes that stimulate normal cell to become cancerous]

15. c [Metastasis is a term for spreading of malignant tumor or cancer cells]

Chapter 17 Bacterial and Viral Genetics

Why It Matters [pp. 351–353]

1. Theodore Escherich; 2. *Bacterium coli*; 3. *Escherichia coli*; 4. bacteriophages 5. phages; 6. eukaryotic

17.1 Gene Transfer and Genetic Recombination in Bacteria [pp. 353–360]

7. sexually; 8. meiosis; 9. asexually; 10. DNA; 11. DNA; 12. minimal; 13. sugar; 14. salt; 15. clones; 16. Lederberg; 17. Tatum; 18. mutagens; 19. mutations; 20. auxotrophs; 21. minimal; 22. minimal; 23. circular; 24. pilus; 25. donor; 26. recipient; 27. conjugation; 28. F+ ; 29. plasmid ; 30. F; 31. F-; 32. plasmid; 33. F-; 34. F; 35. F+; 36. sexual; 37. recombination; 38. Hfr; 39. chromosome; 40. F; 41. pilus; 42. F-; 43. Jacob; 44. Wollman; 45. recombination; 46. genetic map; 47. circular; 48. R; 49. recombination; 50. Griffith; 51. pathogenic; 52. Avery, McCarty, and McLeod; 53. DNA; 54. capsule; 55. DNA; 56. DNA; 57. DNA; 58. recombination; 59. DNA; 60. DNA; 61. artificial transformation; 62. calcium; 63. electroporation; 64. plasmids; 65. Lederberg; 66. Zinder; 67. DNA; 68. DNA; 69.

DNA; 70. recombination; 71. Lederberg; 72. Lederberg; 73. replica; 74. complete; 75. auxotrophs; 76. velveteen; 77. velveteen; 78. auxotrophs , 79. B; 80. C; 81. A; 82. F; 83. D; 84. G; 85. E; 86. D; 87. A; 88. E; 89. H; 90. G; 91. B; 92. C; 93. F; 94. L; 95. I; 96. J; 97. K; 98. N; 99. M; 100. P; 101. O

102.

Mechanism of Recombination	Brief Description
A. Transformation	Where bacteria absorb DNA released by killed bacteria
B. Conjugation	Where bacteria pass DNA from donor cell to the recipient cell through a pilus
C. Transduction	Where a bacteriophage transfers DNA from one bacterium they attack to the next bacteria

103. True. , 104. False. Only some bacteria are known to be able to absorb DNA in nature although other varieties can be made to absorb with the help of special lab treatments., 105. False. When transfer of DNA takes place, complete transfer of F gene is rare.

17.2 Viruses and Viral Recombination [pp. 360–362]

106. core; 107. coat; 108. envelope; 109. nucleic; 110. replication; 111. DNA; 112. RNA; 113. coat; 114. recognition; 115. viruses; 116. bacteria; 117. virulent; 118. temperate; 119. virulent; 120. DNA; 121. recognition; 122. DNA; 123. coat; 124. chromosome; 125. polymerase; 126. DNA; 127. heads; 128. tails; 129. DNA; 130. viruses; 131. lytic; 132. temperate; 133. DNA; 134. DNA; 135. chromosome; 136. prophage; 137. lysogenic; 138. prophage; 139. lytic , 140. C; 141. D; 142. A; 143. F; 144. G; 145. B; 146. E

147.

Bacteriophage type	Type of Multiplication	Brief Description
A. Virulent	Lytic cycle	Where a bacteriophage multiplies inside the host bacterium and new viruses are released by rupturing of the host cell
B. Temperate	Lysogenic cycle	Where the genetic material of the bacteriophage becomes incorporated into the bacterial chromosome and multiplies with the host cell division

148. Lytic cycle is where a phage injects its genetic material into the bacterium. Viral DNA then directs synthesis of 100–200 new viruses. These viruses are released by rupturing of the bacterium. Lysogenic cycle, on the other hand, involves phage injecting its DNA which becomes incorporated into the bacterial chromosome as a prophage. As the bacterium multiplies, the incorporated viral DNA also multiplies. ,149. Phage is active form of a bacterial virus whereas prophage is inactive, dormant form of a virus that become incorporated in bacterial chromosome.

17.3 Transposable Elements [pp. 362–368]

150. transposable; 151. jumping; 152. target; 153. transposition; 154. genetic; 155. gene; 156. transposed; 157. plasmids; 158. Insertion; 159. transposase; 160. transposons; 161. antibiotic; 162. Barbara McClintock; 163. transposable; 164. eukaryotes; 165. transposons; 166. retrotransposons; 167. complementary; 168. reverse transcriptase; 169. complementary; 170. retrotransposon; 171. retroviruses; 172. reverse transcriptase; 173. DNA; 174. provirus, 175. C; 176. D; 177. A; 178. H; 179. B; 180. E; 181. F; 182. G, 183. False. Certain viruses such as HIV is known to stay dormant in the host cell for a number of years., 184. True., 185. False. Barbara McClintock was the first scientist to discover transposable elements in corn.,186. True.

Self-Test

1. a [Binary fission is where a bacterium divides into two genetically identical cells.]
2. d [Auxotrophs was a term used for bacterial mutants that were no longer able to grow on minimal medium.]

3. c [A pilus is a tubelike structure that is coded by the F gene and serves as an anchor as well as a passageway for DNA transfer.]
4. b [Hfr has the F gene as part of the main chromosome.]
5. a [Hfr has the F gene and acts as a donor.]
6. b [Transduction is where a virus picks up a segment of host DNA during its assembly and inserts into the new host as it infects.]
7. b [Transformation has been demonstrated only in some bacteria, although experimentally a number of varieties can be forced to absorb DNA.]
8. b [Only certain animal viruses have a membrane that they receive while exiting the host.]
9. b [Viruses depend upon the host cell machinery for their replication and very few viruses have some enzyme molecules that they may carry.]
10. b [During lysogenic cycle, the DNA of a bacteriophage becomes incorporated into the host DNA and multiplies with it.]
11. a [During lytic cycle, a virus multiplies inside the host cell and uses an enzyme from the host to lyse the cell and exit.]
12. c [Reverse transcriptase copies RNA to make viral DNA; an enzyme that is usually absent in the host cell.]
13. c [Provirus is a stage of virus where its DNA is incorporated into the host cell.]
14. d [Transposable elements in bacteria can move DNA between two locations in the main chromosome or between main chromosome and plasmid or between two plasmids.]
15. b [Transposable elements were first discovered in corn and later found in bacteria and other eukaryotes, including humans.]

Chapter 18 DNA Technologies and Genomics

Why It Matters [pp. 371–372]

1. biotechnology; 2. technologies; 3. engineering.

18.1 DNA Cloning [pp. 372–379]

4. cloning; 5. function; 6. proteins; 7. gene; 8. restriction; 9. sticky; 10. gene; 11. bacteria; 12. restriction; 13. gene; 14. ligase; 15. recombinant; 16. transformation; 17. recombinant; 18. hybridization; 19. proteins; 20. library; 21. mRNA; 22. introns; 23. reverse transcriptase; 24. mRNA; 25. polymerase; 26. complementary; 27. vector; 28. cDNA; 29. polymerase; 30. PCR; 31. separate; 32. primers; 33. polymerase; 34. complementary; 35. DNA; 36. DNA; 37. gel electrophoresis., 38. C; 39. K; 40. L; 41. A; 42. G; 43. D; 44. M; 45. B; 46. P; 47. E; 48. O; 49. J; 50. F; 51. I; 52. N; 53. H., 54. Restriction enzymes cut DNA at specific sites. EcoRI is special because it cuts DNA leaving short, single strands at the two ends that allow for easy insertion into the plasmid., 55. Reverse transcriptase is a viral enzyme that uses RNA to make a complementary DNA, a process that is reverse of transcription. Eukaryotic genes have non-coding sequences (introns) interspersed between protein coding sequences (exons). Once the gene is copied, pre-mRNA is made, which undergoes processing to remove introns In order to clone specific eukaryotic genes, mRNA , after removing introns from its pre-mRNA, can be copied to make the cDNA. This process allows bacterium to make eukaryotic proteins.

18.2 Applications of DNA Technologies [pp. 379–390], **18.3 Genome Analysis** [pp. 390–398]

56. DNA; 57. mutated; 58. restriction; 59. restriction fragment length polymorphisms; 60. Southern blot; 61. DNA; 62. short tandem repeat; 63. PCR; 64. gel electrophoresis; 65. plasmid; 66. transgenic; 67. protein; 68. somatic;

69. germ-line; 70. nucleus; 71. transgenic; 72. egg; 73. embryo; 74. uterus; 75. resistance; 76. tolerance; 77. nutrient; 78. Ti; 79. genes; 80. National Institutes of Health; 81. DNA; 82. structural; 83. functional; 84. bioinformatics; 85. genome; 86. cloned; 87. base-pairs; 88. protein-coding; 89. 2; 90. 50; 91. proteome. , 92. C; 93. D; 94. F; 95. A; 96. E; 97. B; 98. B; 99. E; 100. A; 101. F; 102.D ; 103. C; 104. H; 105. G; 106. I., 107. Cheek cells samples are collected from parents and child → DNA is isolated from each sample → each sample is amplified using PCR → restriction enzyme is used to chop DNA into fragments → DNA fragments are separated on gel electrophoresis → bands in all the samples are compared., 108. mRNA for insulin is isolated from pancreatic cells → reverse transcriptase is used to make the cDNA → cDNA for insulin is inserted into a bacterial plasmid → bacteria are allowed to absorb the recombinant plasmid → bacteria now makes the hormone., 109. False. Genes introduced in the somatic cells affect that individual. Only genes introduced in the gametic cells are passed on to the next generation., 110. True. An organism is called transgenic in it has genes from another species., 111. False. Scientists predicted 100,000 protein coding genes but Human Genome Project showed that there are only 30,000 protein coding genes., 112. False. Only some restriction enzymes leave sticky ends on DNA fragments., 113. False. Enzyme reverse transcriptase is isolated from retroviruses

Self-Test

1. d [Transgenic organisms have genes from other species]
2. a [cDNA is made as a complementary copy of mRNA]
3. b [During the PCR procedure, small DNA sequences are used as the primer to start copying DNA to make multiple copies]
4. b [DNA polymerase is used to make multiple copies of DNA]
5. d [Gel electrophoresis can be used to separate DNA, RNA or proteins]
6. c [When special restriction enzymes are used, DNA is cut to leave short, single stranded ends of the fragments]
7. a [Human Genome Project showed that over 50% of the human genome consists of repeated sequences that have no apparent function]
8. a [Human Genome Project showed that only about 2% of the human genome has the protein-coding sequences]
9. b [The goal of the Human Genome Project was to sequence the DNA of several organisms, in addition to human genome]
10. a [The diploid nucleus from mammary cell of an adult sheep was injected into an enucleated egg cell]
11. a [The enzyme reverse transcriptase uses single stranded RNA as a template]
12. b [Reverse transcriptase is isolated from retroviruses]
13. a [Kary Mullis received Nobel Prize in 1993 for the development of PCR technique]
14. c [Ian Wilmut and Keith Campbell were the first ones to clone sheep]
15. b [Gene therapy has had some success in humans as well].

Chapter 19 The Historical Development of Evolutionary Thought

Why It Matters [pp.401-402]

1. Alfred Russel Wallace

19.1 The Recognition of Evolutionary Change[pp. 402-405]

2. natural history; 3. natural theology; 4. taxonomy; 5. biogeography; 6. comparative morphologists; 7. vestigial structures; 8. fossils; 9. stratification; 10. paleobiology; 11. catastrophism; 12. gradualism; 13. uniformitarianism; 14. d; 15. e; 16. c; 17. b; 18. a

19.2 Darwin's Journeys [pp. 405-411]

19. fossils; 20. common ancestor; 21. beaks; 22. Malthus 23. artificial selection; 24. natural selection; 25. adaptive; 26. evolutionary divergence; 27. descent with modification; 28. f; 29. d; 30. a; 31. c; 32. b; 33. e

19.3 Evolutionary Biology Since Darwin [pp. 411-415]

34. population biology; 35. the modern synthesis; 36. microevolution; 37. macroevolution; 38. orthogenesis; 39. c; 40. b; 41. a; 42. e; 43. d; 44. macro-; 45. bio-; 46. paleo-; 47. ortho-; 48. T; 49. F—homologous; 50. T; 51. F—the modern synthesis

Self-Test

1. b [The other choices are incorrect because they were Greek philosophers but not mainly concerned with natural history.]
2. b [The fly's wing is analogous to the bat's wing because they have the same function (flight) but are structurally unrelated; the bat's wing, human arm and whale's flipper all are derived from the same basic bone structure.]
3. a [Fossils collected from the same stratum may be very different in terms of appearance, size, and structure.]
4. d [DNA and its importance to evolution was unknown in Lamarck's day.]
5. a [Inheritance of acquired characteristics and spontaneous generation have been disproven by experimentation; special creation is an idea that cannot be tested by the scientific method.]
6. b [Selection will favor different traits in different environments and, thus, cause evolutionary divergence.]
7. a [Microevolution is subtle change in the genetic makeup of a population leading to increased adaptation]
8. d [Comparative embryology, comparative molecular biology, and comparative anatomy may all yield information about evolutionary relationships.]
9. a [a is incorrect because evolution is better represented by a tree rather than a linear progression with a single species at the top. The other statements are true.]
10. a [b is incorrect because evolution is better represented by a tree rather than a ladder; c is incorrect because evolution by natural selection is not goal directed and, thus, is unpredictable; d evolutionary research is an ongoing, active branch of biology.]

Chapter 20 Microevolution: Genetic Changes within Populations

Why It Matters [pp. 419-420]

1. Penicillin; 2. resistant

20.1 The Recognition of Evolutionary Change[pp. 420-423]

3. microevolution; 4. species; 5. phenotypic; 6. qualitative; 7. quantitative; 8. polymorphism; 9. T; 10. T; 11. F—genetically; 12. F—sometimes; 13. F—10^{600}; 14. T; 15. Discrete variants of a character; 16. A heritable change in the genetic structure of a population; 17. Gel electrophoresis; 18. A group of individuals of the same species living and interacting with each other; 19. gene pool; 20. allele frequency; 21. genotype frequency; 22. Hardy-Weinberg; 23. genetic equilibrium; 24. null

20.2 Population Genetics [pp. 423-425]

25. No mutations; 26. No immigration or emigration; 27. Infinite population size; 28. All genotypes reproduce equally well; 29. Individuals choose mates randomly

20.3 The Agents of Microevolution [pp. 425-434]

30. mutation; 31. gene flow; 32. genetic drift; 33. genetic bottleneck; 34. founder effect; 35. natural selection; 36. relative fitness; 37. stabilizing selection; 38. directional selection; 39. disruptive selection; 40. sexual; 41. sexual dimorphism; 42. a; 43. e; 44. b; 45. c; 46. d

20.4 Maintaining Genetic and Phenotypic Variation [pp. 435-437]

20.5 Adaptation and Evolutionary Constraints [pp. 437-440]

47. diploidy; 48. Balanced polymorphisms; 49. heterozygote advantage; 50. frequency dependent selection; 51. selectively neutral; 52. relative fitness; 53. microevolution; 54. a; 55. d; 56. e; 57. c; 58. b

Self-Test

1. b [Diploidy will not necessarily result in increased gametogenesis, mutation rates, or genetic drift.]
2. d [An individual's phenotype may vary quantitatively, but it is produced by both genetic and environmental factors.]
3. c [Acquired or environmentally produced traits cannot be passed to offspring; physiological, behavioral, developmental, and other traits are subject to evolutionary change, not just anatomical ones.]
4. d
5. b [Bottlenecks will only result in drift; it has no direct effect on gene flow, mutation rate, or polymorphism.]
6. d [Fitness is a measure of reproductive success; the other answers are not.]
7. d [Nonrandom mating will affect genotype frequencies but not allele frequencies.]
8. d
9. c
10. c [some mutations may have no measurable fitness consequences; selection does not produce perfect organisms regardless of human intervention; many traits that evolved for one purpose in an ancestor get co-opted for other purposes in descendants.]

Chapter 21 Speciation

Why It Matters [pp. 443-444]

1. speciation

21.1 What is a Species? [pp. 444-447]

2. morphological; 3. biological; 4. phylogenetic; 5. subspecies; 6. ring; 7. cline; 8. b; 9. c; 10. a

21.2 Maintaining Reproductive Isolation [pp. 447-449]

11. reproductive isolation mechanism; 12. prezygotic; 13. postzygotic; 14. ecological; 15. temporal; 16. behavioral; 17. mechanical; 18. gametic; 19. hybrid inviability; 20. hybrid sterility; 21. hybrid breakdown; 22. ecological isolation 23. individuals reproduce at different times; 24. courtship displays and vocalizations; 25. anatomy of sex organs is incompatible; 26. gametic isolation; 27. hybrid inviability; 28. hybrid sterility; 29. F_2 offspring are sterile or exhibit other abnormalities

21.3 The Geography of Speciation [pp. 449-454]

21.4 Genetic Mechanisms of Speciation [pp. 454-460]

30. allopatric; 31. species cluster; 32. hybrid zones; 33. reinforcement; 34. parapatric; 35. sympatric; 36. host race; 37. polyploidy; 38. autopolyploidy; 39. unreduced gametes; 40. allopolyploidy; 41. para-; 42. auto-; 43. allo-; 44. sym-; 45. allo-

Self-Test

1. a [The biological species concept is based on an individual's ability to interbreed in nature; the morphological species concept is based on anatomical traits; the phylogenetic species concept is based on analysis of a phlyogenetic tree.]

2. b

3. a [Some biologists use "race" and "subspecies" synonymously.]

4. a

5. a

6. d

7. a [A prezygotic isolating mechanism would not permit fertilization of an egg.]

8. a [Allopatric speciation requires geographical isolation of populations. The other choices do not.]

9. d

10. a [b, c and d are true. Autopolyploidy is rare in animals because most animals are incapable of self-fertiliation.]

Chapter 22 Paleobiology and Macroevolution

Why It Matters [pp. 463 – 464]

1. fossils; 2. comparative morphology 3. paleobiology; 4. macroevolution

22.1 The Fossil Record [pp. 464-469]

22.2 Earth History, Biogeography and Convergent Evolution [pp. 469-471]

5. fossils; 6. radiometric dating 7. half-life; 8. plate tectonics; 9. continental drift 10. continuous distribution; 11. disjunct distribution; 12. dispersal; 13. vicariance; 14. biota; 15. biogeographical realms; 16. endemic species; 17. convergent evolution 18. F—oxygen poor; 19. F—incomplete; 20. T; 21. F—divergent; 22. F—11,200 years old; 23. T; 24. T; 25. F—continuous

22.3 Interpreting Evolutionary Lineages [pp. 473-477]

22.4 Macroevolutionary Trends in Morphology [pp. 477-480]

26. anagenesis; 27. cladogenesis; 28. gradualist; 29. punctuated equilibrium; 30. species selection; 31. preadaptation; 32. allometric growth; 33. heterochrony; 34. paedo-; 35. hetero-; 36. allo-; 37. clado-; 38. a; 39. b; 40. c; 41. d; 42. e

22.5 Macroevolutionary Trends in Biodiversity [pp. 480-483]

43. biodiversity; 44. adaptive raditation; 45. adaptive zone; 46. extinction; 47. background extinction rate; 48. mass extinction; 49. T; 50. F- sudden environmental change; 51. T; 52. F- varying rates

22.6 Evolutionary Developmental Biology [pp. 483-488]

53. Evolutionary developmental biology; 54. homeotic; 55. *Hox*; 56. homeobox; 57. homeodomain; 58. b; 59. c; 60. a

Self-Test

1. b [Aerobic conditions cause decomposition before fossilization can occur. Hard parts fossilize. Lava and volcanic ash would probably destroy tissues before they could fossilize.]

2. a

3. b [Continental drift makes it difficult for populations to maintain contact with each other, leading to disjunct distributions.]

4. d

5. c [The phylogeny of horses is highly branched, thus is cladogenetic. Abiogenesis is the creation of life from nonliving materials. Anagenesis does not create a branched phylogenetic tree. Pangenesis is an erroneous theory of inheritance proposed by Darwin.]

6. d

7.c

8. a

9. b

10. c

Chapter 23 Systematic Biology: Phylogeny and Classification

Why It Matters [pp. 491-492]

1. systematics

23.1 Systematic Biology: An Overview [pp. 492-493], **23.2 The Linnaean System of Taxonomy**[pp. 493-494], **23.3 Organismal Traits as Systematic Characters** [pp. 494-495]

2. phylogeny; 3. phylogenetic trees; 4. taxonomy; 5. classification; 6. binomial nomenclature; 7. binomial; 8. genus; 9. specific epithet; 10. taxonomic hierarchy; 11. family; 12. orders; 13. classes; 14. phyla; 15. kingdoms; 16. domains; 17. taxon; 18. c; 19. a; 20. d; 21. b; 22. b; 23. g; 24. e; 25. the species is *Oncorhynchus mykiss*, genus is *Oncorhynchus*, the specific epithet is *mykiss*; 26. the species is *Xenopus laevis*, genus is *Xenopus*, the specific epithet is *laevis*; 27. the species is *Homo sapiens*, genus is *Homo*, the specific epithet is *sapiens*; 28. d; 29. h; 30. f; 31. a; 32. c; 33. g; 34. b; 35. e

23.4 Evaluating Systematic Characters [pp. 495-497], **23.5 Phylogenetic Inference and Classification** [pp. 497-501], **23.6 Molecular Phylogenetics** [pp. 501-508].

36. homologies; 37. homoplasies; 38. mosaic evolutions; 39. ancestral characters; 40. derived characters; 41. outgroup comparison; 42. monophyletic taxa; 43. polyphyletic taxa; 44. paraphyletic taxa; 45. principle of monophyly; 46. principle of parsimony; 47. traditional evolutionary systematics; 48. cladistics; 49. clade; 50. cladograms; 51. PhyloCode; 52. molecular clock; 53. domains; 54. monophyletic; 55. paraphyletic; 56. polyphyletic; 57. b; 58. a; 59. a; 60. b; 61. b; 62. a; 63. a; 64. b; 65. d; 66. b; 67. f; 68. a; 69. c; 70. e

Self-Test

1. b

2. b, d

3. c

4. d

5. a, b, d [a is correct because when compared to all animals, only one group (the vertebrates) possess a vertebral column; therefore, a vertebral column is a derived character; b is correct because among vertebrates, the two-chambered heart of fish is ancestral and the four-chambered heart of mammals is derived; among insects,

most modern insects have six walking legs while only two groups have four walking legs; therefore, 6 walking legs are ancestral]

6. b, d

7. a, b, c

8. b, c, e

9. c

10. b [b is correct because in monophyletic taxa, species arise from a common ancestor, and the common ancestor for species C and D is species F]

11. b [b is correct because species B and C arise from different evolutionary lineages, and taxa that arise from different evolutionary lineages are considered polyphyletic]

Chapter 24 The Origin of Life

Why It Matters [pp. 511–512]

1. George Lemaitre; 2. gas; 3. dust; 4. earth; 5. 4.6; 6. 3.5.

24.1 The Formation of Molecules Necessary for Life [pp. 512–515]

7. membrane; 8. nucleic; 9. proteins; 10. energy; 11. spontaneously; 12. 4.6; 13. metallic; 14. metallic; 15. heat; 16. dust; 17. gas; 18. water; 19. water; 20. volcanoes; 21. organic; 22. atmosphere; 23. energy; 24. oxygen; 25. organic; 26. prebiotic; 27. Stanley Miller; 28. Harold Urey; 29. reducing; 30. water vapors; 31. organic; 32. organic; 33. reducing; 34. Koichiro Matsuno; 35. polypeptides; 36. meteor; 37. comet; 38. B; 39. C; 40. A; 41. D; 42. C; 43. D; 44. B; 45. A
46.

Hypotheses	*Short description*
A. Oparin-Haldane	Organic chemicals developed in a reducing environment of the ancient earth and atmosphere that was subjected to heat and lightning
B. Hydrothermal Vents	Organic chemicals developed in the submarine volcanoes in the ocean floor
C. Extraterrestrial Origin	Organic chemicals arrived at earth through a meteor or a comet

24.2 The Origin of Cells [pp. 515–519]

47. evaporation; 48. dehydration; 49. membrane; 50. protocells; 51. clay; 52. organic; 53. Noam Lahay; 54. Sherwood Chang; 55. Goldacre; 56. lipid; 57. David Deamer; 58. protocells; 59. oxidation; 60. reduction; 61. electron; 62. electron; 63. ATP; 64. RNA; 65. enzymes; 66. proteins; 67. RNA; 68. ribozymes; 69. RNA; 70. DNA; 71. proteins; 72. DNA; 73. RNA; 74. organic; 75. lipids; 76. protocells; 77. protocells; 78. Richard Dickerson; 79. H_2S; 80. oxygen; 81. water; 82. oxygen; 83. oxygen; 84. stromatolites; 85. B; 86. C; 87. A; 88. D; 89. C; 90. A; 91. D; 92. B; 93. F; 94. E; 95. G; 96. Miller and Urey were able to imitate ancient environment in the lab and form macromolecules and phospholipid bilayer vesicles.; 97. Present day ribozymes support the fact that RNA may have been first molecule to store information and catalyze chemical reactions.; 98. If the atmosphere was oxidative, chemicals would have broken down as they were being formed.; 99. B C A D; 100. False—The ancient photosynthetic bacteria probably used H_2S as the electron donor.; 101. False—The earth's atmosphere was probably reducing and not oxidizing.

24.3 The Origins of Eukaryotic Cells [pp. 519–523]

102. cytoplasm; 103. organelles; 104. Lynn Margulis; 105. prokaryotic; 106. aerobic; 107. endosymbiosis; 108. mitochondria; 109. acceptor; 110. oxygen; 111. prokaryotic; 112. photosynthetic; 113. chloroplast; 114. prokaryotic; 115. DNA; 116. ribosomal; 117. size; 118. 2.2; 119. plasma; 120. Archaea; 121. eukaryotic;

122. unicellular; 123. 800–1000; 124. colonies; 125. differentiation; 126. Common characteristics between prokaryotic cells and organelles like mitochondria and chloroplasts.; 127. Domain Archaea includes organisms that are structurally like common bacteria but share some processes with eukaryotes. ; 128. False—ER, Golgi, and nuclear membranes developed by invagination of plasma membrane.; 129. False—Scientists suspect that multicellularity evolved along several lineages.; 130. True.

Self-Test

1. c [To be able to replicate the genetic information and translate the code to make proteins was critical for the origination of a living cell.]

2. a [Presence of oxygen would make the environment oxidizing while reducing environment was a key to Oparin-Haldane hypothesis.]

3. a [Presence of oxygen would cause quick breakdown of new chemicals.]

4. c [Even today, the hydrothermal vents maintain a reducing environment conducive to making organic chemicals.]

5. b [Ribozymes are known to be self-replicating and catalyzing molecules—a combination of two key characteristics.]

6. b [Prokaryotic cells lack membrane bound organelles.]

7. b [Mitochondria and chloroplast share several characteristics with a prokaryotic cell.]

8. b [ER and nuclear membrane may have formed by the invagination of plasma membrane.]

9. a [The Big Bang Theory explains how earth and other planets may have formed and Earth is estimated to have formed about 4.6 billion years ago.]

10. c [The biggest gap in evolution was probably the transition from prokaryotes to eukaryotes—making eukaryotes only 2.2 billion years old.]

11. b [The earliest fossil of prokaryote was found in the stromatolites of Australia.]

12. a [Reducing environment eliminates the possible breakdown of chemicals by oxygen.]

13. a [Archaeans do have basic prokaryotic structure but share genetic processing of eukaryotes.]

14. d [The DNA sequence in mitochondria and chloroplast does not code for proteins coded by nuclear DNA.]

15. a [Miller-Urey were able to make organic chemicals by subjecting a reducing chemical environment to high temperatures.]

Chapter 25 Prokaryotes and Viruses

Why It Matters [pp. 525–526]

1. prokaryotes; 2. Bacteria; 3. Archaea; 4. diseases; 5. metabolic; 6. smaller; 7. non

25.1 Prokaryotic Structure and Function [pp. 526–534]

8. organelles; 9. mutation; 10. transformation; 11. transduction; 12. conjugation; 13. wall;14. membrane; 15. circular; 16. chromosome; 17. nucleoid; 18. plasmids; 19. coccus; 20. bacillus; 21. vibrio; 22. spirilla; 23. peptidoglycan; 24. Hans Christian Gram; 25. Gram; 26. +; 27. -; 28. cell wall; 29. +; 30. peptidoglycan; 31. -; 32. peptidoglycan; 33. membrane; 34. lipopolysaccharide; 35. -; 36. polysaccharide; 37. capsule; 38. slime layer; 39. flagella; 40. move; 41. pili; 42. DNA; 43. energy; 44. chemoautotrophs; 45. chemoheterotrophs; 46. photoautotrophs; 47. photoheterotrophs; 48. aerobes; 49. obligate; 50. energy; 51. anaerobes; 52. obligate; 53. energy; 54. oxygen; 55. facultative; 56. bacteria; 57. fixation; 58. amino; 59. binary; 60. identical; 61. conjugation; 62. pilus; 63. endospores.; 64. C; 65. A; 66. D; 67. E; 68. G; 69. B; 70. F; 71. I; 72. H; 73. C; 74. D; 75. B; 76. A; 77. B; 78. C; 79. D; 80. A; 81. B; 82. C; 83. E; 84. A; 85. D; 86. C; 87. D; 88. B; 89. A; 90. C; 91. D; 92. B; 93. A; 94. E; 95. a; 96. b; 97. b; 98. a

25.2 The Domain Bacteria [pp. 534-537], **25.3 The Domain Archaea** [pp. 537–540]

99. DNA; 100. RNA; 101. protein; 102. Bacteria; 103. Archaea; 104. Proteobacteria; 105. Green bacteria; 106. Cyanobacteria; 107. Gram-positive; 108. Spirochete; 109. Chlamydia; 110. Woese; 111. extremophiles; 112. mesophiles; 113. plasma; 114. wall; 115. methanogens; 116. halophiles; 117. thermophiles; 118. psychrophiles.; 119. C; 120. B; 121. F; 122. D; 123. A; 124. E; 125. In Archaea, the plasma membrane has chemical bonds between the hydrocarbon chains and glycerol, and the cell wall is made of proteins, polysaccharides, or chemicals related to peptidoglycan.; 126. Thermophiles have DNA polymerase that is thermostable and very useful for PCR.

25.4 Viruses, Viroids, and Prions [pp. 540–546]

127. DNA; 128. RNA; 129. single; 130. double; 131. capsid; 132. enzymes; 133. non; 134. helical; 135. polyhedral; 136. enveloped; 137. complex; 138. bacteriophages; 139. lytic; 140. lysogenic; 141. prophage; 142. capsid; 143. genome; 144. capsid; 145. genome; 146. endocytosis; 147. genome; 148. membrane; 149. injuries; 150. bites; 151. plasmodesmata; 152. bacteria; 153. entry; 154. multiplication; 155. RNA; 156. plants; 157. protein; 158. E; 159. M; 160. F; 161. A; 162. K; 163. J; 164. B; 165. L; 166. N; 167. I; 168. C; 169. H; 170. D; 171. G

172. Complete the following table with the chemical composition of virus, viroids, and prions.

Structural Entity	Genome	Protein	Envelope
Virus	**DNA or RNA**	**Capsid**	**Present or absent**
Viroid	**RNA**	**None**	**None**
Prion	**None**	**Protein particle**	**None**

173. False—Viruses lack the properties of living things such as cell structure, metabolism, and response to external and internal environment.; 174. False—Viruses use the host's chemicals, energy, and enzymes to replicate. Therefore they cannot multiply outside a living host cell.; 175. False—Antibiotics are developed to treat only bacterial infections.

Self-Test

1. a [Viruses, viroids, and prions do not share properties of living things.]
2. d [Bacteria lack a nucleus; their DNA is located in the nucleoid region of the cytoplasm.]
3. d [Bacterial DNA is located in the cytoplasm and not in the nucleus.]
4. d [Gram negative bacteria stain pink with Gram staining.]
5. d [Some bacteria have capsule to protect them from unfavorable environment.]
6. b [Eukaryotic flagella is made of microtubules whereas prokaryotic flagella are made of protein fibers.]
7. c [Facultative anaerobes can produce energy aerobically or anaerobically.]
8. a [Binary fission in bacteria produces clones.]
9. c [Outer membrane in Gram negative bacteria have lipopolysaccharides.]
10. c [Photoautotrophs use light and CO_2 to produce energy and organic chemicals.]
11. b [Chemoheterotrophs need organic chemicals to produce energy and make other organic chemicals.]
12. a [Psychrophiles are found in freezing temperatures.]
13. b [Halophiles live in high-salt environments.]
14. b [Viroids are naked RNA molecules that are known to infect plants.]
15. d [Bacteria are made of simple prokaryotic cells.]

Chapter 26 Protists

Why It Matters [pp. 549–550]

1. aquatic; 2. Protoctista; 3. eukaryotic

26.1 What Is a Protist? [pp. 550–553]

4. diverse; 5. water; 6. unicellular; 7. eukaryotic; 8. organelles; 9. asexually; 10. sexually; 11. fungi, plants, and animals; 12. wall; 13. plants; 14. animals; 15. mitochondria; 16. anaerobically; 17. heterotrophic; 18. food; 19. enzymes; 20. autotrophic; 21. plants; 22. contractile; 23. flagella and cilia; 24. pseudopodia; 25. phytoplankton; 26. zooplankton; 27. D; 28. E; 29. A; 30. B; 31. C;

32.

Major Characteristics	*Description*
Habitat	All aquatic or in moist soil
Cell Structure	Eukaryotic; mostly unicellular, or some are multicellular with no or simple differentiation
Metabolism	All produce ATP aerobically; excavates that lack mitochondria produce ATP anaerobically
Nutrients	Autotrophs produce organic molecules, have pigments, and photosynthesize; heterotrophs either ingest or absorb
Reproduction	Asexual by mitosis; sexual by meiosis and formation of gametes

26.2 The Protist Group [pp. 553–572]

33. kingdom; 34. mitochondria; 35. anaerobic; 36. disc; 37. alveoli; 38. bioluminescent; 39. autotrophic; 40. gametes; 41. hyphae; 42. silica; 43. plants; 44. fucoxanthin; 45. blades; 46. stipes; 47. holdfast; 48. pseudopods; 49. tests; 50. axopods; 51. calcium carbonate; 52. engulf; 53. pseudopodia; 54. mitosis; 55. fruiting; 56. plants; 57. phycobilins; 58. differentiation; 59. plants; 60. fungi and animals; 61. flagellum. 62. D; 63. E; 64. A; 65. F; 66. B; 67. G; 68. C; 69. Protists are diverse with respect to structure, metabolism and reproduction; 70. Small photosynthetic protists that live in large bodies of water are called phytoplanktons. They provide nutrients and oxygen to the zooplanktons, microorganisms, and animals that also live in that water.; 71. Primary endosymbiosis is where a nonphotosynthetic eukaryotic cell engulfs a photosynthetic cyanobacterium that becomes a permanent resident and transforms into a plastid, as in red-green algae and land plants. Secondary endosymbiosis is where a nonphotosynthetic eukaryotic cell engulfs a small photosynthetic eukaryote that becomes a plastid in Euglena.

Self-Test

1. b [Protists are eukaryotic.]
2. a [Most protists are aquatic.]
3. b [Some protists such as Excavates lack mitochondria and therefore produce ATP anaerobically.]
4. b [Contractile vacuole pumps excess water out of the cell.]
5. b [Fucoxanthin gives brown and golden algae their color.]
6. a [Phycobilins gives red algae their color.]
7. a [Animals and fungi probably evolved from the choanoflagellates that are part of Opisthokonts.]
8. c [Excavates lack mitochondria.]
9. d [Amoeba and slime molds belong to the group Amoebozoa.]
10. d [Slime molds that belong to Amoebozoa have been studied extensively to understand the process of differentiation because slime molds can go from single-celled amoeba to a differentiated fruiting body.]
11. b [Primary endosymbiosis is where a nonphotosynthetic eukaryotic cell engulfs a photosynthetic prokaryotic cell that becomes a permanent resident and transforms into a plastid, as in red-green algae and land plants]
12. d [Secondary endosymbiosis is where a nonphotosynthetic eukaryotic cell engulfs a photosynthetic eukaryotic cell that becomes a permanent resident and transforms into a plastid in *Euglena.*]
13. e [Alveolates have complex cytoplasmic structures and move with the help of cilia.]
14. a [Protoctista is a kingdom where organisms that are not plants, animals, fungi, or prokaryotes are just combined.]
15. a [Protoctista includes organisms that reproduce asexually by mitosis or sexually by meiosis.]

Chapter 27 Plants

Why It Matters [pp. 575–576]

1. kingdom Plantae

27.1 The Transition to Life on Land [pp. 576-581]

2. sporopollenin; 3. cuticle; 4. stomata; 5. lignin; 6. apical meristem; 7. embryophytes; 8. xylem; 9. phloem; 10. roots; 11. rhizomes; 12. root systems; 13. shoot systems; 14. spores; 15. alteration of generations; 16. sporophyte; 17. gametophyte; 18. sporangia; 19. homosporous; 20. heterosporous; 21. gametangium; 22. sporangium; 23. gametophyte; 24. sporophyte; 25. homosporous; 26. heterosporous; 27. a; 28. b; 29. b; 30. a; 31. b; 32. a; 33. b; 34. a; 35. a; 36. b.; 37. a; 38. b; 39. a; 40. g; 41. h; 42. d; 43. c; 44. i; 45. f; 46. e; 47. b

27.2 Bryophytes: Nonvascular Land Plants [pp. 581–584], **27.3 Seedless Vascular Plants** [pp. 585–590]

48. bryophytes; 49. epiphytes; 50. gametangium; 51. archegonia; 52. antheridia; 53. Hepatophyta; 54. thallus; 55. gemmae; 56. Anthocerophyta; 57. Bryophyta; 58. protonema; 59. Lycophyta; 60. Pterophyta; 61. sporophylls; 62. cone; 63. strobilus; 64. nodes; 65. sorus; 66. annulus; 67. a; 68. b; 69. a; 70. b; 71. b; 72. a; 73. Hepatophyta; 74. presence of algal-like protein bodies called pyrenoids; 75. Bryophyta; 76. club mosses; 77. sperm require water in order to reach eggs; 78. ferns, horsetails; 79.a; 80. f; 81. b; 82. d; 83. c; 84. e

27.4 Gymnosperms: The First Seed Plants [pp. 590–594], **27.5 Angiosperms: Flowering Plants** [pp. 594–601]

85. Gymnosperms; 86. pollen grain; 87. Pollination; 87. ovule; 89. seed; 90. Cycadophytea; 91. Ginkgophyta; 92. Coniferophyta; 93. microsporers; 94. megaspores; 95. angiosperms; 96. flowers; 97. fruit; 98. Anthophyta; 99. monocots; 100. eudicots; 101. magnoliid; 102. basal angiosperms; 103. double fertilization; 104. ovary; 105. coevolved; 106; a; 107. b; 108. b; 109. a; 110. a; 111. b; 112. b; 113. a; 114. resemble palms; restricted to warmer climates; 115. Ginkgophyta; 116. cone-bearing group with 80% of gymnosperm species; 117. grasses, palms, lilies, orchids; 118. flowering shrubs and trees, nonwoody plants, cacti; 119. magnoliids; 120. star anise, water lilies, *Ambroella*; 121. a; 122. h; 123. e; 124. f; 125. d; 126. c; 127. g; 128. b

Self-Test

1. a, b, c [a is correct because angiosperms have two water-conducting cell type which enable water to flow more rapidly to different parts of the plant; b is correct because double fertilization increase nutrient availability for developing embryo and because the ovary protects seeds from desiccation as well as from attack by herbivores and pathogens; c is correct because fruit aids in the dispersal of seeds; d is not correct; this requirement applies to nonvascular plants]

2. a, b, c, d, e

3. b, d

4. a, b, c

5. b, c

6. b, c, d

7. a

8. a, d

9. b

10. a, c, d

Chapter 28 Fungi

Why It Matters [pp. 605–606]

1. recycling; 2. carbon dioxide; 3. decomposers

28.1 General Characteristics of Fungi [pp. 606–610]

4. multicellular; 5. eukaryotic; 6. chitin; 7. hyphe; 8. mycelium; 9. saprobes; 10. parasites; 11. symbiosis; 12. enzymes; 13. spores; 14. plasmogamy; 15. dikaryon; 16. karyogamy; 17. spores., 18. B; 19. D; 20. A; 21. C; 22. G; 23. E; 24. H; 25. I; 26. F

27.

Terms	Definitions
Saprobes	Organisms that live on dead plant and animal materials
Parasites	Organisms that live on other live organisms and cause harm to them
Symbiosis	Organisms that live in partnerships benefiting each other

28.

Terms	Characteristic of Fungi
Cell Type	Eukaryotic
Cell Wall	Chitin
Number of Cells	Multicellular
Mode of Nutrition	Saprobes/Parasites/Symbionts
Reproduction	Asexual and Sexual Spores

28.2 Major Groups of Fungi [pp. 610–620]

29. 5; 30. sexual; 31. flagellated; 32. sporangia; 33. aquatic; 34. aseptate; 35. coenocytic; 36. +; 37. -; 38. gametangia; 39. haploid; 40. plasmogamy; 41. karyogamy; 42. zygospore; 43. meiosis; 44. roots; 45. mycorrhizae; 46. arbuscules; 47. sugars; 48. minerals; 49. conidia; 50. conidiophores; 51. ascogonium; 52. anthredium; 53. dikaryotic; 54. diploid; 55. ascospores; 56. ascus; 57. ascocarp; 58. basidiocarp; 59. basidia; 60. meiosis; 61. basidiospores; 62. sexual., 63. D; 64. G; 65. C; 66. H; 67. A; 68. I; 69. J; 70. B; 71. F; 72. E

73.

Phyla/Group	Major Characteristics
Ascomycota	Form chains of asexual conidia at the tip of specialized conidiophores; 4–8 sexual ascospores formed inside the asci that cluster in a reproductive body called an ascocarp
Zygomycota	Coenocytic, aseptate hyphe; asexual spores formed inside a sporangium; a thick walled, dormant sexual zygospore formed by fusion of + and – mating type gametangia
Basidiomycota	Form mushroom/basidiocarp as a sexual reproductive body that has numerous basidia with 4 externally borne basidiospores
Chytridiomycota	The only aquatic fungi that bear flagellated spores formed inside a sporangium
Glomeromycota	Mycorrhizae that are symbiotically associated with the roots of 80–90% plants; develop arbuscules to get sugars from plants and give dissolved minerals in exchange
Imperfect Fungi	Fungi whose sexual reproduction has not been demonstrated; a temporary housing group

74.

Phyla/Group	Major Characteristics
Ascomycota	Disease in crops; yeast; penicillin producing species; edible truffles and morels
Zygomycota	Mold on bread, fruits, and leather
Basidiomycota	Mushrooms and puffballs
Chytridiomycota	Chytrids that cause diseases in frogs and fish
Glomeromycota	Mycorrhizae associated with roots of plants

75. Fungi are classified on the basis of their sexual reproduction and the type of reproductive body
76. Imperfect Fungi is a temporary housing group for those fungal species whose sexual reproduction has not been observed. Once the type of sexual reproduction has been identified, the species is moved to its respective phylum.
77. Fungi secrete digestive enzymes into the substrate, dead or living, and then absorb the nutrients through mechanisms in their plasma membrane.

28.3 Fungal Associations [pp. 620–623]

78. symbiotic; 79. mycobiont; 80. photobiont; 81. Ascomycetes; 82. Basidiomycetes; 83. thallus; 84. asexual; 85. soredia; 86. ascocarp; 87. basidiocarp; 88. B; 89. A; 90. D; 91. C

Self-Test

1. b [Fungi do have membrane-bound organelles.]
2. b [Fungi lack photosynthetic pigments and totally depend upon living and dead plants and animals for their nutrients.]
3. b [Sexual reproduction has not been observed in some fungal species.]
4. b [Most fungi are multicellular. However, some can be unicellular such as yeast or aseptate where the hypha is not divided into definite cells.]
5. c [Fungi secrete enzymes to digest the nutrients in their substrate and then absorb.]
6. a [Chytridiomycota produce flagellated spores in their sporangia.]
7. c [Phylum Basidiomycota includes the mushrooms.]
8. d [Molds are included in the phylum Zygomycota.]
9. b [Ascomycota produce ascospores inside their asci.]
10. e [Glomeromycota includes the mycorrhizae that are associated with plant roots.]
11. c [Mycorrhizae get sugars from the plants and provide dissolved minerals to the plants.]
12. d [Imperfect Fungi are temporarily housed under this group due to lack of information on their sexual reproduction.]
13. a [Mycobiont is the fungal partner of a lichen.]
14. c [Soredia are clusters of fungal and algal cells that are used by the lichens for asexual reproduction.]
15. b [Depending upon the type of mycobiont, lichens do reproduce via ascospores or basidiospores.]

Chapter 29 Animal Phylogeny, Acoelomates, and Protostomes

Why It Matters [pp. 627–628]

1. Animalia; 2. species; 3. invertebrates; 4. backbone; 5. vertebrate; 6. backbone

29.1 What Is an Animal? [pp. 628–629]

29.2 Key Innovations in Animal Evolution [pp. 629–633]

7. eukaryotic; 8. one; 9. multicellular; 10. lack; 11. walls; 12. energy; 13. nutrients; 14. animals; 15. heterotrophs;

16. motile; 17. sessile; 18. asexual; 19. sexual; 20. body; 21. tissues; 22. Eumetazoans; 23. lacking tissues; 24. Parazoans; 25. tissues; 26. diploblastic; 27. endoderm; 28. ectoderm; 29. triploblastic; 30. mesoderm; 31. symmetry; 32. radial; 33. bilateral; 34. Parazoans; 35. shape; 36. asymmetrical; 37. body; 38. Acoelomates; 39. coelomate; 40. true; 41. coelom; 42. K; 43. G; 44. M; 45. N; 46. H; 47. A; 48. O; 49. B; 50. F; 51. C; 52. P; 53. L; 54. D; 55. J; 56. E; 57. I

29.3 An Overview of Animal Phylogeny and Classification [pp. 633–635]

58. morphological; 59. embryological; 60. nucleotide; 61. ribosomal; 62. mitochondrial; 63. hypothesis

29.4 Animals without Tissues: Parazoa [pp. 635–636]

29.5 Eumetazoans with Radial Symmetry [pp. 636–639]

64. Parazoan; 65. sponges; 66. Porifera; 67. Eumetazoans; 68. Cnidaria; 69. Ctenophora; 70. tissues; 71. sessile; 72. tissues; 73. radial; 74. gastrovascular; 75. one; 76. blastopore; 77. mouth; 78. anus; 79. diploblastic; 80. mesoglea; 81. ectoderm; 82. endoderm; 83. motile; 84. sessile; 85. motile; 86. B; 87. D; 88. A; 89. C; 90. F; 91. B; 92. E; 93. C; 94. B; 95. C; 96. B; 97. A; 98. B; 99. A; 100. C

29.6 Lophotrochozoan Protostomes [pp. 641–653]

101. Protostomes; 102. Lophotrochozoans; 103. Ecdysozoans; 104. 3; 105. lophophore; 106. Platyhelminthes; 107. flatworms; 108. Acoelomates; 109. triploblastic; 110. 4; 111. free-living; 112. parasitic; 113. Rotifera; 114. corona; 115. parthenogenesis; 116. females; 117. unfertilized eggs; 118. Mollusca; 119. Polyplacophora; 120. Gastropoda; 121. Bivalvia; 122. Cephalopoda; 123. 3; 124. visceral mass; 125. head-foot; 126. mantle; 127. shell; 128. Annelida; 129. segmented; 130. 3; 131. Polychaeta; 132. Oligochaeta; 133. Hirudinea; 134. septa; 135. setae; 136. leeches or Hirudinea; 137. Q; 138. R; 139. P; 140. J; 141. O; 142. S; 143. K. 144. M; 145. L; 146. N; 147. C; 148. A; 149. I; 150. B; 151. E; 152. G; 153. F; 154. H; 155. D

29.7 Ecdysozoan Protostomes [pp. 653–663]

156. external; 157. 3; 158. Ecdysozoan; 159. large; 160. Nematoda; 161. roundworms; 162. Onychophora; 163. southern; 164. Arthropoda; 165. segmented; 166. exoskeleton; 167. grows; 168. 3; 169. head; 170. thorax; 171. abdomen; 172. jointed; 173. arthropods; 174. A significant disadvantage of shedding the exoskeleton is loss of protection and support of the exoskeleton. The animals during molting are extremely vulnerable to predation; 175. Nematodes have sexual reproduction, with a male and female organism. The large number of eggs and sperm play a major role in the success of these worms; 176. A; 177. G; 178. E; 179. F; 180. A; 181. B; 182. D; 183. C; 184.C; 185. B; 186. C; 187. B; 188. D; 189. E; 190. D

Self-Test

1. c [other selections are associated with plants; being heterotrophic is most unique to animals]

2. d [all pairs are appropriately matched except for d—Platyhelminthes do not have a body cavity, they are Acoelomates]

3. b [the origin of mesoderm is different between protostomes and deuterostomes, the other selections are present in both groups]

4. a [these are characteristics that are associated with protostomes]

5. b [having both sets of reproductive systems, male and female is hermaphroditic]

6. b [a radula is a scraping device, mollusks typically have a radula and are often bottom- or rock-dwelling animals]

7. d [cephalopods move rapidly and require a high level of oxygen to support the increased mobility, thus a closed circulatory system will increase delivery of oxygen delivery to tissues.]

8. b [selection a is an incorrect statement, and selections c and d are associated with insects]

9. b [animals with repeating units of internal systems are typically Annelida, the segmented worms]

10. c [occupying different habitats and eating different food make insects that undergo complete metamorphosis very successful]

---

30 Deuterostomes: Vertebrates and Their Closest Relatives

Why It Matters [pp. 667–668]

1. Deuterostomes; 2. embryological; 3. molecular sequence; 4. three; 5. mouth; 6. 2nd; 7. Deuterostome

30.1 Invertebrate Deuterostomes [pp. 668–671]

30.2 Overview of the Phylum Chordata [pp. 671–674]

8. Echinodermata; 9. Hemichordata; 10. Chordata; 11. deuterostomes; 12. echinoderms; 13. bilaterally; 14. larval; 15. radial; 16. 5; 17. Hemichordata; 18. acorn; 19. proboscis; 20. gill slits; 21. Chordata; 22. three; 23. invertebrates; 24. vertebrata; 25. chordates; 26. notochord; 27. segmented; 28. dorsal hollow nerve; 29. pharynx; 30. D; 31. J; 32. G; 33. E; 34. I; 35. B; 36. K; 37. F; 38. A; 39. C; 40. H; 41. E; 42. C; 43. A; 44. B; 45. F; 46. D; 47. A; 48. B; 49. F; 50. A

30.3 The Origin and Diversification of Vertebrates [pp. 674–675]

30.4 Agnathans: Hagfishes and Lampreys, Conodonts, and Ostracoderms [pp. 676–678]

51. three-dimensional; 52. *Hox*; 53. simple 54. fewer; 55. complex; 56. jaw; 57. Agnathans; 58. jaws; 59. gnathostomata; 60. Tetrapods; 61. Tetrapods ; 62. four; 63. locomotion; 64. amniotes; 65. eggs; 66. terrestrial; 67. jawless; 68. hagfish; 69. lampreys; 70. fossils; 71. Conodonts; 72. Ostracoderms

30.5 Jawed Fishes [pp. 678–683]

73. G; 74. H; 75. I; 76. F; 77. J; 78. K; 79. L; 80. A; 81. C; 82. D; 83. E; 84. B

30.6 Early Tetrapods and Modern Amphibians [pp. 683–685]

85. air; 86. sound; 87. vibrations; 88. amphibians; 89. aquatic; 90. larval; 91. three; 92. anurans; 93. frogs; 94. toads; 95. Urodela; 96. salamanders; 97. Gymnophiona; 98. caecelians; 99. G; 100. I; 101. A; 102. C; 103. J; 104. H; 105. D; 106. B; 107. F; 108. E

30.7 The Origin and Mesozoic Radiations of Amniotes [pp. 686–688]

109. amniotes; 110. sac; 111. embryo; 112. amnion; 113. dry; 114. skin; 115. dehydrate; 116. keratin; 117. lipid; 118. amniotic; 119. shell; 120. amniotic; 121. membranes; 122. yolk; 123. albumin; 124. energy; 125. nutrients; 126. water; 127. uric acid; 128. ammonium; 129. water; 130. toxicity; 131. three; 132. synapsida; 133. anapsida; 134. diapsida

30.8 Testudines: Turtles [pp. 688–689]

30.9 Living Nonfeathered Diapsids: Sphenodontids, Squamates, and Crocodilians [pp. 689–692]

30.10 Aves: Birds [pp. 692–694]

135. Testudines; 136. turtles; 137. anapsida; 138. no temporal; 139. turtles; 140. shell; 141. diapsida; 142. two temporal; 143. Sphenodontids; 144. Squamates; 145. lizards; 146. snakes; 147. crocodilian; 148. alligators; 149. crocodiles; 150. aves; 151. birds; 152. E; 153. F; 154. G; 155. H; 156. J; 157. A; 158. C; 159. B; 160. K; 161. I; 162. D

30.11 Mammalia: Monotremes, Marsupials, and Placentals [pp. 695–697]

30.12 Nonhuman Primates [pp. 697–702]

30.13 The Evolution of Humans [pp. 702–707]

163. mammals; 164. synapsida; 165. one temporal; 166. high metabolic; 167. body temperature; 168. teeth; 169. jaws; 170. parental care; 171. brain; 172. reproduction; 173. lay eggs; 174. prototheria; 175. Monotremes; 176. theria; 177. live-bearing; 178. marsupials; 179. eutheria; 180. Placentals; 181. whales; 182. dolphins; 183. humans; 184. apes; 185.

monkeys; 186. arboreal; 187. ground; 188. erect; 189. flexible; 190. grasp; 191. hands 192. feet; 193. cortex; 194. integration; 195. bipedal; 196. upright (erect); 197. bipedal; 198. power; 199. precision; 200. larger; 201. H; 202. E; 203. F; 204. G; 205. K; 206. A; 207. J; 208. D; 209. C; 210. B; 211. I

Self-Test

1. b [the bilaterally symmetrical larval form is characteristic of the phylum Echinodermata, pedicellariae are found in the asteroidea]
2. d [echinoderms have tube feet and acorn worms have gill slits, the dorsal hollow nerve chord is found in chordates]
3. a [the perforated pharynx is maintained with gill arches for gas exchange in an aquatic environment]
4. d [all except for d are Agnathans]
5. d [the swim bladder is a hydrostatic organ which increases buoyancy; if destroyed, the fish would not be as buoyant]
6. a [of the selections, dehydration is a major problem in the terrestrial environment]
7. a [of the selections, only keratin and lipid in skin will be advantageous to terrestrial life]
8. c [hollow limb bones significantly reduces the weight of the skeleton of birds]
9. d [all selections are either other names or examples of prototheria]
10. b [the precision grip allows us to manipulate objects with fine movements]

Chapter 31 The Plant Body

Why It Matters [pp. 711–712]

1. 11,000; 2. Anthophyta

31.1 Plant Structure and Growth: An Overview [pp. 712–715]

3. organs; 4. Tissues; 5. protoplast; 6. lignification; 7. shoot system; 8. root system; 9. determinate growth; 10. indeterminate growth; 11. meristem; 12. apical meristem; 13. primary tissues; 14. primary plant body; 15. primary growth; 16. secondary tissues; 17. lateral meristems; 18. secondary growth; 19. secondary plant body; 20. monocots; 21. eudicots; 22. annuals; 23. biennials; 24. perennials; 25. F—lignin; 26. T; 27. F—shoot apical meristem; 28. F—secondary ; 29. T; 30. Multiples of 3; 31. Arranged randomly; 32. Network; 33. Three

31.2 The Three Plant Tissue Systems [pp. 715-721]

34. ground tissue; 35. vascular tissue; 36. dermal tissue; 37. Parenchyma; 38. Collenchyma; 39. Sclerenchyma; 40. Xylem; 41. tracheids; 42. Vessel members; 43. xylem vessel; 44. phloem; 45. sieve elements; 46. sieve tube; 47. companion cells; 48. Epidermis; 49. cuticle; 50. guard cells; 51. stoma; 52. trichomes; 53. root hairs; 54. c; 55. a; 56. b

31.3 Primary Shoot Systems [pp. 721–727]

57. node; 58. internode; 59. axil; 60. terminal buds; 61. blade; 62. lateral buds; 63. apical dominance; 64. initial; 65. derivative; 66. primary meristems; 67. protoderm; 68. ground meristem; 69. procambium; 70. vascular bundles; 71. stele; 72. cortex; 73. pith; 74. leaf primordial; 75. mesophyll; 76. veins; 77. C; 78. B; 79. A

31.4 Root Systems [pp. 727–730]

31.5 Secondary Growth [pp.730–734]

80. taproot; 81. lateral roots; 82. fibrous root; 83. adventitious; 84. root cap; 85. quiescent center; 86. zone of cell division; 87. zone of elongation; 88. zone of maturation; 89. exodermis; 90. endodermis; 91. pericycle; 92. root primordium; 93. vascular cambium; 94. cork cambium; 95. cork; 96. Fusiform intitials; 97. Ray initials; 98. wood; 99. heartwood; 100. sapwood; 101. bark; 102. periderm; 103. F—Periderm; 104. T; 105. F—fusiform initials

Self-Test

1. a [The secondary cell wall is laid down inside the primary wall.]

2. c

3. c

4. a [Shoot and root apical meristems are responsible for growth in length.]

5. b

6. c

7. b [Leave attach to the stem at the node. The distance between two nodes in the internode.]

8. d

9. c

10. d

Chapter 32 Transport in Plants

Why It Matters [pp. 737–738]

1. cohesion; 2. transpiration

32.1 Principles of Water and Solute Movement in Plants [pp. 738–742]

3. passive transport; 4. active transport; 5. transport proteins; 6. membrane potential; 7. symport; 8. antiport; 9. bulk flow; 10. xylem sap; 11. osmosis; 12. water potential; 13. turgor pressure; 14. megapascals; 15. central vacuole; 16. tonoplasts; 17. aquaporins; 18. wilting; 19. T; 20. F—Symport; 21. T; 22. F—active; 23. T

32.2 Transport in Roots [pp. 742–745]

32.3 Transport of Water and Minerals in the Xylem [pp. 745–750]

24. apoplast pathway; 25. symplast pathway; 26. transmembrane pathway; 27. Casparian strip; 28. transpiration; 29. cohesion-tension mechanism of water transport; 30. root pressure; 31. guttation; 32. crassulacean acid metabolism (CAM); 33. C; 34. E; 35. D; 36. A; 37. B

32.4 Root Systems [pp. 750–754]

38. translocation; 39. phloem sap; 40. source; 41. sink; 42. pressure flow; 43. transfer cells; 44. F—sucrose; 45. F—in any direction; 46. T; 47. T; 48. T

Self-Test

1. d

2. c

3. d

4. d

5. b

6. c [Adhesion is the attraction of water molecules to other charged or polar molecules. Cohesion is the attraction of water molecules to each other. Viscosity is the friction of liquid molecules as the flow past each other or over a solid surface.]

7. b

8. c [via symport]

9. c [CAM photosynthesis occurs most often in hot, dry habitats like those where *Sedum* is found.]

10. a [As long as a pressure gradient exists, phloem sap will move toward the region of lower pressure.]

Chapter 33 Plant Nutrition

Why It Matters [pp. 757–758]

1. water and minerals; 2. light and air; 3. structural; 4. physiological

33.1 Plant Nutritional Requirements [pp.758–762]

33.2 Soil [pp.762–765]

5. 99; 6. Hydroponics; 7. minerals; 8. Essential; 9. metabolic; 10. gradient; 11. macronutrients; 12. carbon, hydrogen, oxygen, nitrogen, phosphorus, potassium, calcium, sulfur, magnesium;13. micronutrients; 14. copper, chlorine, nickel; 15. stunted; 16. chlorosis; 17. water; 18. air; 19. air; 20. water; 21. humus; 22. water; 23. decompose; 24. air; 25. cations; 26. anions; 27. pH; 28. horizons.; 29. B; 30. D; 31. A; 32. E; 33. C

34.

Soil Ingredients	*Value in Soil*
Particles	Hold water and air
Humus	Source of nutrients, holds water and air
Living organisms	Provide organic chemicals and aerates the soil
Minerals	Nutrients for the plant

33.3 Obtaining and Absorbing Nutrients [pp. 765–772]

35. nitrogen, phosphorus and potassium; 36. root hair; 37. transport; 38. mycorrhizae; 39. sugars; 40. nitrogenous; 41. vacuole; 42. cytoplasm; 43. xylem; 44. 80; 45. enzymes; 46. nitrogen-fixing; 47. ammonifying; 48. ammonium; 49. nitrifying; 50. ammonium; 51. amino; 52. nitrogen; 53. nodules; 54. flavenoids; 55. nod; 56. nodules; 57. bacteroids; 58. nitrogenase; 59. leghemoglobin; 60. carnivores; 61. parasites; 62. epiphytes; 63. B; 64. A; 65. C; 66. D; 67. E; 68. D; 69. A; 70. C; 71. B.

72.

Adaptations	Description
Carnivorous	Plants get their nitrogen by luring insects into an enzyme rich structure and digesting them
Parasite	Plants develop special structures to penetrate other plants and tap their nitrogen, sugars, and water
Epiphyte	Plants anchor on other plants to collect rainwater, dissolved nutrients, and light
Symbiotic	Plants and certain bacteria or fungi live in partnership

Self-Test

1. c [Plants absorb nitrogen mostly in the nitrate form.]
2. b [Plants mostly use nitrogen in the ammonium form.]
3. c [nitrogen-phosphorus-potassium]
4. b [Humus adds organic chemicals to the soil.]
5. b [pH directly affects the absorption of minerals by the plants.]
6. b [Absorption of minerals and water takes place just above the root tip where root hair are present.]
7. d [Most orchids are epiphytes and anchor on the tree to reach light and catch raindrops.]

8. c [Nitrifying bacteria convert ammonium to nitrates, a form that plants absorb.]
9. b [Bacteroids are bacteria that live in root nodules.]
10. b [Nod genes stimulate plant root cells to make leghemoglobin.]
11. a [Hydroponics was developed to learn about the mineral requirements of plants.]
12. a [Some of the minerals act as cofactors for the plant enzymes.]
13. b [They are both important, just the amount needed is different.]
14. a [C, H, and O are macronutrients that form the basic macromolecules.]
15. b [Plants store most of their water and minerals in the central vacuole.]

Chapter 34 Reproduction and Development in Flowering Plants

Why It Matters [pp. 775–776]

1. flowers; 2. pollination; 3. fertilization; 4. seeds

34.1 Overview of Flowering Plant Reproduction [pp. 776–777]

34.2 The Formation of Flowers and Gametes [pp. 778–780]

34.3 Pollination, Fertilization and Germination [pp. 781–787]

5. sporophyte; 6. haploid; 7. gametophyte; 8. gametophyte; 9. gametes; 10. diploid; 11. sporophyte; 12. floral shoot; 13. inflorescence; 14. calyx; 15. sepals; 16. corolla; 17. petals; 18. stamens; 19. filament; 20. anther; 21. pollen; 22. carpels; 23. ovary; 24. style; 25. stigma; 26. Complete; 27. Incomplete; 28. monoecious; 29. dioecious; 30. anther; 31. cell wall; 32. gametophyte; 33. sperm; 34. tube; 35. ovary; 36. one; 37. gametophyte; 38. embryo; 39. one; 40. two; 41. three; 42. central; 43. micropyle; 44. anther; 45. stigma; 46. pollination; 47. pollen; 48. style; 49. ovary; 50. fertilization; 51. seed; 52. allele; 53. allele; 54. stigma; 55. style; 56. ovule; 57. sperm; 58. zygote; 59. triploid; 60. endosperm; 61. embryo; 62. cotyledons; 63. cotyledon; 64. endosperm; 65. radicle; 66. plumule; 67. coat; 68. coleorhiza; 69. coleoptile; 70. ovary; 71. pollen; 72. pericarp; 73. ovary; 74. Simple ; 75. Aggregate; 76. Multiple; 77. dormancy; 78. imbibition; 79. embryo; 80. hydrolytic; 81. radicle; 82. root; 83. plumule; 84. E; 85. C; 86. D; 87. G; 88. F; 89. A; 90. H; 91. B; 92. H; 93. E; 94. A; 95. G; 96. B; 97. I; 98. L; 99. C; 100. J; 101. D; 102. K; 103. F; 104. B; 105. C; 106. A; 107.F ; 108. D; 109. E; 110. E; 111. A; 112. B; 113. F; 114. C; 115. G; 116. D

117.

Events	*Description*
A. Pollination	The transfer of pollen from an anther to the stigma of the carpel
B. Fertilization	The fusion of male sperm cell and the female egg cell
C. Germination	The seed imbibes water to come out of dormancy and grow into a seedling

118.

Fruits	*Description*
A. Simple fruits	A fruit that develops from a single ovary of a flower
B. Aggregate fruits	A fruit that develops from multiple ovaries of a single flower
C. Multiple fruits	A fruit that develops from several ovaries of multiple flowers

34.4 Asexual Reproduction of Flowering Plants [pp. 787–789]

34.5 Early Development of Plant Form and Function [pp. 789–798]

119. vegetative; 120. totipotency; 121. Fragmentation; 122. Apomixis; 123. stock; 124. callus; 125. hormones; 126. somaclonal; 127. protoplasts; 128. root-shoot; 129. apical; 130. basal; 131. morphogenesis; 132. oriented; 133. expansion; 134. homeotic; 135. C; 136. D; 137. A; 138. G; 139. H; 140. B; 141. I; 142. E; 143. K; 144. F; 145. L; 146. J

SELF-TEST

1. a [A flowering plant is a sporophyte because it bears microspores and megaspores.]
2. b [A pollen grain is a 3 celled gametophyte that contains 2 haploid sperm cells and a pollen tube cell.]
3. b [An embryo sac represents the female gametophyte of a flowering plant.]
4. d [A pollen contains two sperm cells and one pollen tube cell.]
5. c [Ovule becomes the seed.]
6. d [The diploid central cell fuses with the sperm cell and eventually forms the endosperm.]
7. c [Haploid sperm cell fuses with diploid central cell to form a triploid cell that forms the endosperm.]
8. b [Some eudicots have fleshy cotyledons while others have fleshy endosperm.]
9. b [Strawberry is an aggregate fruit that is derived from many ovaries of the same flower.]
10. c [A pineapple is derived from the ovaries of many flowers that stay together.]
11. a [Garden peas and green beans are simple fruits because they are derived from one ovary of a single flower.]
12. b [Mutation often occur in the cells of a cultured callus.]
13. d [Radicle forms primary root of the seedling.]
14. a [Pollination is the process of transfer of pollen grain from the anther to the stigma.]
15. b [Dedifferentiation is when specialized cells become unspecialized]

Chapter 35 Control of Plant Growth and Development

Why It Matters [pp. 801–802]

1. light, temperature, moisture; 2. seed/fruit

35.1 Plant Hormones [pp. 802–812]

3. chemical; 4. environmental; 5. hormones; 6. organic; 7. vascular; 8. meristem; 9. stimulate; 10. light; 11. light; 12. gravity; 13. promote; 14. tips; 15. stimulate; 16. dormancy; 17. bolting; 18. tips; 19. xylem; 20. division; 21. senescence; 22. abscission; 23. fruits; 24. hormones; 25. stimulate; 26. carotenoids; 27. inhibits; 28. germinate; 29. fatty; 30. pathogens; 31. predators; 32. D; 33. A; 34. B; 35. C; 36. D; 37. A; 38. B; 39. C;

40.

Hormone	*Action*
Auxins	Promote growth of the stem and lateral roots; promote fruit development; help plant in responding to light and gravity
Gibberallins	Promote cell division, seed germination, and bolting
Cytokinins	Promote cell division and inhibit senescence
Ethylene	Promote senescence, abscission, and fruit ripening
Brassinosteroids	Promote stem elongation, vascular development, and growth of pollen tube
Abscisic acid	Inhibits stem growth and promotes abscission
Jasmonates	Protect plants from pathogens and predators

35.2 Plant Chemical Defenses [pp. 812–816]

35.3 Plant Responses to the Environment: Movements [817–821]

35.4 Plant Responses to the Environment: Biological Clocks [821–825]

35.5 Signal Responses at the Cellular Level [826–828]

41. bacteria, viruses, fungi and insects; 42. Jasmonates and ethylene; 43. salicylic; 44. inhibitors; 45. proteins; 46. peroxide; 47. hypersensitive; 48. secondary; 49. phytoalexins; 50. tropism; 51. Phototropism; 52. phototropin; 53. Gravitropism; 54. otoliths; 55. downward; 56. upward; 57. Thigmotropism; 58. Nastic; 59. circadian; 60. photoperiodism; 61. phytochrome; 62. Pr; 63. Pfr; 64. long-day; 65. short-day; 66. vernalization; 67. dormancy; 68. gibberallins; 69. receptor; 70. chemical; 71. proteins; 72. C; 73. G; 74. A; 75. H; 76. I; 77. B; 78. J; 79. D; 80. E; 81. F; 82. K

SELF-TEST

1. a [Auxins were the first plant hormone to be identified.]
2. a [Auxins exhibits polar transport and move away from unidirectional light.]
3. b [Gibberallins are involved in breaking of seed and bud dormancy.]
4. b [Gibberallins are involved in bolting seen in rosette plants.]
5. c [Cytokinins coordinate growth of roots and shoots in concert with the auxins.]
6. d [The diploid central cell fuses with the sperm cell and eventually forms the endosperm.]
7. b [Abscission is the process of dropping of flowers, fruits, and leaves.]
8. d [Bolting is used for extension of the floral stem in rosette plants.]
9. c [Thigmotropism movement or growth of a plant in response to contact with an object.]
10. d [Nastic movement refers to temporary, reversible response to a unidirectional stimulus.]
11. e [Photoperiodism refers to response of a plant due to changes in the length of light and dark periods during each 24-hour period.]
12. e [Vernalization is low temperature stimulation of flowering.]
13. a [Senescence is aging in plants.]
14. e [Phytochrome is involved in photoperiodism.]
15. a [Gibberallins are made by plants and fungi.]

Chapter 36 Introduction to Animal Organization and Physiology

Why It Matters [pp. 831–832]

1. Homeostasis; 2. internal; 3. maintenance; 4. structure; 5. function

36.1 Organization of the Animal Body [p. 832]

6. internal; 7. external; 8. water; 9. cell; 10. tissues; 11. organ; 12. tissues; 13. function; 14. organ system; 15. organs

36.2 Animal Tissues [pp. 833–840]

16. D; 17. A; 18. B; 19. C; 20. D; 21. A; 22. B; 23. E; 24. C; 25. C; 26. G; 27. A; 28. F; 29. B; 30. E; 31. D; 32. epithelial; 33. connective; 34. muscular; 35. nervous; 36. Epithelial; 37. cell layers; 38. shape; 39. 6; 40. connective; 41. loose CT; 42. fibrous CT; 43. cartilage; 44. bone; 45. adipose; 46. blood; 47. cells; 48. fibers; 49. extracellular matrix; 50. 3; 51. muscle; 52. skeletal; 53. smooth; 54. cardiac; 55. 2; 56. nervous; 57. neurons; 58. glial; 59A. absorption, secretion, protection, diffusion; 59B. lines body cavities, covers surfaces; 59C. Connective; 59D. Forcibly shortens; 59E. skeletal mass, walls of organs and tubes, heart; 59F. responds to stimuli, conducts electrical activity for communication

36.3 Coordination of Tissues in Organs and Organ Systems [pp. 840–841]

60. cell; 61. tissues; 62. organs; 63. organ system; 64. 11; 65. responding to stimuli; 66. movement; 67. acquisition of nutrients and production of wastes; 68. metabolism; 69. protection against foreign substances; 70. C; 71. D; 72. F; 73. E; 74. K; 75. A; 76. I; 77. J; 78. B; 79. E; 80. G; 81. F; 82. A, I; 83. I; 84. C; 85. H

36.4 Homeostasis [pp. 841–843]

86. Homeostasis is a dynamic equilibrium. Certain body functions must be maintained in a range that is compatible with life, such as temperature, blood pressure, or pH of body fluids. Control mechanisms detect changes in these variables and adjust the range to meet the demand. When the new demand is changed, a modification of the variable will occur to maintain an internal environment that is compatible with life. 87. A positive feedback mechanism will produce more product. This is a type of amplification process. The product will often result in the production of more product. A negative feedback mechanism maintains the output in a range. If the output falls, the feedback results in more output to maintain the output in an acceptable range. If the output increases, the feedback will result in a decrease in output to maintain the output in the acceptable range.

88. C; 89. D; 90. E. 91. B; 92. A

Self-Test

1. a [a cell must maintain homeostasis between the intracellular and extracellular environments]
2. c [the mitochondria (organelle) is within the hepatic cell which is within the epithelium (major tissue type) which is within the liver (organ)]
3. d [epithelial tissue is characterized by lining body cavities with little or no extracellular matrix]
4. c [these characteristics are of blood, a connective tissue; red blood cells are involved in oxygen delivery to cells; and white blood cells are involved in immunity]
5. c [connective tissue is the most varied and is characterized by specific cell types associated with the tissue type, fibers, and matrix]
6. b [the excretory system is critical in maintaining osmotic balances, ions or electrolytes, and pH of body fluids]
7. d [the endocrine system is a major component of homeostatic mechanisms; hormones are chemical mediators that are transported to cells via the circulation]
8. b[sweat glands are derivatives of the skin (integument) which is a major protective coat between the internal and external environment of organisms]
9. b [positive feedback systems are characterized by an amplification of the response or product]
10. a [sweat glands are a way to reduce body temperature by loss of water and the associated heat of that water. When body temperature increases, homeostatic mechanisms result in increased blood flow to the skin and increased activity of sweat glands.]

Chapter 37 Systematic Biology: Phylogeny and Classification

Why It Matters [pp. 847–848]

1. nervous system

37.1 Neurons and their Organization in Nervous System: An Overview [pp. 848–851]

2. neural signaling; 3. neurons; 4. reception; 5. transmission; 6. integration; 7. response; 8. afferent neurons; 9. sensory neurons; 10. interneurons; 11. efferent neurons; 12. effectors; 13. cell body; 14. dendrites; 15. axons; 16. axon hillock; 17. axon terminals; 18. neural circuits; 19. glial cells; 20. astrocytes; 21. oligodendrocytes; 22. Schwann cells; 23. nodes of Ranvier; 24. synapse; 25. presynaptic cell; 26. postsynaptic cell; 27. electrical synapse; 28. chemical synapse;

29. neurotransmitter; 30. synaptic cleft; 31. pre-; 32. post-; 33. interneuron; 34. b; 35. a; 36. a; 37. b; 38. b; 39. c; 40. a; 41. Efferent neurons carry impulses away from the interneuron/network to effectors in general (could be a gland or muscle). Motor neurons are specific efferent neurons that carry impulses to muscle (could be smooth muscle, skeletal muscle, or cardiac muscle); 42. b; 43. d; 44. a; 45. c; 46. neural support cells; 47. astrocytes; 48. wrap around axons of neurons in the central nervous system; 49. wrap around axons of neurons in a peripheral nervous system; 50. neurons; 51. e; 52. c; 53. f; 54. d; 55. b; 56. g; 57. a; 58. dendrites; 59. axon; 60. axon hillock; 61. axon terminal; 62. cell body (soma)

37.2 Signal Conduction by Neurons [pp. 851–857], **37.3 Conduction across Chemical Synapses** [pp. 858–862], **37.4 Integration of Incoming Signals by Neurons** [pp. 862–863]

63. membrane potential; 64. resting potential; 65. polarized; 66. action potential; 67. depolarized; 68. threshold potential; 69. hyperpolarized; 70. all-or-nothing principle; 71. refractory period; 72. voltage-gated ion channels; 73. propagation; 74. salutatory conduction; 75. presynaptic membrane; 76. ligand-gated ion channels; 77. postsynaptic membrane; 78. synaptic vesicles; 79. exocytosis; 80. direct neurotransmitters; 81. indirect neurotransmitters; 82. excitatory postsynaptic potential (EPSP); 83. inhibitory postsynaptic potential (IPSP); 84. graded potential; 85. temporal summation; 86. spatial summation; 87. neuron-; 88. de-; 89. hyper-; 90. a; 91. b; 92. b; 93. a; 94. a; 95. b; 96. b; 97. a; 98. b; 99. a; 100. All cells display a separation of positive and negative charges across their membrane, but this potential (membrane potential) remains unchanged. Some cells, such as neurons and muscles, possess membranes (excitable membranes) that are capable of changing their potential. The resting potential is the membrane potential of an unstimulated nerve or muscle cell; 101. e; 102. b; 103. a; 104. d; 105. c; 106. f; 107. resting potential; 108. threshold potential; 109. depolarization; 110. repolarization; 111. refractory period; 112. hyperpolarization (undershoot).

Self-Test

1. b, c, d, e

2. a, c

3. a, c, d

4. a, c, e

5. a, c

6. a, c

7. b, e

8. a, d [a is correct because the membrane cannot be stimulated if the influx of sodium is blocked; therefore, the inactivation of voltage-gated sodium channels marks the onset of the refractory period. The membrane will remain refractory until the inactivation of the voltage-gated sodium channel is lifted; this corresponds with the reestablishment of the resting potential, making d also correct.]

9. c, d [c is correct because as sodium influx occurs and the membrane becomes depolarized in a specific region, adjacent voltage-gated sodium channels are prompted to open; however, because voltage-gated sodium channels become inactivated at the peak of an action potential, only those channels that are in front of the wave (which have not recently opened and are not inactive) are capable of responding and opening, while those behind the wave (which have just recently opened) are inactive and not capable of responding. Thus, the impulse travels only in one direction in a wavelike manner; d also is correct because of the all-or-nothing nature of action potentials; once threshold is reached, the voltage-gated sodium channels open, allowing rapid (initially) sodium influx and depolarization of the membrane in that specific region. Because the amplitude of each action potential is the same, the impulse is constantly "refreshed" as the wave of depolarization moves along the axon, assuring that it will reach the axon terminus. (This is important because some motor neurons can be several meters in length!)]

10. a, c, d

11. b, d [b is correct because an inhibitory neuron releases neurotransmitters that bind to the postsynaptic cell and can open chloride and potassium channels. Because of the asymmetric distribution of ions across the membrane (high sodium and chloride outside, high potassium inside), the opening of chloride channels causes chloride influx (down its concentration gradient), while the opening of potassium channels results in potassium efflux (down its concentration gradient). The net result is that the inside becomes more negative relative to the outside than when the membrane was at resting potential; therefore, the membrane is said to be hyperpolarized; d also is correct because the flux of potassium and chloride is proportion to the strength of the inhibitory signal (based on how many neurotransmitters where bound and how many channels were affected).]

12. a

Chapter 38 The Cell: An Overview

Why It Matters [pp. 867–868]

1. nervous system

38.1 Invertebrate and Vertebrate Nervous Systems Compared [pp. 868–871]

2. nerve nets; 3. ganglia; 4. brain; 5. nerve cords; 6. central nervous system (CNS); 7. peripheral nervous system (PNS); 8. neural tube; 9. spinal cord; 10. ventricles; 11. central canal; 12. forebrain; 13. midbrain; 14. hindbrain; 15. hypothalamus; 16. postganglionic; 17. preganglionic; 18. a; 19. b; 20. loose meshes of neurons in certain animal groups with radial symmetry; 21. ganglion; 22. bundle of nerves that extend from a central ganglion of more complex invertebrates

38.2 The Peripheral Nervous System [pp. 871–872], **38.3 The Central Nervous System and Its Functions** [pp. 872–879]

23. somatic nervous system; 24. autonomic nervous system; 25. sympathetic division; 26. parasympathetic division; 27. meninges; 28. cerebral spinal fluid; 29. gray matter; 30. white matter; 31. reflex; 32. brainstem; 33. cerebral cortex; 34. blood-brain barrier; 35. reticular formation; 36. cerebellum; 37. thalamus; 38. hypothalamus; 39. basal nuclei; 40. limbic system; 41. amygdale; 42. hippocampus; 43. olfactory bulbs; 44. corpus callosum; 45. primary somatosensory area; 46. association areas; 47. primary motor area; 48. lateralization; 49. b; 50. a; 51. a; 52. b; 53. a; 54. b; 55. control body movement (mostly voluntary); 56. autonomic; 57. sympathetic; 58. para-sympathetic; 59. connects the two cerebral hemispheres; 60. primary somatosensory area; 61. primary motor area; 62. integrates sensory information and formulates responses; 63. a; 64. j; 65. l; 66. e; 67. g; 68. c; 69. d; 70. i; 71. k; 72. n; 73. h; 74. m; 75. b; 76. f

38.4 Memory, Learning, and Consciousness [pp. 879–883]

77. memory; 78. learning; 79. consciousness; 80. short-term memory; 81. long-term memory; 82. long-term potentiation; 83. sensitization; 84. electroencephalogram; 85. rapid eye-movement (REM) sleep; 86. a; 87. b; 88. a; 89. e; 90. d; 91. f; 92. c; 93. b

Self-Test

1. d [d is correct because the hypothalamus is responsible for coordination various growth, development, reproductive, osmoregulatory and other processes; neurons in the hypothalamus release hormones that affect these processes (see chapter 40); a is not correct because this region is involved in sensory integration and motor control; b is not correct because this region is involved in high functions such as emotions, memory, etc.; c is not correct because this region is the region that receives sensory information; and e is not correct because this region integrates and sorts sensory information]

2. b, d [b is correct because the autonomic nervous system has primary responsibility for the viscera and blood vascular system; d is correct because most arteriole are innervated only with branches of the sympathetic division]

3. b, c [b is correct because the autonomic nervous system has primary responsibility for the viscera and blood vascular system; d is correct because the parasympathetic division is responsible for coordinated most processes associated with feeding and digestion (e.g., saliva release, gut peristalysis, etc.)]

4. a, b, c, e

5. b

6. a, b

7. b

8. a, c

9. c

10. a, d, e

Chapter 39 Sensory Systems

Why It Matters [pp. 885–886]

1. sensory systems

39.1 Overview of Sensory Receptors and Pathways [pp. 886–888], **39.2 Mechanoreceptors and the Tactile and Spatial Senses** [pp. 888–891], **39.3 Mechanoreception and Hearing** [pp. 891–894]

2. sensory receptors; 3. sensory transduction; 4. mechanoreceptors; 5. photoreceptors; 6. chemoreceptors; 7. thermoreceptors; 8. nociceptors; 9. frequency of action potentials; 10. number of neurons activiated; 11. sensory adaptation; 12. Pacinian corpuscles; 13. proprioceptors; 14. statocysts; 15. statoliths; 16. sensory hair cells; 17. lateral line system; 18. neuromasts; 19. stereocilia; 20. cupula; 21. vestibular apparatus; 22. semicircular canals; 23. utricle; 24. saccule; 25. otoliths; 26. stretch receptors; 27. muscle spindles; 28. Golgi tendon organ; 29. tympanum; 30. pinna; 31. outer ear; 32. tympanic membrane; 33. middle ear; 34. malleus; 35. incus; 36. stapes; 37. oval window; 38. inner ear; 39. semicircular canal; 40 utricle; 41. saccule; 42. cochlea; 43. organ of Corti; 44. round window; 45. echolocation; 46. chemoreceptor; 47. thermoreceptor; 48. otolith; 49. proprioceptor; 50. photoreceptor; 51. nociceptor; 52. a; 53. b; 54. lateral line system; 55. detect position and orientation; used for equilibrium in invertebrates; 56. perceives position and motion of head; 57. muscle spindle; 58. organ of Corti; 59. Proprioceptors that detect stretch and compression of tendon; 60. neuromast; 61. a; 62. h; 63. i; 64. e; 65. c; 66. g; 67. b; 68. f; 69. d; 70. pinna; 71. Eustachian tube; 72. stapes; 73. incus; 74. malleus; 75. semicircular canals; 76. oval window; 77. auditory canal; 78. tympanic membrane; 79. round window; 80. cochlea; 81. outer ear; 82. middle ear; 83. inner ear

39.4 Photoreceptors and Vision [pp. 894–899], **39.5 Chemoreceptors** [pp. 899–902], **39.6 Thermoreceptors and Nociceptors** [pp. 902–904], **39.7 Electroreceptors and Magnoreceptors** [pp. 904–906]

84. ocellus; 85. compound eye; 86. ommatidia; 87. cornea; 88. photopigment; 89. single-lens eye; 90. lens; 91. retina; 92. iris; 93 pupil; 94. accommodation; 95. aqueous humor; 96. vitreous humor; 97. ciliary body; 98. rods; 99. cones; 100. fovea; 101. peripheral vision; 102. photopigments; 103. retinal; 104. opsins; 105. rhodopsin; 106. bipolar cells; 107. ganglion cells; 108. horizontal cells; 109. amacrine cells; 110. lateral inhibition; 111. photopsin; 112. optic chiasm; 113. lateral geniculate nuclei; 114. sensilla; 115. taste buds; 116. olfactory hairs; 117. transient receptor potential (TRP); 118. magnoreceptors; 119. A compound eye contain 100s to 1000s of individual visual units. A single-lens eye has one lens and operates like a camera; 120. Photo-pigments consist of a covalent complex of retinal and one of several different proteins know as opsins. The photopigment in rod cells is rhodopsin. Cone cells contain

different types of photopsins based upon different opsin forms; humans have three photopsins; 121. Accommodation is the process of focusing and image by moving the lens back and forth relative to the retina; 122. bipolar cell; 123. extend over entire retina, and axons come together to form optic nerve; 124. amacrine cell; 125. connect with different photoreceptor cells and bipolar cells; 126. photoreceptor; 127. radiant energy; 128. distinguish tastes of sweet, sour, salty, and umami; 129. olfactory hair; 130. chemicals; 131. electroreceptor; 132. communication; locate objects (including prey); 133. magnetic field; 134. nociceptor; 135. tissue damage; noxious chemicals; 136. a; 137. i; 138. f; 139. c; 140. d; 141. b; 142. e; 143. g; 144. h; 145. ciliary body; 146. iris; 147. lens; 148. pupil; 149. cornea; 150. aqueous humor; 151. vitreous humor; 152. retina; 153. fovea; 154. optic nerve

Self-Test

1. c

2. a, c, e [a is correct because these are used by invertebrates to detect position; c is correct because this structure is used for maintaining equilibrium in vertebrates; e is correct because some fish have these structures to provide information about orientation; b and d are not correct because these structures detect vibration (sound)]

3. b

4. a, b, c

5. a

6. b, d [invertebrates use b to focus images, whereas vertebrates use d; a is not correct because this process sharpens images by enhancing contrast; c is not correct because this structure is the visual unit of a compound eye]

7. d

8. a

9. d [d is correct because these compounds bind to opioid receptors and block the release of substance P, which makes b not correct because this compound is released from axons and conveys the sensation of pain to the CNS; a is not correct because this chemical gives the "hot" taste in food; c is not correct because insulin stimulates the uptake of glucose and other nutrients into cells and does not affect pain perception]

10. a, b, c

---

40 The Endocrine System

Why It Matters [pp. 909–910]

1. hormones; 2. endocrine systems

40.1 Hormones and Their Secretions [pp. 910–912], **40.2 Mechanisms of Hormone Action** [pp. 912–919]

3. endocrine glands; 4. neurosecretory cells; 5. amine; 6. peptide; 7. steroid; 8. fatty acid derivative; 9. growth factors; 10. prostaglandins; 11. neurohormone; 12. hyposecretion; 13. hypersecretion; 14. hyperglycemic; 15. endocrine; 16. neuronsecretion; 17. the release of a hormone from an epithelial cell in a gland that is transported in the blood and generally effective at a distance from its site of secretion; 18. the release of a chemical signal into extracellular fluid that regulated the activity of a neighboring cell; 19. autocrine (or autoregulation); 20. epinephrine; 21. insulin; 22. cortisol; 23. fatty acid derivative; 24. c; 25. d; 26. b; 27. a

40.3 The Hypothalamus and Pituitary [919–922]

28. pituitary gland; 29. posterior pituitary; 30. anterior pituitary; 31. tropic hormone; 32. releasing hormones; 33. inhibiting hormones; 34. antidiuretic hormone (ADH); 35. oxytocin; 36. prolactin; 37. growth hormone; 38. thyroid stimulating hormone (TSH); 39. adrenocorticotropic hormone (ACTH); 40. follicle stimulating hormone (FSH); 41. luteinizing hormone (LH); 42. gonadotropins; 43. melanocyte stimulating hormone (MSH); 44.

endorphins; 45. The anterior pituitary is a distinct lobe that produces and secretes many hormones into the general circulation (e.g., growth hormone). The posterior pituitary does not produce any hormone. It is the site where axons from the hypothalamus terminate to release tropic hormones into the portal system that regulate the anterior pituitary or to release peptides such as ADH or oxytocin directly into the general circulation; 46. anterior pituitary; 47. stimulates growth and regulates metabolism; 48. prolactin; 49. peptide; 50. anterior pituitary; 51. TSH; 52. peptide; 53. stimulates adrenal cortex to produce cortisol and aldosterone; 54. peptide; 55. anterior pituitary; 56. promotes gamete development; 57. peptide; 58. anterior pituitary; 59. promotes sex steroid production; 60. MSH; 61. peptide; 62. anterior pituitary (intermediate lobe, when present); 63. peptide; 64. anterior pituitary (intermediate lobe, when present); 65. inhibit perception of pain; 66. peptide; 67. posterior pituitary (produced in hypothalamus); 68. stimulates water reabsorption (conservation); 69. oxytocin; 70. peptide; 71. posterior pituitary (produced in hypothalamus); 72. a; 73. b; 74. neurosecretory neuron (some produce releasing hormones; some produce inhibiting hormones); 75. hypothalamus; 76. protein vein; 77. posterior pituitary; 78. anterior pituitary; 79. neurosecretory neuron (some produce ADH, some produce oxytocin); 80. hypothalamus; 81. anterior pituitary; 82. posterior pituitary; 83. a; 84. c; 85. b

40.4 Other Major Endocrine Glands of Vertebrates [pp. 922–928],

86. thyroid gland; 87. thyroxine (T4); 88. triiodothyronine (T3); 89. metamorphosis; 90. calcitonin; 91. parathyroid hormone; 92. parathyroid gland; 93. vitamin D; 94. adrenal medulla; 95. adrenal cortex; 96. catecholamines; 97. epinephrine; 98. norepinphrine; 99. glucocorticoids; 100. mineralocorticoids; 101. cortisol; 102. aldosterone; 103. testes; 104. ovaries; 105. androgens; 106. estrogens; 107. progestins; 108. testosterone; 109. 17β-estradiol; 110. progesterone; 111. Islets of Langerhans; 112. pancreas; 113. insulin; 114. glucagon; 115. diabetes mellitus; 116. pineal gland; 117. melatonin; 118. a; 119. c; 120. e; 121. g; 122. i; 123. j; 124. f; 125. h; 126. b; 127. d; 128. both are steroid hormones produced and secreted by the adrenal cortex; glucocorticoids regulate carbohydrate metabolism, whereas mineralocorticoids regulate ion and water balance; 129. Both are "thyroid hormones" derived from the amino acid tyrosine; T4 has four iodine atoms, whereas T3 has three iodine atoms; 130. amine; 131. regulate basal metabolic rate; triggers metamorphosis in amphibians; 132. calcitonin; 133. thyroid gland; 134. increases blood calcium; 135. epinephrine; 136. adrenal medulla; 137. cortisol; 138. steroid; 139. steroid; 140. adrenal cortex; 141. increase sodium and water reabsorption; 142. testosterone; 143. estrdiol; 144. steroid; 145. promotes development and maintenance of secondary sex characteristics; 146. prepares and maintains uterus for implantation; 147. gonadotropin releasing hormone; 148. posterior pituitary (produced in hypothalamus); 149. peptide; 150. islets of Langerhans; 151. anabolic; stimulates nutrient uptake into cells and macromolecule synthesis; 152. glucagon; 153. peptide; 154. islets of Langerhans; 155. melatonin; 156. peptide; 157. helps maintain daily biorhythms

40.5 Endocrine Systems in Invertebrates [p. 928–930]

158. brain hormone; 159. ecdysone; 160. juvenile hormone; 161. molt-inhibiting hormone; 162. a; 163. d; 164. b; 165. c

Self-Test

1. a

2. b, c [b is correct because the secretion of the hormone from the source cell is reduced in negative feedback; c is correct because homeostasis is maintained with negative feedback]

3. a, c, e [a and c are correct because steroids and thyroid hormones are soluble in the cell membrane and enter the cytoplasm where they bind to receptors inside the cell; e is correct because the hormone-receptor complexes formed after hormone binding affect the transcription of genes; b and d are not correct because these hormones bind to membrane-associated receptors and initiate rapid responses inside cells without altering gene expression]

4. b [b is correct because glucagon acts on the liver to promote the breakdown of stored glycogen to glucose as well as the formation of glucose (i.e., gluconeogensis) from noncarbohydrate sources (e.g., amino acids); these actions result in the elevation of glucose levels in the blood]

5. c, f

6. b, c

7. a, d, e

8. b, d

9. b, d

10. a, c

Chapter 41 Muscle, Bones, and Body Movements

Why It Matters [pp. 933–934]

1. skeletal; 2. cardiac; 3. smooth; 4. skeletal muscle

41.1 Vertebrate Skeletal Muscle [pp. 934–941]

5. muscle fibers; 6. myofibrils; 7. thick filaments; 8. thin filaments; 9. sarcomere; 10. T-tubules; 11. sarcoplasmic reticulum; 12. neuromuscular junction; 13. acetylcholine; 14. sliding filament mechanism; 15. muscle twitch; 16. tetanus; 17. slow-muscle twitch; 18. fast-muscle twitch; 19. motor units; 20. sarcomere; 21. myofibril; 22. myoglobin; 23. Slow fibers contract slowly with low intensity, whereas fast fibers contract rapidly with high intensity. Differences are due different rates of ATP hydrolysis by myosin crossbridges; 24. a; 25. b; 26. a; 27. h; 28. g; 29. c; 30. b; 31. d; 32. I; 33. f; 34. e; 35. neuromuscular junction; 36. T-tubule; 37. sarcoplasmic reticulum; 38. myofibril; 39. sarcomere

41.2 Skeletal Systems [pp. 941–943], **41.3 Vertebrate Movement: The Interactions between Muscles and Bones** [pp. 943–946]

40. hydrostatic skeleton; 41. exoskeleton; 42. endoskeleton; 43. axial skeleton; 44. appendicular skeleton; 45. synocial; 46. cartilaginous; 47. fibrous; 48. agonist; 49. antagonistic pairs; 50. extensor muscles; 51. flexor muscles; 52. b; 53. c; 54. a; 55. a; 56. b; 57. b; 58. a; 59. a; 60. d; 61. e; 62. f; 63. b; 64. c

Self-Test

1. c [c is correct because only actin and myosin filaments move relative to one another; b is not correct because actin and myosin are arranged parallel to one another; d is not correct because actin does not normally dissociate during sarcomere shortening]

2. c, d

3. b

4. c, e

5. d, e

6. a, b, c

7. c, e

8. d

9. a

10. d

Chapter 42 The Circulatory System

Why It Matters [pp. 949–950]

1. circulatory system

42.1 Animal Circulatory Systems: An Introduction [pp. 950–953], **42.2 Blood and Its Components** [pp. 953–956]

2. open circulatory system; 3. hemolymph; 4. sinuses; 5. closed circulatory system; 6. arteries; 7. capillaries; 8. veins; 9. atria; 10. ventricles; 11. systemic circuit; 12. pulmonocutaneous circuit; 13. pulmonary circuit; 14. plasma; 15. albumins; 16. globulins; 17. fibrinogen; 18. erythrocytes; 19. red blood cells (RBC); 20. erythropoietin; 21. leukocytes; 22. platelets; 23. fibrin; 24. erythrocyte; 25. leukocyte; 26. hemoglobin; 27. a; 28. c; 29. b; 30. b; 31. a; 32. a; 33. b; 34. leukocytes; 35. specialized to transport O_2; 36. platelets; 37. Fibrinogen is a precursor to fibrin. The large soluble fibrinogen protein is converted by enzymes released from platelets to the insoluble fibrin to help form clots; 38. e; 39. a; 40. b; 41. I; 42. G; 43. c; 44. H; 45. F; 46. d

42.3 The Heart [pp. 956–961] **42.4 Blood Vessels of the Circulatory System** [pp. 961–965], **42.5 Maintaining Blood Flow and Pressure** [pp. 965–966] **42.6 The Lymphatic System** [pp. 966–967]

47. aorta; 48. systole; 49. diastole; 50. atrioventricular valves; 51. neurogenic heart; 52. myogenic heart; 53. sinoatrial node; 54. pacemaker cells; 55. atrioventricular node; 56. electrocardiogram; 57. hypertension; 58. arterioles; 59. precapillary sphincter; 60. venules; 61. cardiac output; 62. lymphatic system; 63. lymph; 64. lymph nodes; 65. a; 66. b; 67. b; 68. a; 69. b; 70. a; 71. a; 72. b; 73. extensive network of vessels that collects excess interstitial fluid; 74. lymph node; 75. lymph; 76. a record reflecting the electrical activity of the heart by attaching electrodes to the surface of the body; 77. a; 78. g; 79. d; 80. h; 81. f; 82. b; 83. e; 84. c

Self-Test

1. d

2. a, c, e [a is correct because these cells are red in appearance due to the iron complexed to hemoglobin, which binds O_2, making e correct; c is correct because RBCs are derived from stem cells in bone marrow; b is not correct because RBS typically have a biconcave shape; d is not correct because erythrocytes are not white blood cells]

3. b, c

4. a, d, e [a and e are correct because the pulmonary artery carries deoxygenated blood from the right ventricle to the lungs to get oxygenated; d is correct because this blood also carries relatively high levels of CO_2 to the lungs where it can be released into the air]

5. a

6. c, d, e [see answer to question 37]

7. e

8. c, d

9. a

10. a

Chapter 43 Defenses against Disease

Why It Matters [pp. 971–972]

1. system

43.1 Three Lines of Defense against Invasion [pp. 972–973], **43.2 Nonspecific Defenses: Innate Immunity** [pp. 973–976]

2. innate immune system; 3. adaptive immune system; 4. immune response; 5. inflammation; 6. macrophages; 7. cytokines; 8. mast cells; 9. neutophils; 10. chemokines; 11. eosinophils; 12. complement system; 13. membrane attach complexes; 14. interferons; 15. natural killer cells; 16. apoptosis; 17. lymphocyte; 18. a specific protein that binds to pathogens to mark them for elimination; 19. a drug that blocks the effects of histamine; 20. a white blood cell located in lymphoid tissue; 21. an enzyme that attaches to and destroys pathogens; 22. a; 23. b; 24. c; 25. a; 26. b; 27. first to recognize pathogens; engulfs pathogen and kills it; secretes signal to initiate other immune reponses; 28. mast cell; 29. attracted to infected site by chemokines; engulfs pathogen and kills it; usually dies itself afterward; 30. eosinophils; 31. natural killer cells; 32. f; 33. g; 34. b; 35. c; 36. e; 37. a; 38. d

43.3 Specific Defenses: Adaptive Immunity [pp. 976–989]

39. antigen; 40. B cells; 41. T cells; 42. thymus gland; 43. antibody-mediated immunity; 44. cell-mediated immunity; 45. plasma cells; 46. antibodies; 47. memory cells; 48. B cell receptors; 49. T cell receptors; 50. epitopes; 51. immunoglobins; 52. light chains; 53. heavy chains; 54. dendritic cell; 55. class II major histocompatibility complex; 56. antigen presenting cells (APC); 57. major histocompatibility complex; 58. $CD4^{+}$ cells; 59. clonal expression; 60. $CD4^{+}$ cells; 61. helper T cells; 62. effector T cell; 63. plasma cells; 64. memory cells; 65. clonal selection; 66. memory helper T cells; 67. immunological memory; 68. primary immune response; 69. secondary immune response; 70. active immunity; 71. passive immunity; 72. cell-mediated immunity; 73. CD8 receptors; 74. $CD8^{+}$ cells; 75. cytotoxic T cells; 76. monoclonal antibodies; 77. hybridomas; 78. The MHC is derived form a large cluster of genes expressed in a few immune cells types (dendritic cells; macrophages, B cells). MHC proteins bind to antigen molecules inside the cell, then translocates it to the surface of the cell, making the cell an antigen-presenting cell; 79. b; 80. a; 81. a; 82. b; 83. b; 84. a; 85. lymphocytes that arise and mature in the bone marrow; derivatives contribute to antibody-mediated immunity; 86. T cells; 87. B cell derivative that produces antibodies; 88. cell types derived from T cells and B cells and are responsible for initiating a rapid immune response upon reexposure to an antigen; 89. phagocytic cell that initiates adaptive immunity by engulfing foreign cell; 90. $CD4^{+}$ cell; 91. derived from activated CD4+ cells and leads to antibody-mediated immunity; 92. $CD8^{+}$ cell; 93. Derived from activated $CD8^{+}$ cells and destroys infected cells; 94. a; 95. k; 96. I; 97. f; 98. g; 99. c; 100. d; 101. e; 102. h; 103. j; 104. b

43.4 Malfunction and Failure of the Immune System [pp. 989–992], **43.5 Defenses in Other Animals** [pp. 992–994]

105. immunological tolerance; 106. autoimmune reaction; 107. allergens; 108. anaphylactic shock; 109. hemolin; 110. c; 111. e; 112. a; 113. d; 114. b

Self-Test

1. a, c, e

2. b, c, d [d is correct because allergens induce B cells to make an overabundance of IgE, which, in turn, stimulates mast cells to overproduce histamine, making b and c correct; a and e are not correct because allergens do not appear to be taken into cells, interact with MHC, or induce clonal production of killer T cells]

3. b, d

4. c

5. a, d [a and d are correct because complement is a complex of proteins in the plasma; some of these, when activated, attach to the surface of pathogens, resulting in the perforation of their membrane and destruction; b and c are not correct because the activation of compliment is nonspecific and the proteins are not antibodies.]

6. b

7. a, b, c, d

8. b, c, e [a is correct because B cells are lymphocytes that arise from stem cells in the bone marrow; c is correct because when a cell encounters an antigen, it incorporates it into the cell and presents it on its surface. The B cells are induced to proliferate when they encounter and bind to a helper T cell presenting the same antigen on its surface, producing a clone of B cells with identical B cell receptors. Some of these clones differentiate into plasma cells which secrete the same antibody that was originally displayed on the parent B cell's surface; e is correct because each unique antigen would result in a unique B cell clone; a is not correct because B cells are not granular; d is not correct because B cells do not derive from plasma cells; rather, plasma cells derive from B cells]

9. a, d

10. b

Chapter 44 Gas Exchange: The Respiratory System

Why It Matters [pp. 997–998]

1. respiratory system

44.1 The Function of Gas Exchange [pp. 998–1000], **44.2 Adaptations for Respiration** [pp. 1000–1003]

2. physiological respiration; 3. respiratory medium; 4. breathing; 5. respiratory surface; 6. tracheal system; 7. gills; 8. lungs; 9. ventilation; 10. perfusion; 11. external gills; 12. internal gills; 13. counter-current exchange; 14. tracheae; 15. spiracles; 16. positive pressure breathing; 17. negative pressure breathing; 18. alveoli; 19. one-way; 20. ventilation; 21. hyperventilation; 22. hypoventilation; 23. Counter-current exchange occurs when the respiratory medium (in this case, water) flows in the direction opposite to that which the respiratory surface (in this case, lamellae) is perfused with blood. The adaptive significance is that the diffusion gradient is maintained across the entire length of the respiratory surface, increasing the O_2 extraction efficiency; 24. a; 25. b; 26. a; 27. b; 28. a; 29. b. 30. b; 31. a; 32. aie; 33. tracheal system; 34. water; 35. gills; 36. air; 37. lungs; 38. air; 39. lungs; 40. a; 41. g; 42. d; 43. b; 44. h; 45. c; 46. e; 47. f

44.3 The Mammalian Respiratory System [pp. 1004–1007]

48. pharynx; 49. larynx; 50. trachea; 51. bronchi; 52. bronchioles; 53. pleura; 54. diaphragm; 55. external intercostals muscles; 56. internal intercostals muscles; 57; tidal volume; 58. vital capacity; 59. residual volume; 60. carotid bodies; 61. aortic bodies; 62. b; 63 a; 64. b; 65. a; 66. c; 67. nasal passages; 68. pharynx; 69. epiglottis; 70. larynx; 71. trachea; 72. lung; 73. bronchi; 74. mouth; 75. pleura; 76. intercostals muscles; 77. diaphragm; 78. bronchiole; 79. alveoli

44.4 Mechanisms of Gas Exchange and Transport [pp. 1007–1010], **44.5 Respiration and High Altitude and in Ocean Depths** [pp. 1010–1012]

80. partial pressure; 81. hemoglobin; 82. oxygen dissociation curve; 83. carbonic anhydrase; 84. buffer; 85. erythropoietin; 86. myoglobin; 87. The sigmoid shape of the oxygen dissociation curve reflects the cooperate binding characteristics of the four subunits. Initially, O_2 binds to the first subunit with some difficulty (lag phase). After O_2 binds to the first subunit, it changes the shape of the Hb molecule such that the binding affinity of the second Hb subunit is increased and O_2 binds with greater ease; similarly, binding of the second O_2 changes the shape of the Hb molecule further and results in heightened affinity of the 3rd subunit for O_2 (exponential phase). Finally, as the 4th O_2 binds, the Hb molecule becomes saturated; 88. b; 89. a; 90, b; 91. c; 92. a; 93. d

Self-Test

1. a, b, e

2. e [e is correct because atmospheric pO_2 is typically 150 mmHg near sea level; normal arterial pO_2 is generally 100 mmHg (after extraction from atmospheric air) and tissue pO_2 levels are typically 10 mmHg (where O_2 is being used)]

3. a, b, c

4. d

5. c

6. a, b, d [a is correct because counter-current exchange is used in bony fish; d is correct because in counter-current circulation, the respiratory medium (in this case, water) flows in the direction opposite to that which the respiratory surface (in this case, lamellae) is perfused with blood. B is correct because the counter-current circulation results in the pO_2 of water always being higher than the pO_2 of blood, thus favoring the diffusion of O_2 from water to blood across the entire length of the respiratory surface]

7. a, b, c [In tissues, CO_2 is generated as a by-product of metabolism and its levels increase; the more CO_2 produced by metabolism, the more diffuses from the tissues into the blood and, into RBCs (remember the cell membrane is permeable to small molecules like CO_2); a is correct because the more CO_2 that enters the RBC, the more CO_2 is available to react with Hb; b is correct because the higher the CO_2 concentration, the more CO_2 there is to react with water to shift the equilibrium of the reaction of $CO_2 + H_2O \Xi [H_2CO_3] \Xi H^+ + HCO_3^-$ to the right (as written); therefore, the more CO_2 that enters the RBC, the more H^+ there is produced and the more H^+ that can react with Hb-O_2, forcing the dissociation of O_2 from Hb, thereby forming HHb (reduced Hb); c is correct because the RBCs exchange Cl^- with HCO_3^-; with higher levels of HCO_3^- being produced in the RBC (from above carbonic acid reaction), the more Cl^- will be taken into the cell in exchange for HCO_3^-, resulting in higher levels of HCO_3^- in the plasma (making HCO_3^- in the plasma the major form in which CO_2 is transported in the blood).]

8. a, c

9. a, b, c, d

Chapter 45 Animal Nutrition

Why It Matters [pp. 1015–1016],

1. nutrients; 2. Nutrition; 3. digestion; 4. absorption; 5. ingestion; 6. digestion; 7. carbohydrates; 8. proteins; 9. lipids; 10. nucleic acids

45.1 Feeding and Nutrition [pp. 1016–1018]

11. carbohydrates; 12. fats; 13. malnutrition or undernutrition; 14. proteins; 15. muscle; 16. nucleic acid; 17. energy (ATP); 18. synthesize; 19. essential; 20. amino acids.; 21. fatty acids; 22. minerals; 23. vitamins; 24. B; 25. C; 26. D; 27. A

45.2 Digestive Processes [pp. 1018–1020]

28. breakdown; 29. smaller; 30. absorbed; 31. enzymatic hydrolysis; 32. specific; 33. bond; 34. molecule; 35. intracellular; 36. extracellular; 37. B; 38. D; 39. A; 40. C; 41. Intracellular digestion occurs within the cell. In order for the cell not be broken down by digestive enzymes, particles are contained within vacuoles that will fuse with lysosomes containing digestive enzymes. Food particles are taken into the cell by endocytosis. After digestion has occurred and materials absorbed from the vacuole, waste products are eliminated from the cell by exocytosis;

42. Extracellular digestion occurs in a tube or saclike structure, actually outside of the organism. The potential for different types of food sources is increased, since size is not a limiting factor. The primary limiting factor will be the available enzymes to breakdown the food material; 43. A, B; 44. E; 45. C, D; 46. A; 47. C, D; 48. A, B; 49. A, B, C

45.3 Digestion in Humans and Other Mammals [pp. 1020–1031]

50. essential; 51. 8; 52. 2; 53. diet; 54. hydrophilic; 55. hydrophobic; 56. D; 57. skin; 58. K; 59. micro-organisms, bacteria; 60. F; 61. I; 62. H; 63. C; 64. D; 65. B; 66. E; 67. G; 68. A; 69. 5; 70. 4; 71. 14; 72. 12; 73. 11; 74. 6; 75. 3; 76. 10; 77. 8; 78. 13; 79. 1; 80. 7; 81. 2; 82. 15; 83. 9; 84. mouth; 85. teeth; 86. salivary glands; 87. esophagus; 88. gastroesophageal; 89. stomach; 90. 3; 91. Gastric; 92. HCl; 93. pepsinogen; 94. acidicity; 95. mucus; 96. pyloric; 97. small intestine; 98. pancreas; 99. liver; 100. mucosa; 101. digestion; 102. absorption; 103. ileocecal; 104. colon or large intestine; 105. water; 106. rectum; 107. feces; 108. anus

45.4 Regulation of the Digestive Process [pp. 1032–1033]

109. C; 110. E; 111. B; 112. F; 113. D; 114. A; 115. False, autonomic; 116. True; 117. False, hypothalamus; 118. True

45.5 Digestive Specializations in Vertebrates [pp. 1033–1036]

119. diet; 120. herbivores; 121. long; 122. storage; 123. carnivores; 124. short; 125. symbiotic; 126. ruminants; 127. 4; 128. cellulose

Self-Test

1. d [ATP production as well as biological molecule production is dependent on the digestion of organic molecules]
2. d [Kangaroos are herbivores which have long intestinal tracts, only the tiger, a carnivore, has a short intestinal tract]
3. a [fats have the most potential kcal per gram]
4. b [essential amino acids, fatty acids, vitamins and minerals must be obtained in the diet]
5. d [bulk feeders consume chunks or particles of food; teeth and claws would be advantageous adaptations]
6. a [saclike body plans have one opening for both intake of food and removal of waste products]
7. c [all selections except the pancreas are involved in mechanical processing or mixing]
8. b [long term antibiotic treatment will decrease the bacteria that normally inhabit the gut; these bacteria synthesize vitamin K which is necessary for synthesis of clotting factors]
9. c [the pyloric sphincter is a ring of smooth muscle that acts as a valve between the stomach and small intestine; if this valve were blocked, chyme would not be able to move out of the stomach into the small intestine]
10. a [Secretin reduces gastric acid release as well as stimulates the pancreas to release a rich bicarbonate solution; reduction in this hormone would result in a more acidic environment]

Chapter 46 Regulating The Internal Environment

Why It Matters [pp. 1039–1040],

1. concentration; 2. osmoregulation; 3. excretion; 4. water; 5. ions; 6. waste products

46.1 Introduction to Osmoregulation and Excretion [pp. 1040–1043]

7. osmosis; 8. high; 9. water; 10. low; 11. water; 12. semipermeable membrane; 13. passive; 14. concentration; 15. osmolarity; 16. isoosmotic; 17. hyperosmotic; 18.hypoosmotic; 19. osmoregulater; 20. osmoconformer; 21. metabolic; 22. excretion; 23. removal; 24. balance; 25. G; 26. E; 27. D; 28. F; 29. C; 30. A; 31. B

46.2 Osmoregulation and Excretion in Invertebrates [pp. 1043–1045]

32. Marine; 33. osmoconformers; 34. terrestrial; 35. osmoregulaters; 36. external; 37. energy; 38. metabolic (toxic); 39. ammonia; 40. energy; 41. hyperosmotic; 42. varied; 43. osmoconformers; 44. B; 45. C; 46. A; 47. C; 48. A; 49. B; 50. A; 51. B

46.3 Osmoregulation and Excretion in Mammals [pp. 1045–1052]

46.4 Regulation of Mammalian Kidney Function [pp. 1052–1054]

52. kidney; 53. nephron; 54. cortex; 55. medulla; 56. urine; 57. wastes; 58. renal pelvis; 59. urinary bladder; 60. ureter; 61. hyperosmotic; 62. water; 63. nephron; 64. pertiubular capillaries; 65. hyperosmotic; 66. auto-regulatory; 67. filtration; 68. juxtaglomerular; 69. distal; 70. afferent; 71. 2; 72. renin-angiotensin-aldosterone; 73. RAAS; 74. antidiuretic; 75. ADH; 76. B; 77. I; 78. F; 79. H; 80. J; 81. G; 82. D; 83. C; 84. E; 85. A; 86. False, increases; 87. False, increase; 88. True; 89. False, hypothalamus; 90. True; 91. True; 92. Decreased blood pressure results in increased renin, which increases the level of angiotensin. Angiotensin increases blood pressure by direct action on blood vessels. In addition, angiotensin increases aldosterone from the adrenal cortex, which increases sodium reabsorption and water follows in the kidney. Increased water reabsorption increases blood pressure; 93. Decreased osmolarity is detected by the osmoreceptors in the hypothalamus, which causes a decrease in the amount of ADH released from the posterior pituitary. With less ADH, the permeability of the distal convoluted tubules and collecting ducts to water will decrease, and less water will be reabsorbed by the kidney. The osmolarity of blood fluids would increase with less water reabsorption by the kidney.

46.5 Kidney Function in Nonmammalians Vertebrates [pp. 1054–1056]

94. water; 95. salts; 96. hyperosmotic; 97. conserved; 98. excreted; 99. urea; 100. nitrogenous; 101. isoosmotic; 102. hyperosmotic; 103. excrete; 104. conserve; 105. conserve; 106. water-free; 107. covering (adaptations); 108. D; 109. B; 110. B; 111. D; 112. C, E; 113. A; 114. A; 115. C; 116. C

46.6 Introduction to Thermoregulation [pp. 1056–1058]

117. negative; 118. thermoreceptors; 119. set point; 120. gain; 121. loss; 122. Endotherms; 123. Ectotherms; 124. C; 125. D; 126. A; 127. B; 128. A, B

46.7 Ectothermy [pp.1058–1060]

129. external; 130. birds; 131. mammals; 132. behavioral; 133. deep; 134. upper; 135. radiating; 136. thermal acclimatization; 137. enzymes; 138. temperature

46.8 Endothermy [pp. 1060–1066]

139. B; 140. A; 141. A; 142. B; 143. B; 144. B; 145. A; 146. False, large; 147. False, summer, winter; 148. True; 149. True; 150. True

Self-Test

1. b [if the osmolarity of the intracellular and extracellular fluid were equal (isoosmotic), the animal would be an osmoconformer]

2. a [material that is either reabsorbed from the filtrate or secreted into the filtrate is moved either by facilitated or active transport]

3. d [these are filtration systems, so movement of material can either be by carrier mediated transport—activated or facilitated—or by filtration]

4. c [removal of nitrogenous wastes in the form of uric acid crystal would indicate a conservation of water, the environment is most likely terrestrial and very dry]

5. c [the only portion of nephron without aquaporins is the ascending segment]

6. d [hyperosmotic body fluids means an increased solute concentration and less water. ADH would increase which would make the last portion of the nephron permeable to water—water absorption would increase]

7. d [salmon are marine teleosts, and nitrogenous waste removal would be by secretion of ammonia from the gills]

8. b [low blood pressure could be corrected by increased water reabsorption or decreased urine output; the only selection that would accomplish this would be increased aldosterone, which increases sodium reabsorption and water follows; also, aldosterone will increase ADH]

9. b [shivering would produce heat; this would be transferred to the egg by conduction—direct contact]

10. b [shivering would increase heat production, decrease epinephrine, and vasodilation would decrease heat production]

Chapter 47 Animal Reproduction

Why It Matters [pp. 1069–1070],

1. biological; 2. adaptation; 3. egg; 4. sperm; 5. diversity; 6. reproduce or mate

47.1 Animal Reproductive Modes: Asexual and Sexual Reproduction [pp. 1070–1071]

7. sexual; 8. asexual; 9. genetic; 10. clonal; 11. identical; 12. fission; 13. budding; 14. fragmentation; 15. egg; 16. fertilization; 17. meiosis; 18. parthenogenesis; 19. unique; 20. female; 21. genetic diversity; 22. genetic recombination; 23. independent assortment; 24. C; 25. D; 26. A; 27. B

47.2 Cellular Mechanisms of Sexual Reproduction [pp. 1071–1078]

28. gametes; 29. gametogenesis; 30. meiosis; 31. germ; 32. gonads; 33. testes; 34. ovaries; 35. sperm; 36. spermatogenesis; 37. egg, or ovum; 38. oogenesis; 39. D; 40. H; 41. A; 42. G; 43. F; 44. C; 45. B; 46. E; 47. External fertilization occurs outside of the body, typically in aquatic species; synchronization of female and male gametes is critical to success. Internal fertilization occurs within the female reproductive tract in many terrestrial species; typically synchronization of gamete release is not necessary; 48. The acrosome reaction involves release of enzymes that breakdown egg-coating material; fast block is a wave of depolarization that occurs within seconds of fusion of the sperm nucleus into the cytoplasm of the egg, while the slow block occurs within minutes of this reaction. Fast block also initiates the release of calcium ions that release enzymes from cortical granules, thus contributing to a barrier (the slow block) against sperm penetration. These are mechanisms to prevent multiple sperm from fusing with the egg.

47.3 Sexual Reproduction in Humans [pp. 1078–1087]

49. dual; 50. hormone; 51. estrogens; 52. follicular; 53. progesterone; 54. inhibin; 55. ovulation 56. Leydig; 57. seminiferous; 58. androgens; 59. follicle-stimulating (FSH); 60. luteinizing (LH); 61. gonadotropin-releasing (GnRH); 62. oocytes; 63. estrogens; 64. uterine; 65. LH; 66. ovulation; 67. progesterone; 68. contractions; 69. Sertoli; 70. spermatogenesis; 71. Leydig; 72. androgens; 73. GnRH; 74. FSH; 75. LH; 76. GnRH; 77. feedback; 78. C; 79. E; 80. F; 81. C; 82. B; 83. A; 84. A; 85. E; 86. F; 87. B; 88.F;

47.4 Methods for Preventing Pregnancy: Contraception [pp. 1087-1090]

89. contraception; 90. ovulation; 91. sperm; 92. fertilization; 93. implantation; 94. vasectomy; 95. tubal ligation; 96. D; 97. A; 98. E; 99. C; 100. B; 101. True; 102. False, high levels; 103. True; 104. False, prevent;

Self-Test

1. d [fragmentation occurs when there are separate pieces of the parent]

2. d [The only advantage of asexual reproduction listed would be: not necessary to find a mate; all the other options are actually advantages of sexual reproduction]

3. d [if the fast block were inhibited, then polyspermy could occur; if this happened, the number of chromosomes would be greater than the diploid number]

4. b [this animal is a monotreme, which means that it is a mammal but it lays eggs]

5. b [after ovulation, the follicular cells become the corpus luteum, which produce progesterone]

6. c [if GnRH were not inhibited, another cycle could start prior to the end of the first, thus additional FSH and LH would be released from the pituitary and another oocyte could be stimulated]

7. b [the prostate produces a very alkaline (basic) fluid, thus if this were inhibited, the semen would be more acidic]

8. a [without a pituitary, the individual would be sterile, there would be no FSH or LH]

9. d [human chorionic gonadotrophin]

10. a [IUD is the only option that doesn't prevent fertilization, rather, IUDs prevent implanation]

Chapter 48 Animal Development

Why It Matters [pp. 1093–1094],

1. fertilized egg; 2. embryo; 3. adult; 4. morphology; 5. differentiation; 6. DNA; 7. developmental

48.1 Mechanisms of Embryonic Development [pp. 1094–1097]

48.2 Major Patterns of Cleavage and Gastrulation [pp. 1097–1101]

8. zygote; 9. genetic (nuclear); 10. egg; 11. cytoplasmic; 12. yolk; 13. animal; 14. vegetal; 15. polarity; 16. axes; 17. cleavage; 18. cleavage; 19. gastrulation; 20. organogenesis; 21. morula; 22. blastula; 23. gastrula; 24. gastrulation; 25. ectoderm; 26. endoderm; 27. mesoderm; 28. archenteron; 29. blastopore; 30. anus; 31. mouth; 32. protostomes; 33. mouth; 34. deuterostomes; 35. anus; 36. selective cell adhesions; 37. induction; 38. determinations; 39. differentiation; 40. C; 41. B; 42. A; 43. A, B, C; 44. C; 45. A; 46. B; 47. C; 48. C; 49. C; 50. B; 51. E; 52. C; 53. G; 54. B; 55. F; 56. A; 57. D; 58. B; 59. B; 60. A; 61. C; 62. B; 63. C; 64. A; 65. C

48.3 From Gastrulation to Adult Body Structures: Organogenesis [pp. 1101–1104]

66. ectoderm; 67. mesoderm; 68. endoderm; 69. organs; 70. organogenesis; 71. apoptosis; 72. notochord; 73. mesoderm; 74. nervous; 75. neurulation; 76. induction; 77. neural plate; 78. mesoderm; 79. somites; 80A. 5; 80B. 2; 80C. 3; 80D. 1; 80E. 4; 80F. 7; 80G. 6

48.4 Embryonic Development of Humans and Other Mammals [pp. 1104–1109]

81. three trimesters; 82. cleavage; 83. gastrulation; 84. organogenesis; 85. fetus; 86. fertilization; 87. 1st (upper) third; 88. implantation; 89. blastocyst; 90. blastocoel; 91. inner cell mass; 92. embryo; 93. trophoblast; 94. trophoblast; 95. implantation; 96. endometrium (uterine wall); 97. embryonic disc; 98. epiblast; 99. hypoblast; 100. embryo; 101. epiblast; 102. hypoblast; 103. gastrulation; 104. neurulation; 105. reptile-bird; 106. D; 107. A; 108. B; 109. E; 110. C; 111. The Y chromosome and the sex-determining region encode a protein that initiates development of fetal testes which secrete testosterone and anti-Mullerian hormone. These two hormones cause the Wolffian ducts to become male reproductive organs and inhibit the development of Mullerian ducts.; 112. The extra-embryonic membranes, the chorion, amnion, and placenta, are derived from the trophoblast—a single cell layer that surrounds the embryo at the blastocyst stage and the hypoblast cell layer of the embryonic disc. Both cell layers are derived from the developing embryo.

48.5 The Cellular and Molecular Basis of Development [pp. 1109–1115]

48.6 The Genetic and Molecular Control of Development [pp. 1115–1122]

113. orientation; 114. rate; 115. furrow (axes); 116. G1; 117. microtubules; 118. microfilaments; 119. adhesions; 120. molecular; 121. False, gap genes; 122. True; 123. True; 124. False, pair-rule; 125. False, and

Self-Test

1. c [cytoplasmic determinants are from the egg having their effect throughout development but primarily during early cleavage of the zygote]

2. b [the blastopore becomes the mouth in protostomes and the anus in deuterostomes]

3. d [endoderm is the innermost layer and will form the linings of major organ systems]

4. b [a few cells from the hypoblast develop into the germ cells, which migrate to the developing gonads of the embryo]

5. b [the primitive streak defines the axis of the embryo, providing organizational cues for right and left or bilateral symmetry]

6. a [neural crest cells are unique to vertebrates; they form when the neural tube closes; cranial nerves are derived from these cells]

7. c[amniocentesis is evaluation of amniotic fluid that surrounds the fetus]

8. d [the SRY protein is a product of a gene on the Y chromosome, so without this protein, the Mullerian ducts would develop into female reproductive structures]

9. a [microtubules and microfilaments appear to play a significant role in the orientation or axes of the development of cleavage furrows that determine cell orientation in the embryo]

10. c [segmentation genes subdivide the embryo into regions or segments (somites) of the embryo]

Chapter 49 Population Ecology

Why It Matters [pp. 1125 -1126]

1. biotic; 2. abiotic

49.1 The Recognition of Evolutionary Change [pp. 1126 – 1127]

3. organismal ecology; 4. population ecology; 5. community ecology; 6. ecosystem ecology; 7. biosphere; 8. e; 9. c; 10. b; 11. d; 12. a

49.2 Population Characteristics [pp. 1127-1129]

13. geographic range; 14. habitat; 15. population size; 16. Population density; 17. dispersal pattern; 18. clumped; 19. uniform; 20. random; 21. age structure; 22. generation time; 23. sex ratio; 24. uniform; 25. random; 26. clumped

49.3 Demography [pp. 1129-1132]

27. immigration; 28. emmigration; 29. demography; 30. life table; 31. cohort; 32. age-specific survivorship; 33. age-specific mortality; 34. age-specific fecundity; 35. survivorship curve; 36. a; 37. b 38. d; 39. c

49.4 Evolution of Life Histories [pp. 1132-1133]

40. life history; 41. energy budget; 42. passive parental care; 43. active parental care; 44. b; 45. b; 46. a; 47. b

49.5 Models of Population Growth [pp. 1133-1139]

48. per capita growth rate; 49. exponential; 50. intrinsic growth rate; 51. logistic; 52. carrying capacity; 53. intraspecific competition; 54. time lag; 55. F—population size does not change; 56. T; 57. F—decreases; 58. F—intraspecific competition

49.6 Population Regulation [pp. 1139-1145]

59. density dependent; 60. density independent; 61. r-selected; 62. K-selected; 63. a; 64. b; 65. a; 66. b; 67. b

49.7 Human Population Growth [pp. 1145-1149]

68. demographic transition model; 69. family planning programs; 70. stable; 71. decreasing; 72. increasing

Self-Test

1. a [All branches of biology, including ecology, involve the study of life. The word "abiotic" implies the absence of life.]

2. a [In the U.S., population density is high in cities, and there are also large areas where few people live.]

3. b

4. a

5. b [Death rate may be higher than birthrate in both models, resulting in negative population growth; density dependent factors and K-selected species are more typical of logistic growth.]

6. d

7. c [The effects of temperature on a population are usually unrelated to population density.]

8. b

9. d

10. a [Age structure can predict the proportion of a population made up of different age classes in the future.]

Chapter 50 Population Interactions and Community Ecology

Why It Matters [pp.1151-1152]

1. brood parasites; 2. ecological community

50.1 Population Interactions [pp. 1152-1160]

3. coevolution; 4. predation; 5. herbivory; 6. Optimal foraging theory; 7. crypsis; 8. aposematic; 9. Batesian mimicry; 10. model; 11. mimic; 12. Mullerian mimicry; 13. interspecific competition; 14. interference competition; 15. exploitative competition; 16. Competitive exclusion principle; 17. niche; 18. fundamental niche; 19. realized niche; 20. Resource partitioning; 21. character displacement; 22. symbiosis; 23. mutualism; 24. commensalism; 25. parasitism; 26. host; 27. ectoparasite; 28. endoparasite; 29. Parasitoids; 30. b; 31. d; 32. a; 33. c; 34. a; 35. a; 36. Probably beneficial to both species. The greater the number of distasteful individuals of either species, the faster predators will learn to avoid them; 37. Probably beneficial to both species. Since the majority of individuals are models, predators are most likely to sample distasteful individuals; 38. Detrimental to both species. Predators are most likely to encounter palatable mimics and will be less apt to avoid both species.

50.2 The Nature of Ecological Communities [pp. 1160 – 1163]

39. species composition; 40. ecotone; 41. b; 42. a; 43. c

50.3 Community Characteristics [pp. 1163 – 1166]

50.4 Effects of Population Interactions on Community Characteristics [pp. 1166-1167]

50.5 Effects of Disturbance on Community Characteristics [pp. 1167-1170]

44. richness; 45. relative abundance; 46. species diversity; 47. trophic levels; 48. primary producers; 49. autotrophs; 50. consumers; 51. primary consumers; 52. secondary consumers; 53. tertiary consumers; 54. omnivores; 55. detritivores; 56; decomposers; 57. heterotrophs; 58. food chain; 59. food web; 60. stability; 61. keystone species; 62. intermediate disturbance hypothesis; 63. D; 64. A, B and C; 65. E; 66. E; 67. D

50.6 Ecological Succession: Responses to Disturbances [pp. 1170 – 1174]

50.7 Variations in Species Richness among Communities [pp. 1174 – 1176]

68. ecological succession; 69. primary succession; 70. climax community; 71. secondary succession; 72. aquatic succession; 73. facilitation; 74. inhibition; 75. tolerance; 76. disclimax; 77. equilibrium theory of island biogeography; 78. D; 79. B; 80. C; 81. A

Self-Test

1.b [a and c are simply adaptations that do not necessarily result from interaction with other organisms; d results from intraspecific interactions; coevolution is based on interactions between different species]

2. b [Symbioses are interactions between individuals of different species.]

3. c

4. d

5. d

6. a

7. d

8. d [a, b and c had established communities before the disturbance, therefore the recovery is secondary succession]

9. a

10. d [The assumption is that extinction rates are inversely correlated with island size; larger islands have more potential niches and competition between immigrants is less likely.]

Chapter 51 Ecosystems

Why It Matters [pp.1181-1182]

1. ecosystem

50.1 Energy Flow and Ecosystem Energetics [pp. 1182 – 1190]

2. gross primary productivity 3. respiration; 4. net primary production; 5. biomass; 6. standing crop biomass; 7. limiting nutrient; 8. secondary productivity; 9. ecological efficiency; 10. ecological pryamids; 11. pyramid of biomass; 12. turnover rate; 13. pyramid of numbers; 14. pyramid of energy; 15. trophic cascade; 16. a; 17. b; 18. c; 19. T; 20. T; 21. F—large

51.2 Nutrient Cycling in Ecosystems [pp. 1191 - 1199]

51.3 Community Characteristics [pp. 1199 – 1200]

22. biogeochemical cycle; 23. generalized compartment; 24. hydrologic; 25. nitrogen; 26. nitrogen fixation; 27. ammonification; 28. nitrification; 29. denitrification; 30. carbon; 31. phosphorus; 32. simulation; 33. c; 34. a; 35. c; 36. a; 37. c; 38. d; 39. b

Self-Test

1. b

2. a

3. d [Biomass and numbers are the basis for two of the three ecological pyramids.]

4. d

5. a

6. b

7. d [Photosynthesis is the only one the four choices that removes carbon from the atmosphere.]

8. b

9. a

10. b

Chapter 52 The Biosphere

Why It Matters [pp.1203-1204]

1. biosphere; 2. hydrosphere; 3. lithosphere; 4. atmosphere

52.1 Environmental Diversity of the Biosphere [pp. 1205-1209]

5. biome 6. tropics; 7. adiabatic; 8. maritime; 9. continental; 10. monsoon cycles; 11. rain shadow; 12. microclimate; 13. A; 14. B; 15. C; 16. c; 17. a; 18. b

52.2 Organismal Response to Environmental Variation [pp. 1209-1211]

52.3 Terrestrial Biomes [pp. 1211-1219]

19. climograph; 20. tropical rain forest; 21. tropical deciduous forest; 22. tropical montane forest; 23. savanna; 24. thorn forest; 25. deserts; 26. chaparral; 27. temperate grasslands; 28; temperate deciduous forest; 29. boreal forest; 30. taiga; 31. arctic tundra; 32. permafrost 33. alpine tundra; 34. a; 35. b; 36. c; 37. g; 38. d; 39. e; 40. f

52.4 Freshwater Biomes [p. 1219-1221]

41. photic zone; 42. aphotic zone; 43. wetlands; 44. littoral zone; 45. limnetic zone; 46. profundal zone; 47. spring overturn; 48. autumn overturn; 49. epilimnion; 50. hypolimnion; 51. thermocline; 52. oligotrophic; 53. eutrophic; 54. c; 55. b; 56. a; 57. b; 58. c

52.5 Marine Biomes [1221-1225]

59. benthic; 60. intertidal; 61. neritic; 62. pelagic; 63. oceanic; 64. abyssal; 65. estuaries; 66. salt marshes; 67. coral reefs; 68. nekton; 69. benthos; 70. B; 71. E; 72. D; 73. A; 74. C

Self-Test

1. d [The tilt of the earth on its axis results in parts of its surface receiving differing amounts of solar radiation at different times of the year]
2. a
3. c
4. d [Tropical rainforests have the greatest species richness of all terrestrial biomes.]
5. a [Although days are very long during the arctic summer, temperatures are warm enough for only a short period.]
6. c
7. d [Oxygen rich water sinks, nutrient rich water rises, and the thermocline disappears.]
8. c
9. d
10. c

Chapter 53 Biodiversity and Conservation Biology

Why It Matters [pp. 1229-1230]

1. biodiversity

53.1 The Benefits of Biodiversity [pp. 1230-1232]

53.2 The Biodiversity Crisis [pp. 1232-1239]

2. ecosystem services; 3. habitat fragmentation; 4. edge effects; 5. desertification; 6. pollution; 7. acid precipitation; 8. overexploitation; 9. B; 10. A; 11. C

53.3 Biodiversity Hotspots [pp. 1239-1241]

53.4 Conservation Biology: Principles and Theory [pp. 1241-1247]

53.5 Conservation Biology: Practical Strategies and Economic Tools [pp. 1247-1251]

12. biodiversity hotspots; 13. endemic; 14. Endangered Species; 15. Conservation Biology; 16. population viability analysis; 17. minimum viable population size; 18. metapopulations; 19. source populations; 20. sink populations; 21. landscape ecology; 22. Single Large Or Several Small; 23. ecotourism; 24. ecosystem valuation; 25. c; 26. b; 27. a; 28. b; 29. a; 30.c; 31. A; 32. C; 33. B

Self-Test

1. d [Biodiversity impacts all humans, perhaps especially those listed.]

2. c

3. c [Desertification concentrates salts in the soil, a process called salinization.]

4. d [Sulfur dioxide combines with water vapor in the atmosphere to produce sulfuric acid.]

5. d

6. c

7. c

8. d

9. b [All of the listed scientists would probably have some interest in SLOSS, but probably landscape ecologists more so than the others.]

10. d

Chapter 54 The Physiology and Genetics of Animal Behavior

Why It Matters [pp. 1253 -1254]

1. behavioral repertoire; 2. animal behavior; 3. ethology; 4. neuroscience

54.1 Genetic and Environmental Contributions to Behavior [pp. 1254-1255]

54.2 Instinctive Behaviors [pp. 1255-1257]

54.3 Learned Behaviors [pp. 1257-1259]

5. instinctive; 6. learned; 7. fixed action patterns; 8. sign stimuli; 9. learning; 10. imprinting; 11. critical period; 12. classical conditioning; 13. operant; 14. insight learning; 15. habituation; 16. b; 17. b; 18. a; 19. b; 20. a; 21. A; 22. F; 23. B; 24. C; 25. D; 26. E; 27. F, unconditioned; 28. F, operant; 29. F, habituation; 30. T

54.4 The Neurophysiological Control of Behavior [pp. 1259 – 1260]

54.5 Hormones and Behavior [pp. 1260-1263]

54.6 Nervous System Anatomy and Behavior [pp. 1263-1267]

31. territory; 32. E; 33. A; 34. C; 35. D; 36. B

Self-Test

1. d

2. b

3. a

4. c

5. b

6. c [The operant is the desired behavioral response from the subject. The reinforcement is the reward for performing the operant.]

7. a

8. b

9. b

10. d

Chapter 55 The Ecology and Evolution of Animal Behavior

Why It Matters [pp.1269–1270]

1. proximal; 2. ultimate; 3. migration

55.1 Migration and Wayfinding [pp. 1270–1274]

4. piloting; 5. compass orientation; 6. navigation; 7. A; 8. C, D; 9. B; 10. C

55.2 Habitat Selection and Territoriality [pp. 1274–1276]

11. kinesis; 12. taxis; 13. territoriality; 14. c; 15. b; 16. a

55.3 The Evolution of Communication [pp. 1276–1278]

17. acoustic signaling; 18. visual signaling; 19. chemical; 20. pheromones; 21. tactile; 22. electrical; 23. b; 24. a, b, c, d; 25. e; 26. b

55.4 The Evolution of Reproductive Systems and Mating Behavior [pp. 1279–1281]

27. reproductive strategies; 28. parental investment; 29. sexual selection; 30. courtship displays; 31. lek; 32. mating system; 33. polygyny; 34. polyandry; 35. polygamy; 36. monogamy; 37. promiscuity; 38. C; 39. B; 40. D; 41. A

55.5 Evolution of Social Behavior [pp. 1281–1285]

55.6 An Evolutionary View of Human Social Behavior [pp. 1285–1286]

42. social behavior; 43. dominance hierarchy; 44. altruism; 45. kin selection; 46. haplodiploidy; 47. reciprocal altruism; 48. 0.5; 49. 0.5; 50. 0.25; 51. 0.125

Self-Test

1. d

2. b

3. c

4. a [The other choices all involve resources that would be difficult to defend.]

5. a and/or b [Acoustical and chemical signals can easily go around obstacles.]

6. c

7. c

8. b

9. d [Sharing of blood meals in vampire bats often occurs between nonrelated individuals.]

10. d